Physics, Fabrication, and Applications of Multilayered Structures

NATO ASI Series

Advanced Science Institutes Series

A series presenting the results of activities sponsored by the NATO Science Committee, which aims at the dissemination of advanced scientific and technological knowledge, with a view to strengthening links between scientific communities.

The series is published by an international board of publishers in conjunction with the NATO Scientific Affairs Division

A	Life Sciences	Plenum Publishing Corporation
B	Physics	New York and London
C	Mathematical and Physical Sciences	Kluwer Academic Publishers Dordrecht, Boston, and London
D	Behavioral and Social Sciences	
E	Applied Sciences	
F	Computer and Systems Sciences	Springer-Verlag
G	Ecological Sciences	Berlin, Heidelberg, New York, London,
H	Cell Biology	Paris, and Tokyo

Recent Volumes in this Series

Series B: Physics

Physics, Fabrication, and Applications of Multilayered Structures

P. Dhez

Paris University
Orsay, France

and

C. Weisbuch

Thomson CSF
Orsay, France

Plenum Press
New York and London
Published in cooperation with NATO Scientific Affairs Division

Proceedings of a NATO Advanced Study Institute on
Physics, Fabrication, and Applications of Multilayered Structures,
held June 22–July 4, 1987,
in Ile de Bendor, Bandol, France

Library of Congress Cataloging in Publication Data

NATO Advanced Study Institute on Physics, Fabrication, and Applications of
Multilayered Structures (1987: Bandol, France)
    Physics, fabrication, and applications of multilayered structures / edited by
P. Dhez and C. Weisbuch.
       p.   cm.—(NATO ASI series. Series B, Physics; v. 182)
    "Proceedings of a NATO Advanced Study Institute on Physics, Fabrication,
and Applications of Multilayered Structures, held June 22–July 4, 1987 in Ile de
Bendol, Bandol, France"—T.p. verso.
    "Published in cooperation with NATO Scientific Affairs Division."
    Includes bibliographies and index.
ISBN 978-1-4757-0093-0   ISBN 978-1-4757-0091-6 (eBook)
DOI 10.1007/978-1-4757-0091-6
    1. Superlattices as materials—Congresses. 2. Sputtering (Physics)—Con-
gresses. I. Dhez, Pierre. II. Weisbuch, C. (Claude), 1945–   . III. North Atlantic
Treaty Organization. Scientific Affairs Division. IV. Title. V. Series.
QC611.8.S86N39   1987   88-22671
621.3815′2—dc19   CIP

© 1988 Plenum Press, New York
Softcover reprint of the hardcover 1st edition 1988

A Division of Plenum Publishing Corporation
233 Spring Street, New York, N.Y. 10013

SPECIAL PROGRAM ON CONDENSED SYSTEMS OF LOW DIMENSIONALITY

This book contains the proceedings of a NATO Advanced Study Institute held within the program of activities of the NATO Special Program on Condensed Systems of Low Dimensionality, running from 1983 to 1988 as part of the activities of the NATO Science Committee.

Other books previously published as a result of the activities of the Special Program are:

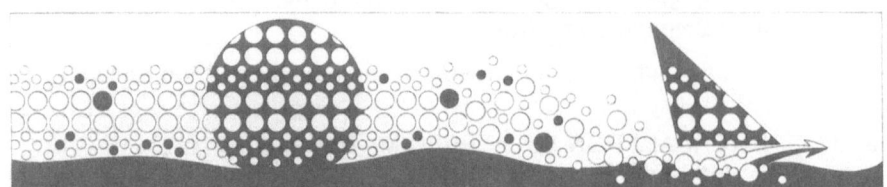

The logo of the School is an attempt to reflect both the environment of the Ile de Bendor near Marseilles, where the School was held, and the imperfect structure of interfaces, one important and intricate aspect of multilayered structures. A boat, the sun, and the sea—which were appreciated by the participants during breaks—appear in stylized form, combined with a representation of a progressively organized structure—which was the goal for all of the multilayered media considered during the School. The concept for the logo was developed by A. Fert and P. Dhez, and it was designed by P. Emery, a young artist living in Gif-sur-Yvette, near Orsay.

ORGANIZATION COMMITTEE

P. DHEZ (Chairman), *Orsay*
A. FERT (Treasurer), *Orsay*
Y. BRUYNSERAEDE, *Leuven*
C. FALCO, *Tucson*
A. L. GREER, *Cambridge*

M. J. KELLY, *Wembley*
F. MEZEI, *Berlin*
M. van der WIEL, *Nieuwegein*
M. VOOS, *Paris*
C. WEISBUCH, *Orsay*

Preface

Low-dimensional materials are of fundamental interest in physics and chemistry and have also found a wide variety of technological applications in fields ranging from microelectronics to optics. Since 1986, several seminars and summer schools devoted to low-dimensional systems have been supported by NATO. The present one, **Physics, Fabrication and Applications of Multilayered Structures,** brought together specialists from different fields in order to review fabrication techniques, characterization methods, physics and applications.

Artificially layered materials are attractive because alternately layering two (or more) elements, by evaporation or sputtering, is a way to obtain new materials with (hopefully) new physical properties that pure materials or alloys do not allow. These new possibilities can be obtained in electronic transport, optics, magnetism or the reflectivity of x-rays and slow neutrons. By changing the components and the thickness of the layers one can track continuously how the new properties appear and follow the importance of the multilayer structure of the materials. In addition, with their large number of interfaces the study of interface properties becomes easier in multilayered structures than in monolayers or bilayers. As a rule, the role of the interface quality, and also the coupling between layers, increases as the thickness of the layer decreases. Several applications at the development stage require layer thicknesses of just a few atomic layers.

The different lectures given during the school review the state of the art of characterization techniques using x-rays, electrons, neutrons, and ion-backscattering, with special attention paid to electron microscopy. A large variety of multilayered media are considered, from the simplest amorphous and crystalline layer to lattice-unmatched semiconductors (so-called strained-layer superlattices). Several types of bilayer components are described, such as metal-semiconductors, metal-insulators and semiconductor-insulators. The novel physics of these materials is often fascinating and is reviewed by experts in the field.

Finally, some applications of these structures are described in the fields of microelectronics, x-rays and neutron mirrors. It should be emphasized that no clearcut distinction between growth, characterization, physical properties and application can be made. The reader will be aware that quite often the physical property under study is the best way to characterize the multilayered structure. This makes the field essentially multidisciplinary and we hope to have conveyed this fact through the assembly of talented lecturers originating from various disciplines.

In partial contradiction to the message carried by the logo of the school, and in spite of the beautiful surroundings of the island, student participation was intense as is very well attested by the poster abstracts reprinted at the end of the lecture notes. The evening sessions were not the least active or poorest attended. Let us thank both the lecturers and students for having made the school a most stimulating event.

We would like to thank the Esprit program of the EEC, the V.A.S., GEC and Thomson CSF companies, and the LURE laboratory at Orsay for their financial support, and Mrs. C. Pendl for typing part of the manuscript and for careful editing and correcting of the whole text.

Orsay, March 1988   P. Dhez

                                                          C. Weisbuch

Contents

PART I
GROWTH AND CHARACTERIZATION OF
MULTILAYER STRUCTURE

GROWTH OF METALLIC AND METAL-CONTAINING SUPERLATTICES

Charles M. Falco

Optical Sciences Center; Department of Physics
University of Arizona, Tucson, AZ 85721   U.S.A.

INTRODUCTION

Advances in the technology to produce ultra-high vacuums, and in development of several vapor deposition techniques, have now made possible the sequential monolayer-by-monolayer deposition of more than one material (each of which may be an element, alloy, or compound). Nearly flawless semiconductor superlattices have been synthesized, as described in the chapters in this book by E. Kasper, S. Luryi, J-Y. Marzin, D. B. McWhan, P. M. Petroff, K. Ploog, H. Sakaki, T. Sakamoto, C. Weisbuch, and R. H. Williams.

Within the past few years interest has developed in producing such compositionally modulated materials with at least one of the constituents being a metal [1,2]. Although it might appear that this interest in multilayered metals started recently, as Figure 1 shows, research on metallic multilayers and superlattices has shown a steady increase for over fifteen years. It should be noted that the two curves in Figure 1 are derived from slightly different types of data (e.g. the top curve includes all MBE-grown materials, not just hetero-structures and superlattices). In spite of this useful comparisons can be made. It is interesting to note how closely the fields of semiconductors and of metals have paralleled each other in development since approximately 1970. The sustained expansion of research on MBE-grown semiconductors can largely be attributed to the large number of practical electronic and opto-electronic devices arising from this area research. Recently, various applications of metallic superlattices have started to appear. A few examples are optics for soft-x-rays[3,4] and magneto-optic data recording [5-7]. Although it is dangerous to extrapolate exponential growth curves, the emergence of such applications should continue to sustain growth of both basic and applied research on metallic multilayers.

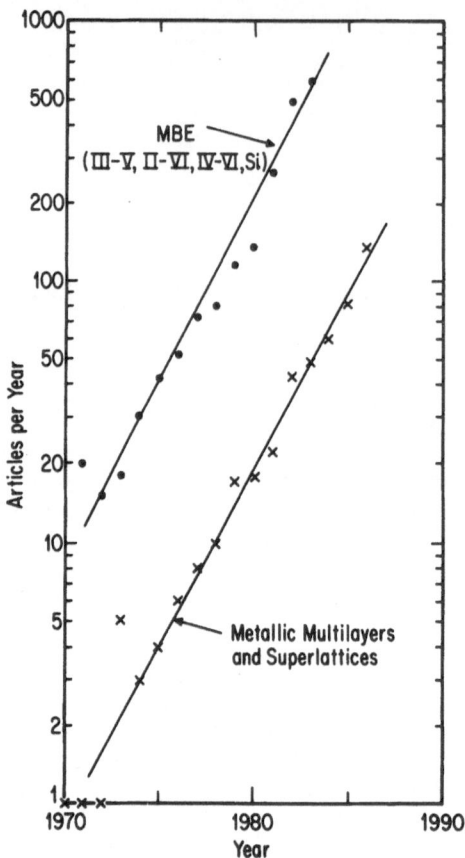

Figure 1. Number of articles published each year since 1970 on
all MBE-grown materials (primarily semiconductors),
and on metallic multilayers and superlattices [8]

ARTIFICIALLY STRUCTURED MATERIALS INVOLVING METALS

        In semiconductors the directional nature of the covalent
bonds aids the formation of high-quality crystals. In metals
the bonding is much less directional, making the incorporation
of defects easier. Thus, single-crystal artificially
structured materials with at least one metallic component are
more difficult to grow than a structure composed completely of
semiconductors. On the other hand, while highly perfect
structures are required for semiconductors to exhibit useful
electronic properties, it is important to realize that metals
can tolerate greater levels of structural and chemical
imperfection and still exhibit phenomena of interest. Metals
are much less sensitive to most impurities and structural
imperfections than are semiconductors. Many physical
properties are not strongly affected in a less than perfect
metal, whereas semiconductors are seriously degraded by
defects, impurities, grain boundaries, etc.

4

The above discussion of atomic bonding mechanisms suggests a reason why it is more difficult to grow a "perfect" metal crystal than it is a perfect semiconductor crystal. It also indicates why usually it is less important. However, certain physical properties of metallic superlattices can be strongly affected by the growth conditions. For example, the superconducting transition temperature of superlattices containing Nb will be depressed by oxygen contamination. Nb is a strong oxygen-getter, so that ultra-high vacuum and high deposition rates are required to keep oxygen contamination negligible in the growing film. For metal/non-metal samples produced by sputtering, Ar gas inclusion in the growing film can sometimes be a problem.

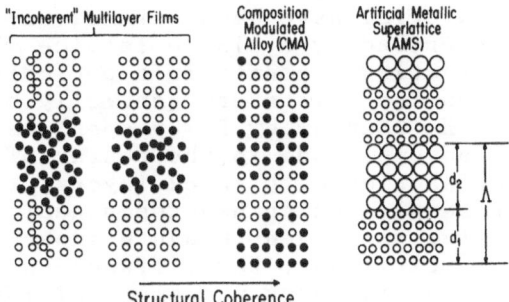

Figure 2. Schematic illustration of the degrees of structural coherence obtained in metallic multilayers and superlattices

STRUCTURAL PERFECTION OF METALLIC SUPERLATTICES

Sputtering,[9,10] evaporation[11,12] and Molecular Beam Epitaxy (MBE)[13,14] techniques have all been used to produce metallic multilayers and superlattices with layering periodicities close to atomic dimensions. This has been successful with a large number of metals. Many of these materials have highly regular periodicity along the layered direction, but varying degrees of perfection within the layers and at interfaces between the layers. These multilayers and superlattices have degrees of structural perfection which can be usefully classified in three categories, as described below and illustrated in Figure 2. It should be emphasized that for many studies or applications ideal single crystal epitaxy as exemplified by the semiconductor GaAs/GaAlAs system is not required. Thus, the preparation and study of metallic materials falling into each category of this scheme is worthwhile.

Referring to Figure 2, "multilayer" films have sharply defined boundaries between successive layers, and highly regular average spacing between layers. However, the structure within either or both layers may be amorphous or polycrystalline, with little or no correlation between the orientation of the crystallites. Thus, no structural information is transferred from a given layer to the next layer. In spite of the low degree of structural coherence of this category of metallic multilayers, these materials have been shown to exhibit interesting physical phenomena (e.g. superconductivity[15,16] as discussed in the Chapter by I. K. Schuller), and to have useful properties (e.g. for x-ray optics[3,4] as described in detail in the Chapter by E. Spiller, or magneto-optic data storage[5-7]).

"Composition Modulated Alloys (CMAs)" are usually formed from when materials with nearly identical lattice constants are used to produce a multilayer. An example is Cu/Ni, with lattice constants differing by less than 2.5%. The magnetic properties of this material have been extensively studied [17-19]. CMAs can be described by a single lattice whose sites are occupied by alternating elements in a periodic manner. The interfaces between the layers are generally broadened by interdiffusion (see the Chapter by F. Spaepen), so that the composition modulation amplitude in the center of a layer can range from $\leq$100% for samples with thick layers, to ~0% for thin layers, or for samples which have been heat treated to enhance the interdiffusion. Even though a superlattice exists for this category of multilayer, the diffuseness of the interfaces prevents studying many phenomena of interest, such as the effect of dimensionality on magnetism.

"Metallic Superlattices" have sharp interfaces between the two components, as do multilayered films. Ideally, the interfaces would be atomically sharp, although, as discussed in the Chapter by D. B. McWhan, it would be difficult to experimentally demonstrate that such perfectly sharp layering existed in a sample. The degree of "sharpness" required in practice would depend upon the phenomenon being investigated. Besides sharp interfaces, superlattices exhibit long range structural coherence which is maintained across many layers of the material, as for a CMA. Thus, one has the roughly natural lattice spacing within the layer modulated by the imposed superlattice layer periodicity. Two examples of superlattices are Mo/Ta (grown by sputtering [10]), and Gd/Y (grown by MBE [13]).

Recently it has been shown that it is possible to grow metallic superlattices consisting of a single monolayer each of the constituents [20]. Figure 3 shows the evolution of the x-ray diffraction spectra for a series of Mo/Ta superlattices ranging in layer thickness from n = 153 atoms per layer to n = 1. These samples were grown by sputtering.

It can be seen from the narrow x-ray linewidths in Figure 3 that, unlike previously reported metallic superlattices, these materials maintain long range structural coherence perpendicular to the layering. This suggests that, at least for certain materials combinations, it should be possible to grow other "monolayer superlattices" with metallic constituents. The structural coherence length can be obtained

from the x-ray linewidths by deconvoluting the instrumental
resolution function and applying the Scherrer equation [21].
This is shown in Figure 4.

FABRICATION OF METALLIC SUPERLATTICES

Introduction

Metallic multilayers and superlattices have been made by
sputtering, evaporation and MBE techniques. In this chapter
the distinction between evaporation and MBE is made depending on
whether pressures in the 10^{-10} torr range are involved, and
in situ analysis techniques such as High Energy Electron
Diffraction are used.

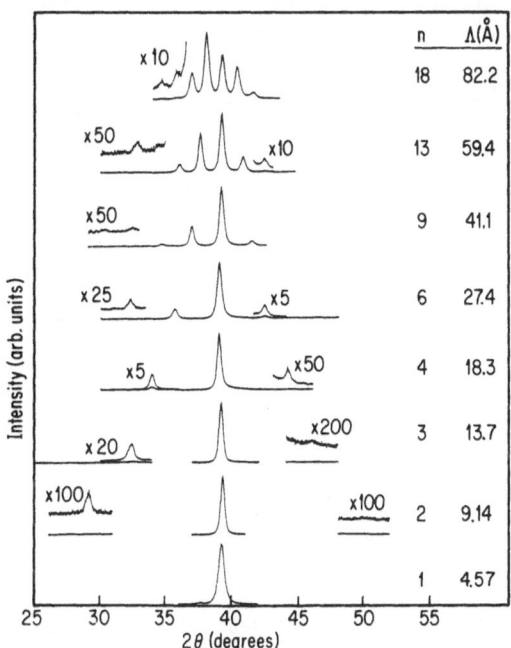

Figure 3. X-Ray diffraction spectra for a series of Mo/Ta
superlattices of modulation wavelengths varying
from n = 18 to n = 1 atomic layers per superlattice
layer

Sputtering

A schematic diagram of a sputtering system used for
the preparation of metallic superlattices is shown in Figure
5 [22]. The base pressure of this diffusion pumped system is
<7 x 10^{-8} torr. We have shown that such a system is capable
of depositing many metals with purity equivalent to the
starting target material [23].

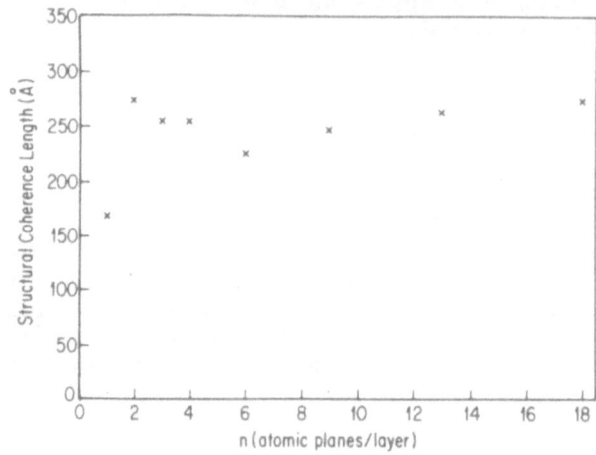

Figure 4. Structural coherence length vs. number of atomic
planes n of each material in a superlattice layer

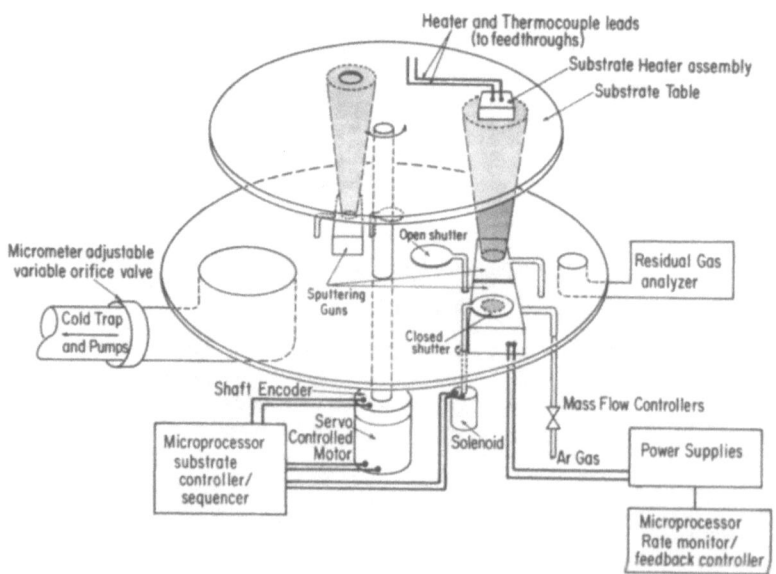

Figure 5. Schematic diagram of a sputtering system used for
depositing metallic superlattices [22]

During sputtering the Ar sputtering gas pressure is usually <3 x 10^{-3} torr, and is held constant to ±0.2% using gas flow controllers. The sputtering rate is only weakly pressure dependent in this range, so that this is adequate to reduce pressure-induced sputtering rate variations to a negligible level. A microprocessor is programmed to operate a servo-controlled motor which rotates the substrates alternately above the sputtering guns at controlled angular rotation rates, or to hold them in place above each gun for a determined length of time. Variations in layer thickness due to rotation rate or to timing errors are negligible.

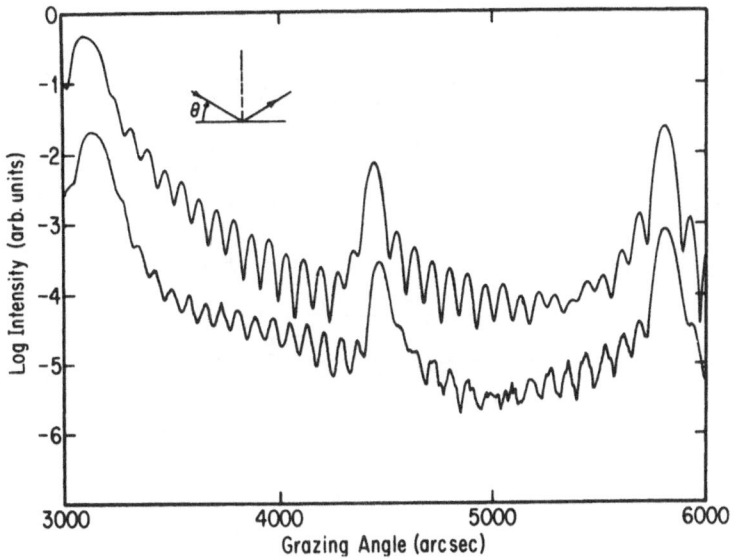

Figure 6. Expanded portion of measured (bottom curve) and calculated (top curve) low-angle x-ray diffraction spectra from a W/Si multilayer

There are two sources of variation in superlattice layer thickness which occur for sputtered samples. The first is the variation due to fluctuations or drift in the sputtering rate during a deposition. This is analogous to the problem of temperature regulation of an effusion cell used for MBE. The second is caused by the spatial variation in sputtering rate at the substrate due to a finite size sputtering target. The same problem occurs for an MBE system. Feedback control of the sputtering power supplies is capable of regulating the sputtering rates (typical rates used are 20 - 100 Å/sec) to ±0.1%. This is the dominant source of error in sputtering these materials. Experimentally, it has been shown that an upper limit on superlattice layer thickness variations produced in such a sputtering system is ±0.3% [24]. This can be seen from Figure 6, where an expanded portion of a low-angle x-ray diffraction scan from a sample produced in this system is shown.

The subsidiary maxima between the Bragg peaks in Figure 6 are due to the finite number of layers in the sample, and are very sensitive to layer thickness variations. By modeling these data it has been shown that any variations in layer thickness larger than 0.5% ~would have washed out these subsidiary maxima [25].

Because of the finite size of any sputtering target, absolute uniformity of layer thicknesses to arbitrary accuracy over the entire substrate is impossible to achieve unless a planetary system is used for holding the substrates.

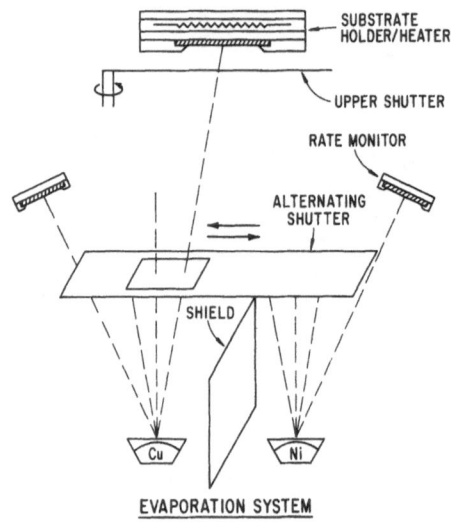

Figure 7. Schematic diagram of an evaporation system used for producing metallic multilayers and superlattices

Evaporation

Evaporation using thermally heated ovens or electron beam guns, at pressures $>10^{-8}$ torr, has been successfully used to prepare many metallic multilayers with interesting physical properties [11,12,17,19]. In this technique, shutters are used to alternately deposit the materials on substrates, or the substrates are rotated over the evaporation sources. A typical evaporation system is shown in Figure 7. In principle, layers down to the monolayer range are possible using evaporation techniques.

A relatively large concentration of impurities can be incorporated in the films during evaporation, due to the background pressure. As a worst case estimate at 1×10^{-7}

torr a monolayer of impurities can form in approximately 10 seconds, assuming unity sticking coefficient. If an evaporation rate of 10 Ångstroms per second were used, this would result in a sample with 1% impurities incorporated. This would be unacceptable for a semiconductor multilayer, but in a metal it still allows many interesting phenomena to be observed.

Molecular Beam Epitaxy (MBE)

Due to the high cost, fewer than ten MBE machines are now in use or on order for the fabrication of metallic superlattices [26]. However, as for semiconductor superlattices, MBE is a powerful technique, and should find increasing use as further applications of these materials are identified.

The requirements for a "metals MBE" machine differ from those for a GaAs machine. Many metals of interest have high melting points, so that high vapor pressures cannot be achieved using effusion cells. As a result, electron beam evaporation sources must be incorporated in the growth chamber in order to obtain desired growth rates. Provision should be made for UHV sputtering sources as well, along with a higher temperature platform (~1200 °C) for epitaxy of refractory metals. These, and other requirements are very similar to those needed for the epitaxy of silicon. Thus, MBE systems based on the new generation of Si MBE machines should find immediate use for producing metallic superlattices. A schematic diagram of such a machine is shown in Figure 8 [27].

DISCUSSION

In spite of the considerable work on the growth of metal and metal-containing multilayers and superlattices over the past fifteen years, as illustrated in Figure 1, the range of systems that can be successfully grown has not yet been thoroughly explored. Although a large number of metal/metal systems have been prepared to date, systematic investigation of the optimum deposition conditions for most of them is lacking. Parameters such as substrate temperature, rate of deposition, base vacuum pressure, sputtering pressure, type of substrate, etc. can greatly influence the growth and quality of these materials. Given this wide range of parameter space, only a few of the metallic systems have been investigated with the required level of detail thus far.

Metal/amorphous-semiconductor multilayers incorporating Ge and Si have been made successfully. Two examples are Nb/Ge [15] and W/Si [25]. Layered structures of metals and crystalline semiconductors (or electrically active amorphous semiconductors) have not yet been produced but could be of great technological importance. Here, the new generation of "high temperature" MBE machines (incorporating electron beam evaporation sources and, possibly, UHV sputtering guns) is likely to lead to important results.

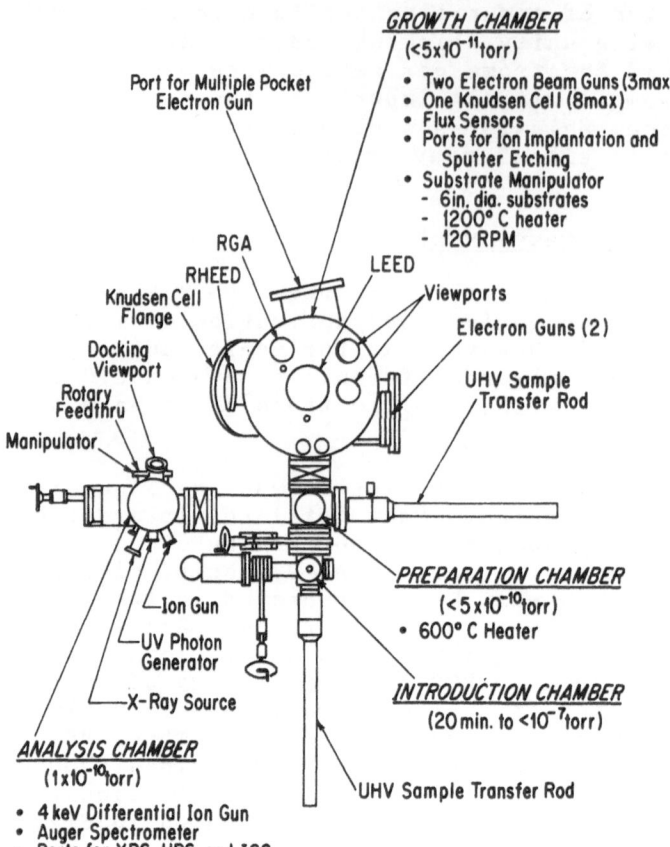

GROWTH CHAMBER
($< 5 \times 10^{-11}$ torr)

- Two Electron Beam Guns (3 max)
- One Knudsen Cell (8 max)
- Flux Sensors
- Ports for Ion Implantation and Sputter Etching
- Substrate Manipulator
  - 6 in. dia. substrates
  - 1200° C heater
  - 120 RPM

Port for Multiple Pocket Electron Gun

RGA
RHEED
Knudsen Cell Flange
Docking Viewport
Rotary Feedthru
Manipulator
LEED
Viewports
Electron Guns (2)
UHV Sample Transfer Rod

Ion Gun
UV Photon Generator
X-Ray Source

PREPARATION CHAMBER
($< 5 \times 10^{-10}$ torr)
- 600° C Heater

INTRODUCTION CHAMBER
(20 min. to $< 10^{-7}$ torr)

UHV Sample Transfer Rod

ANALYSIS CHAMBER
(1×10^{-10} torr)

- 4 keV Differential Ion Gun
- Auger Spectrometer
- Ports for XPS, UPS, and ISS

Figure 8. Schematic diagram of an MBE system for producing metallic multilayers and superlattices [27]

Much of the work on metal/insulator multilayers has been motivated by the possibility of producing efficient optical coatings for the far-UltraViolet and soft X-ray region (the "X-UV" region) [3,4], as described in the chapter by E. Spiller. Both evaporation and sputtering techniques have been used to produce these structures to date. For X-UV optical coatings, work has concentrated on alternating materials of high electron density with those of low electron density, to produce multilayers with the largest differences in optical constants. Amorphous C and B are commonly the low electron density materials of choice, although amorphous Si is very useful for certain X-ray wavelengths. An example of a W/amorphous-Si multilayer mirror designed for operation near 210 Å is shown in Figure 9 [25]. Application of MBE techniques to produce crystalline layers of Si, with improved optical properties in certain regions of the X-UV and the ability to withstand heat loads without recrystallization, should prove important in the future.

To date, no metallic multilayer or superlattice has been found which has the same degree of structural and chemical perfection as their semiconductor counterpart. Crucial

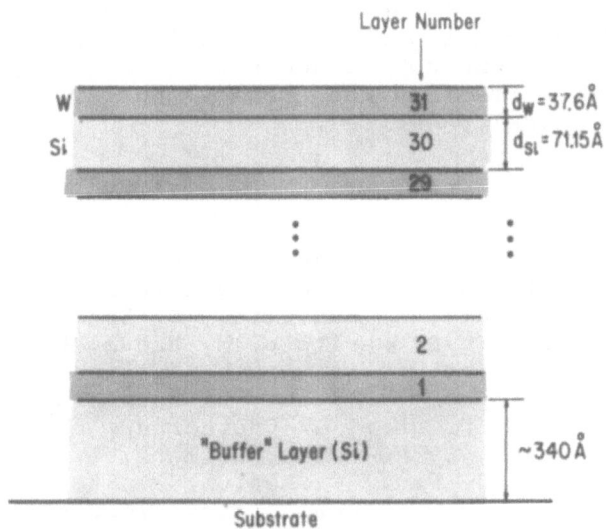

Figure 9. Schematic of a multilayer designed as a mirror for
normal incidence 210 Å X-UV light

factors in determining structural quality are thought to
include the binary phase diagram of the components being
layered (especially their mutual solubility and the presence
or absence of intermetallic compounds), interdiffusion rates,
the existence or absence of lattice matching and compatible
crystalline symmetries, the availability of suitable epitaxial
substrates, ability to control substrate temperature over a
wide range, precise control over the deposition conditions,
etc. Determining which factors affect the formation of
high-quality metallic superlattices or layered structures is
crucial, and is likely to remain an important research problem
for at least the next several years.

SUMMARY

        In summary, several techniques have been developed in
recent years to deposit high-quality metal and metal-
containing superlattices. Highly regular and controllable
layering in a wide variety of metallic systems has been
achieved, and semiconductor-like quality has been approached
in special cases. The precise nature of the interfaces
(lattice match, strains, chemical composition, electronic
structure, etc.) in these metallic systems has not yet been
entirely determined, and will be the subject of future work.
The application of MBE growth techniques, with the possibility
of in situ determination of interfacial structure and
electronic properties, should lead to future improvements in
these metallic materials.

Finally, it should be emphasized that many of the interesting and important effects that will be possible in layered metallic systems do not necessarily require the high degree of crystal perfection and chemical purity characteristic of semiconductors.

ACKNOWLEDGEMENTS

It is a pleasure to acknowledge the contributions of the postdocs and students currently working with me on metallic superlattices: W. Bennett, A. Boufelfel, C. England, F. Fernandez, J. Makous, G. Maritato, M. Nahum, D. Person, D. Schulze and A. Shapiro. This work benefitted greatly from recent collaborations with: J. Aboaf, P. Baumgart, J. Bell, J. Corno, A.M. Cucolo, P. Dhez, G. Güntherodt, B. Hillebrands, A. Khandar, M. Lagally, C. Majkrzak, R. Mock, L. Nevot, F. Nizzoli, B. Pardo, D. Paul, C. Seaton, G. Stegeman, R. Vaglio, B. Vidal, and R. Zanoni. Many discussions over the years with R. C. Dynes, M. Grimsditch, J. B. Ketterson, J. A. Leavitt and I. K. Schuller contributed to this work, and are gratefully acknowledged. Finally, without the technical expertise of Paul Fournier of L. M. Simard and Dr. Peter Chow of Perkin-Elmer in designing state-of-the-art sputtering and MBE systems, none of this research would have been possible.

This research is supported by the AFOSR (URIP Grant 86-0347), AFOSR/ARO/JSOP (F49620-85-C-0039), DOE (DE-FG02-87ER45297), NATO (Travel Grant RG-328/83) and the Optical Data Storage Center at the University of Arizona.

REFERENCES

1. See, for example, various chapters in "Synthetic Modulated Structures," edited by L. L. Chang and B. C. Giessen (Academic Press, New York, 1985).
2. "Interfaces, Superlattices and Thin Films," edited by J. D. Dow and I. K. Schuller (Materials Research Society, Pittsburgh, 1987).
3. See various articles in "Multilayer Structures and Laboratory X-Ray Laser Research," edited by N. M. Ceglio and P. Dhez (SPIE, Bellingham WA, $\underline{688}$, 1986).
4. See various articles in "Soft X-Ray Optics and Technology," edited by E-E. Koch and G. Schmahl (SPIE volume $\underline{733}$, Bellingham WA, 1987).
5. N. J. Sato, Appl. Phys. $\underline{59}$ (1986) 2514.
6. W. R. Bennett, D. C. Person, and C. M. Falco, "Proc. 4th Topical Meeting on Optical Data Storage," (Optical Society of America, in press)
7. M. Tanaka, H. Yuzurihara, and T. Tokita, IEEE Trans. on Magnetics (in press).
8. Data for semiconductors taken from K. Ploog, Chapter 18 in "The Technology and Physics of Molecular Beam Epitaxy," edited by E.H.C. Parker (Plenum, New York, 1985). Data for metals from C. M. Falco (unpublished).
9. W. P. Lowe, T. W. Barbee, T. H. Geballe, and D. B. McWhan, Phys. Rev. B $\underline{24}$ (1981) 6193.

10. J. Makous, C. M. Falco, A. M. Cucolo, and R. Vaglio, "Proc. of 18th Int'l Conf. on Low Temp. Phys." (in press)

11. J. E. Hilliard, in "Modulated Structures – 1979," edited by J. M. Cowley, J. B. Cohen, M. B. Salamon and B. J. Wuensch (American Institute of Physics, New York, 1979), p. 407.

12. N. K. Flevaris, D. Baral, J. E. Hilliard, and J. B. Ketterson, Appl. Phys. Letters $\underline{38}$ (1981) 992.

13. J. Kwo, D. B. McWhan, M. Hong, E. M. Gyorgy, L. C. Feldman, and J. E. Cunningham, in "Layered Structures, Epitaxy and Interfaces," edited by J. M. Gibson and L. R. Dawson (Materials Research Society, Pittsburgh, 1985), p. 509.

14. S. M. Durbin, J. E. Cunningham, M. E. Mochel, and C. P. Flynn, J. Phys. F: Metals Phys. $\underline{L223}$ (1981) 11.

15. S. T. Ruggiero, T. W. Barbee, and M. R. Beasley, Phys. Rev. Lett. $\underline{45}$ (1980) 1299.

16. Q. S. Yang, C. M. Falco, and I. K. Schuller, Phys. Rev. $\underline{27}$ (1983) 3867.

17. B. J. Thaler, J. B. Ketterson, and J. E. Hilliard, Phys. Rev. Lett. $\underline{41}$ (1978) 336.

18. J. Q. Zheng, C. M. Falco, J. B. Ketterson, and I. K. Schuller, Appl. Phys. Lett. $\underline{38}$ (1981) 424.

19. J. F. Dillon, E. M. Gyorgy, L. W. Rupp, Y. Yafet, and L. R. Testardi, J. Appl. Phys. $\underline{52}$ (1981) 2256.

20. J. L. Makous and C. M. Falco, to be published.

21. H. P. Klug and L. E. Alexander, "X-Ray Diffraction Procedures," (John Wiley, New York, 1974).

22. Figure 5 is a schematic diagram based on a commercial sputtering system designed to produce metallic superlattices (L. M. Simard model DH-1).

23. C. M. Falco, J. Appl. Phys. $\underline{65}$ (1984) 1218.

24. W. R. Bennett, J. A. Leavitt, C. M. Falco, Phys. Rev. B (in press).

25. C. M. Falco, F. E. Fernandez, P. Dhez, A. Khandar, L. Nevot, B. Pardo, and J. Corno, Proc. of the SPIE, $\underline{733}$, 343 (1987).

26. I am aware of MBE machines used for growing metallic multilayers at Argonne National Laboratory, University of Arizona, AT&T Bell Laboratories, IBM Almaden and Yorktown, University of Illinois, Johns Hopkins University, and the University of Michigan.

27. Figure 8 is a schematic diagram based on a commercial MBE system designed for silicon epitaxy, with features suitable for producing metallic superlattices (Perkin-Elmer model 433-S).

MULTILAYER STRUCTURES: ATOMIC ENGINEERING IN ITS INFANCY

Troy W. Barbee, Jr.

Chemistry and Materials Science Department
Lawrence Livermore National Laboratory
Livermore, California 94550, USA

ABSTRACT

Thin film deposition processes were understood to be "atom by atom" in nature very early in the development of the appropriate synthesis technologies. This understanding undoubtedly resulted in a leap of imagination to a time when the atomic positions of chemically specific species necessary to achieve a scientific result, or to fulfill a technologically significant need, could be theoretically determined and then experimentally realized--in a word, engineered. Hence "atomic engineering" requires that we know which atoms go in which places for a given reason, and further that we be able to achieve physically the desired result in a systematic and reproducible manner. In this paper a perspective on materials engineering is given as background for a discussion of atomic engineering--now and in the future. The current status of multilayer science and technology will then be reviewed in this context. Important opportunities are discussed and a short summary is presented.

INTRODUCTION

In the traditional fields of metallurgy and materials science as well as chemical engineering, electrical engineering and now applied physics, a primary discipline defined activity has been the development of an understanding of the relationships between properties and microstructure. As used here microstructure refers to the distribution of "groups" of atoms which fill the space occupied by the solid body. The scale of such structures ranges from near atomic to macroscopic, and includes defect microstructures, precipitates in alloys, phase distributions in other multiphase alloy systems, gram size distributions and morphologies in polycrystalline materials, etc. Application areas and properties range over the whole gamut of materials usage in the modern world: mechanical, electrical, magnetic, corrosion resistance, catalysts, and so on.

The understanding we have gained from these disciplinary researches has provided two important perspectives. First, that microstructure control so as to optimize desired properties is possible. Second, that the theoretical models gained from other experimental and theoretical efforts provide a basis for materials design at the microstructure level. These two perspectives form the basis for "materials engineering" in modern technology. Attaining these capabilities, though impressive, has been enabled by equally impressive advances in other fields. A similar relationship will develop in atomic engineering.

Experimental studies of the microstructure/property relationships have been strongly dependent on the tools used to characterize microstructures including x-ray diffraction, transmission electron microscopy, scanning electron microscopy, Auger electron spectroscopy, secondary-ion mass spectroscopy, optical techniques, photoelectron spectroscopies, ESCA, reflection high energy electron diffraction, low energy electron diffraction, positive ion x-ray induced x-ray emission, electron microprobe x-ray fluorescence, high speed digital computational capabilities, and so on. These tools, used on significant samples, have provided a base for the theory from which the engineering design of materials is derived.

In summary, materials engineering has developed as a meaningful activity as the result of a wide ranging array of both scientifically and technologically directed programs. These have generated the prerequisite tools (both theoretical and experimental) needed to establish the data base and methodology we now use. It can be forseen that a similar path will be taken in the development of atomic engineering as an accepted activity and as a viable tool in the advancing science and technology of materials.

ATOMIC ENGINEERING

Engineering in the context used here has two components as stated in the abstract to this paper. Each of these components, design (theory) and fabrication (experiment), are strongly interdependent; at any given moment both benefit from and provide guidance to the other. Atomic engineering is now on the brink of an exciting and productive era based on parallel and correlated theoretical and experimental progress in the study of condensed matter. These activities continue to sustain a clear autonomy as a result of their complexity, of the demands placed on practitioners, and of the current levels of development.

Design (theory) is a developing capability which includes specification of the atomic species to be used, and their positions in a lattice or structure to be formed. Such parameters are determined by the physical property of interest and the synthesis processes chosen. It is expected that the choice of synthesis process will be dictated by the final form of the material desired, the atomic species involved, and the property of interest. These design issues are clearly complex, strongly interactive, and at this time, essentially independent in both their abstract and practical forms.

This methodology is strikingly similar to that implemented in the design of solid state devices and processes necessary to make them physically available. Such devices are based on an understanding of the condensed matter physics of the materials and microstructure engineering to produce devices. It is only in the past decade that coherent design procedures have been computer implemented for such activities. These have met with increasing success and are now important in custom chip or device design and production. What is being proposed here is a step from 500 to 1000 nm level to the 0.2 nm level; a scale change of 2 to 5 x 10^3. This is a formidable task which is already in progress, gaining momentum and direction. I will not comment further on this as a more detailed exposition on theory and its current developmental level in this context is best left to its practitioners.

Fabrication (experiment), as already pointed out, includes many of the same considerations as design (theory) though it is foucussed in a substantially different direction. Generally, the synthesis processes used must be capable of producing thick films of high structural quality and purity. Chemical species distribution on the atomic and microstructural levels which in standard terms are described as uniformity in thickness and composition must also be controllable.

Additionally, if interfaces between either elemental or compound materials are involved, control of interfacial character is desirable. Even more demanding are situations encountered with multilayer structures where multiple interfaces between strongly dissimilar materials only a few tenths to tens of nanometers apart are typical. It is interesting that this characteristic of multilayer structures—multiple "identical" inter-faces—will be a powerful tool in the development of scientific under-standing and design guidelines.

Molecular Beam Epitaxy (M.B.E.) and multilayer structures have resulted in a resurgence of work on thin film growth theory from analyti-cal and computational viewpoints. Also, ranges of behavior not previously explored are being investigated. This new work is providing the initial stages of a framework within which existing data may be organized and which will therefore give guidance to further experimentation. I note that com-putational efforts using molecular dynamic techniques are providing infor-mative results for defects in compounds, surface structures, and the early stages of film nucleation and growth. These studies, when coupled with kinetic models and arguments, give a unique view of thin film formation. It is important to recognize that many of these thin film growth theories include inter-atomic potentials of the two- and three-body types. Chemical specificity and atomic position are inherent in the computational work, and are included in a rather global sense in kinetic approaches. Hence, we now begin to see the basic tools needed for the process design segment of Atomic Engineering in the early stages of development.

This discussion of atomic engineering has been intended to establish a point of view important in what is to follow, rather than as a technical evaluation and background on the subject, thus, the lack of references. Interestingly, this descriptive approach is probably appropriate since an organized area of research and technological development termed "atomic engineering" has not been previously described. The technical discussion in the following will center on multilayer structures, but will wander far afield in an attempt to construct the initial parts of a framework in which future efforts may find both guidance and meaning.

MULTILAYER STRUCTURES

The general concept of multilayer structures is at present well accepted as the ability to synthesize such materials for both science and technology has been demonstrated at an increasingly large number of labor-atories. That such structues are a step toward atomic engineering is not as apparent a concept. These new structures represent control of compo-sition in one dimension, typically parallel to the growth direction. Most work to date has been with layered structures in which the atomic interactions of the individual layer constituents have been primary in determining the structure and properties of those layers. This emphasis has developed since the search has been for physical property changes and structural changes directly attributable to the layering, i.e. a layering effect. A hierarchy of layering structure has now, to a limited extent, developed.

This hierarchy ranges from the deposition of single layers of a given material down to layers only one or two atomic planes thick. The structural region of interest in this paper ranges from layers approximately five atomic planes thick to individual layers only fractions of one atomic plane in average thickness. Thus, layers of submonolayer coverages represent a striking untapped potential at this point. Of course, thicker layers will continue to offer continually broadening opportunity, as the physical property-layering structure relationships are better understood.

I note that the emphasis here joins Molecular Beam Epitaxy technology and general multilayer technology in a manner not previously expected. Much of MBE work is with materials for which the condensation processes are atom specific, i.e. sticking coefficients are zero for coverages greater than one monolayer. This natural selectivity allows atomic layer growth under appropriate conditions by sequential exposure of a surface to different species. For the majority of elements sticking coefficients are not as coverage dependent and physical control of the atoms striking the sample surface is required. Multilayer technology now gives us this capability.

In the following a limited technical review of multilayer synthesis, structure and characterization in the context of atomic engineering is presented. The potential and opportunities offered are then explored and research opportunities discussed. A short summary is then given.

TECHNICAL DISCUSSION: MULTILAYERS

A striking growth of the field of multilayer research has occurred over the past decade. This is reflected first by this conference, the continuing International Conferences "Superlattices and Microstructures" initiated in 1984, the bimonthly journal Superlattices and Microstructures (published by Academic Press), the Symposium titled "Mulitlayers: Synthesis, Properties and Nonelectronic Applications" to be held at the Materials Research Society Annual Meeting in December 1987, the topical symposia on multilayer x-ray optics at the SPIE--The International Society for Optical Engineering--annual meetings of 1985, 1986, and 1987, the 3rd International Conference on Modulated Semiconductor Structures (MSS-III), July (1987), Montpellier, France, and the book Synthetic Modulated Structures, edited by L. Chang and B.C. Giessen (Academic Press, New York, 1985). In this section a discussion of multilayer synthesis is presented first. Structural characterization is then discussed and issues pertinent to this paper considered.

The most successful processes for multilayer synthesis at this time are vapor condensation based and have been reviewed for the general case of metal multilayers by T. W. Barbee, Jr.[1] and for x-ray optics by T. W. Barbee, Jr.[2,3] and E. Spiller[4]. Molecular Beam Epitaxy (MBE) is discussed in detail elsewhere in this volume and will not be pursued further here. Several other techniques including pulsed flow chemical vapor deposition[5], electrochemical deposition[6], primary ion beam deposition[7,8], secondary ion beam deposition[8], and pulsed laser plasma source deposition[9] have been reported. In the following a general outline of multilayer synthesis processes is given and important features noted.

Formation of a multilayer structure by physical vapor deposition requires the sequential deposition of two materials. There is, therefore, a common time sequence to the processes dictated by the structural parameters of the deposit and the synthesis technology used. Two specific procedures have been used. In one, the vapor beams are interrupted by shutters. In the other, the sample is moved through independent vapor beams in a controlled manner.

The time sequence is shown schematically in Fig. 1 with t=0 defined as the start of deposition of the first layer on the substrate. Negative time runs back through an "in-situ" substrate processing, high vacuum pump down, roughing, and mounting of the substrate in the system. Positive time runs through the layered microstructure deposition synthesis process, termination of layering, venting of the system, and removal of the substrate from the system.

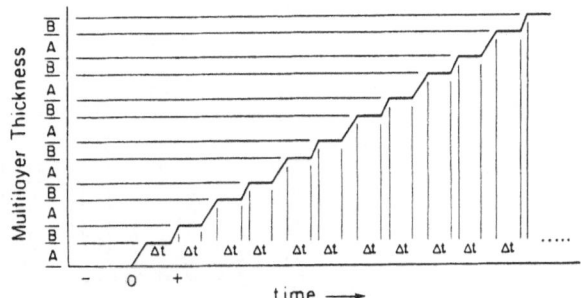

Fig. 1. Schematic of the multilayer thickness-time relationship. The time Δt is that between deposition of layers.

Although only the multilayer deposition synthesis process is discussed in this section, it is important to understand that all these components are integral to the full synthesis process. In Fig. 1 the time when no deposition is occurring is shown as Δt and corresponds to a shuttering effect or to the rotation time between sources for a moving substrate case. This time segment is of potential importance since contamination of the multilayer from the system ambient can occur, thereby affecting the multilayer quality and physical properties.

It is important to note that in the static substrate-shuttered case the deposition rate for any given layer is essentially constant for the period of deposition of the layer. In a rotating sample case the substrate sweeps by the deposition sources, intercepting the sputtered or evaporated atom flux at a range of distances from and orientations relative to the sources. Therefore, the deposition rate varies monotonically from zero to a maximum and back to zero as the substrate passes the source. The angle of incidence of the depositing atoms varies from a value smaller than approximately 30° to the deposition surface at low deposition rates, to normal incidence at the highest deposition rates, and then back to low angles of incidence. This qualitative difference can affect the uniformity and structure of the layers deposited and should be clearly recognized.

An effective framework for the discussion of multilayers can be developed by systematic construction of a hypothetical multilayer. It is then possible to define the range of structural and physical characteristics of each component, specific combinations of components, and the full multilayer structure. Figure 2 is a schematic diagram of the substrate, component layers A and B, interfaces between the substrate and layers A or B, interfaces between layers A and B, interfaces of layers A or B with ambient, and the full multilayer structure. Implicit in this approach is a temporal scale with time equal to zero at the substrate and increasing as the multilayer is synthesized in a manner consistent with the synthesis process.

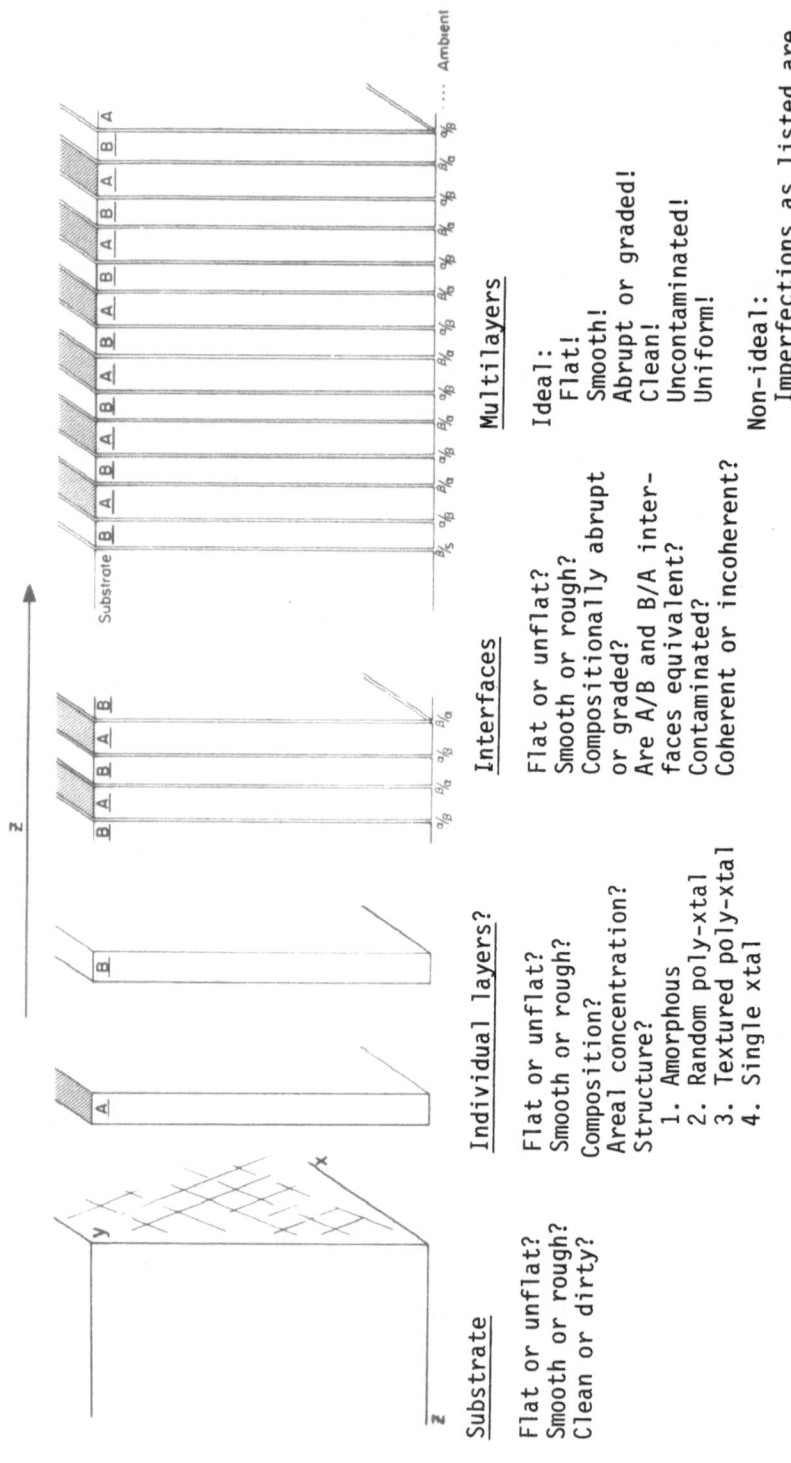

Fig. 2. The components comprising a multilayer film are schematically shown and their interrelationships are listed.

Included in Figure 2 is a listing of statements and questions defining important characteristics of the components of a multilayer. The terms clean and contaminated indicate the presence of undesired and often uncontrollable species on the deposition surface starting with the substrate. Note that the deposition surface is present during the full synthesis process. Such contaminants may therefore be included in the depositing layers or, as a result of the time sequence of layering, be more concentrated at the interfaces between layers.

The individual layers have a specific set of properties directly related to bulk forms of the particular materials (A and B). The primary questions are as follows: What are the compositions and the structures of the layers? Is the number density of atoms in a given layer per unit of surface area the same for all unit areas or in accord with multilayer design? The range of unit area starts at the level of the effective surface area of single atoms and increases to the full area (scale) of the multilayer. The concept of areal atomic concentration in a layer is one way to describe formally nonuniform thickness independent of roughness or flatness. Specific synthesis questions relate to the film nucleation and growth behavior, as deposition of material A onto a substrate or layer B may differ substantially from deposition of B onto a substrate or onto layer A.

Interfaces within the multilayer have the additional characteristic that they may be compositionally abrupt or compositionally graded over some distance in the growth direction. Compositionally abrupt means that composition changes, A to B, occur over one atomic distance normal to the interface. Such abruptness does not preclude roughness, although it may be experimentally difficult to differentiate rough compositionally abrupt from compositionally graded interfaces (rough or smooth). Interfaces are specifically described as layer A onto the substrate or to ambient (α/S or α/A, respectively), layer B onto the substrate or to ambient (β/S or β/A, respectively) and layer A onto layer B (α/B). This level of detail is used because the equivalencies of α/S to β/S, α/A to β/A, or α/β to β/α are not known.

Multilayer structures represent a composite of all the specific characteristics of the individual component elements. It is, in principle, possible to describe the structure of a multilayer by specifying each of the particular features as previously delineated for each appropriate position in the structure. Since this is not easily done, the approach of preference will entail statistical averages over the structure as a whole. Such averages will be correlatable with experiment, as the observed responses average over the structure active in the specific characterization measurement. Additionally, since the observed responses cannot be, in most cases, unambiguously traced to a specific structural feature of a multilayer, care must be exercised when interpreting experimental results.

It is clear from this outline of multilayer synthesis processes that the earliest stages of layer nucleation and growth are of prime importance. This is particularly true when the layering technique is extended so that a macroscopic sample is formed atomic layer by atomic layer. This deposition regime is no longer layering as we have thought of it, but rather a mode dominated by the atomic interactions between what were previously only interfacial atoms. It is under these experimental conditions that atomic engineering to form unique compounds or alloys, which may or may not be compositionally layered, is clearly defined.

Theory for this domain, though not specifically developed, may be gathered in several books and review articles. More classical approaches are described by J.A. Venables and G.L. Price[10], B. Lewis and J.C.

Anderson[11], and R.W. Vook[12]. Selected computer experiments made using Monte Carlo and molecular dynamics techniques which modeled the statistical physics of surface or interface phenomena are reviewed by F.F.Abraham[13]. Although the more classical approaches provide general guidance, I believe the computational simulations will have a dramatic impact in this work. This will require the development of more sophisticated two- and three-body interatomic potentials for most of the elements in the Periodic Table.

It is clear that in spite, of our lack of theoretical guidance we have been able to form multilayer structures containing most of the elements of the Periodic Table with intent. This is shown in Figure 3, where I have indicated the elements included in multilayers with intent--a quite impressive summary.

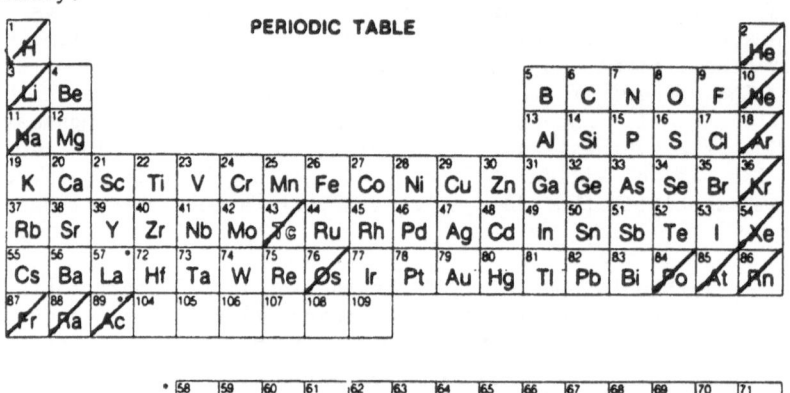

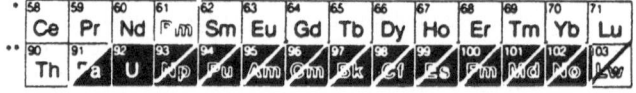

Figure 3. Periodic Table annotated to show the range of elements included in multilayers with intent.

Multilayer characterization may focus on any one of the structural characteristics listed in Figure 2, may be concerned with the parametric relationship between some property and multilayer period, or may be specifically directed to synthesis process-structure-property relationships. All the tools listed as part of the discussion of materials engineering are applicable in multilayer or atomic engineering. Multilayer synthesis is now known to benefit from in-situ continuous monitoring of scattered soft x-ray, optical or electron signals. E. Spiller[4] has demonstrated the effectiveness of monitoring x-ray reflectivity as a multilayer optic is synthesized. This allows corrections for errors in individual layer thickness as well as monitoring growth surface roughness at the 0.1 nm level. More recently optical ellipsometry has been used[13] to monitor layer thickness as multilayer optic structures are grown. Control at the 0.2% per layer is possible with this technique. They also demonstrated conclusively layer smoothing with increasing layer number. It is interesting that removal of the ellipsometric control and good stabilization of sputter source power supplies and mechanical motion, decreased the layer drift from 0.2% to approximately 0.02%. Dynamic control in this case is thus limited by the characterization signal sensitivity.

By way of contrast RHEED has been demonstrated to provide oscillatory signals having periods directly proportional to one monolayer growth of GaAs. The 40 keV electron beam was incident at less than 1° grazing and the (100) and (120) azimuth signals monitored. The signal observed is shown in Figure 4 for $Al_{0.41}Ga_{0.59}As$ through 700 oscillations. This allowed growth of bilayer superlattices which exhibited strong superlattice

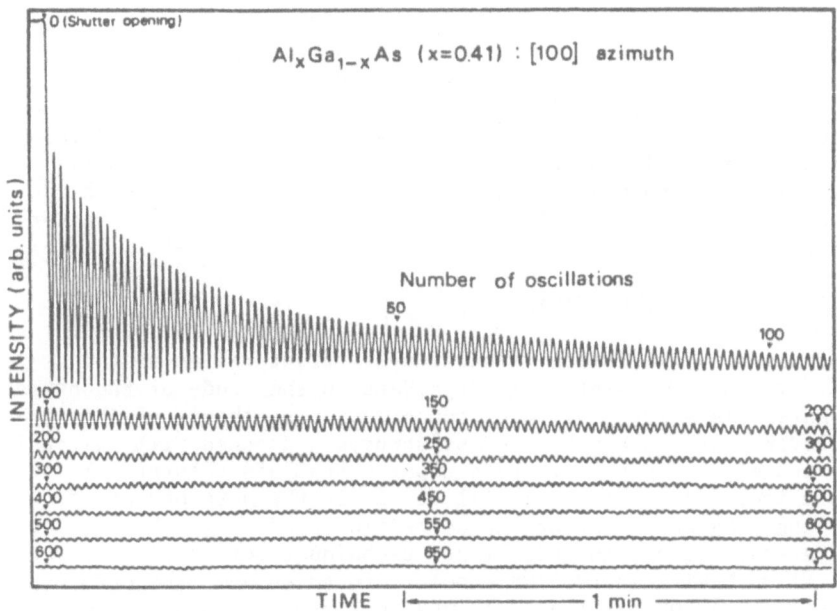

Fig. 4. RHEED intensity oscillations of the specular
beam observed in (100) azimuth of (001) GaAs
substrate during the growth of $Al_{0.41}Ga_{0.59}As$.

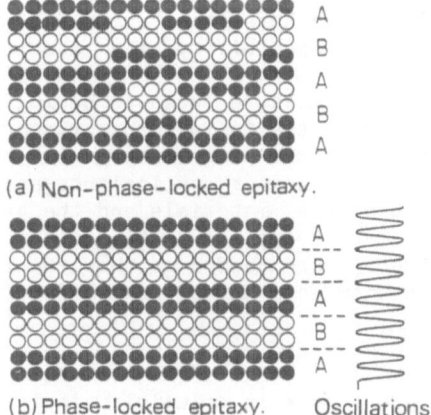

(a) Non-phase-locked epitaxy.

(b) Phase-locked epitaxy.   Oscillations

Fig. 5. Simplified models of the cross-sectional view of superlattice structures
grown by two methods: (a) conventional
non-PLE method (time-measuring), and
(b) PLE method.

Raman scattering effects. A schematic of the structural perfection
attainable is shown in Figure 5. These results are striking and clearly
demonstrate submonolayer sensitivity with what must be very large in plane
coherence.

As is probably apparent from this discussion it is my belief that
the highest priority characterization need is for in-situ monitoring
techniques primarily for structure evaluation. Such techniques should have
broad dynamic range and fractions of a monolayer sensitivity. The development of such a technique is an opportunity: a formidable opportunity.
Additionally in-situ techniques allowing monitoring of parameters characterizing electronic structure would enable engineering of this characteristic of materials.

OPPORTUNITIES AND FUTURE DIRECTIONS

The opportunities and future of multilayer structures as a step
toward atomic engineering is strongly dependent on the study of the basic
physics and materials science of vapor condensation synthesis of thin and
thick films having perfect equilibrium structures and nonequilibrium atomic
arrangements and microstructures. Both experimental and theoretical
components are included. This work will result in the development of
deposition phase diagrams relating the deposition product to the synthesis
process parameters, the development of new techniques for atomic layer
synthesis of films having controlled atomic structures and microstructures,
the development of unique materials for physical property studies and
technological applications, and the ability to design synthesis processes
to produce atomically engineered materials having specified properties.

As previously demonstrated, advances in materials synthesis techniques by vacuum vapor condensation processes have made possible the formation of both highly perfect thin films of equilibrium phases and, more
important, nonequilibrium phases and structures of dramatic diversity.
Such diversity presents a broad ranging set of opportunities for the
materials synthesis practitioner as well as for the general scientific
and technological communities involved in all aspects of the condensed
matter science and technology. Research in this area now has the potential
of providing capabilities which have here been termed "atomic engineering"
and that are based on process reproducibility at an atomic level.

These advances have, in large measure, been technological in nature;
theory and modeling have until recently followed experiment. There have
not been internally consistent research programs in which both experiment
and theory have interacted so as to foster new insights and significant
advances in both the formation of real materials and the predictability
or guidance as to process parameters applied. It is clear that both
activities, experiment and theory, are needed so that the interplay between
current understanding and new knowledge may be most effectively developed.

RESEARCH OPPORTUNITIES

The vapor condensation process is in principle simple, but in practice
complex. The simplicity lies in the limited number of intrinsic process
variables basic to the phenomenon. The complexity originates from a wide
ranging set of extrinsic variables which are largely determined by the
technological approach implemented. Irrespective of such complexity,
several basic questions may be asked that are generic to materials research
as opposed to questions that are answered by means of materials research
but are generic to other disciplines. Materials research, in the broadest
sense, deals with structure-property relationships. Progress is made by
advances in the ability to control structure and thereby properties.

Questions to be addressed include:

1. What are the factors important in determining the nature of the deposition products in vapor deposition experiments?

2. Can a separation of variables be accomplished so that both experimental and theoretical isolation of important effects is possible and improved understanding and control thereby developed?

3. What tools are either needed for or appropriate to research which will be effective in the pursuit of the questions asked above?

Developing answers to the above questions will allow definition of DEPOSITION PHASE DIAGRAMS which are alloy system specific and which result in more general scaling relationships. As an example, directionally bonded low-coordination-number elements can be vapor quenched into the non-crystalline state onto substrates at room or higher temperatures. Several rather complete studies of the effects of substrate temperature, substrate cleanliness, and deposition rate on resultant structure, have been reported for homoepitaxy of sputter deposited germanium. In Figure 6, reproduced from a review of epitaxial film growth by sputtering by Francombe[16], the relationships between deposit structure, deposition rate and substrate temperature are shown. These results can be termed a Deposition Phase Diagram and provide, with sufficient characterization of the process parameters, the information needed for development of basic understanding of the synthesis process structure-property relationships. Experimental results of this nature are very limited and basically exist only for a few elemental materials.

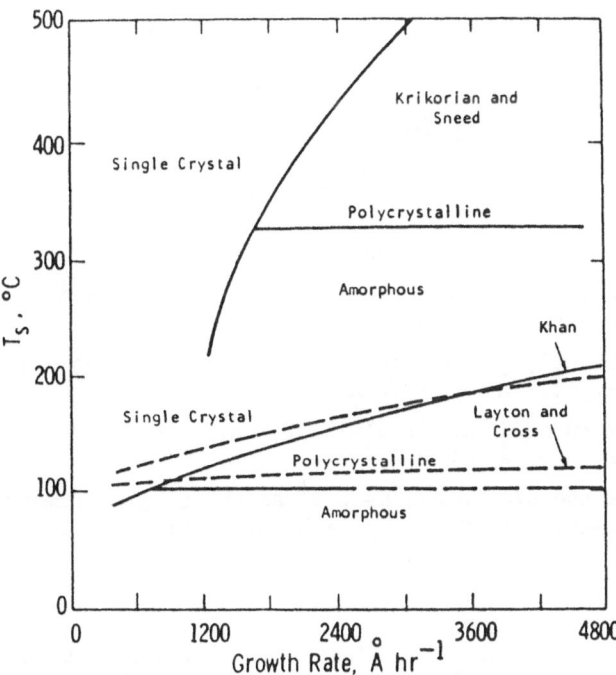

Fig. 6. Effect of film growth rate and substrate temperature T_s on the structure of Ge films sputter-deposited on Ge(111) substrates.

Another pertinent experiment is one[17] on multilayer niobium-zirconium synthesized using sputter deposition onto native oxide coated (100) single crystal silicon substrates. In Figure 7 powder diffraction scans taken using 1.5418 Å radiation are logarithmically plotted over the angular range in two theta where the zirconium (00.2) and niobium (110) reflections lie. Side bands resulting from both the composition and lattice parameter modulation are seen and were analyzed to give the periods shown in Figure 7. Read Camera X-ray photographs taken at a glancing angle

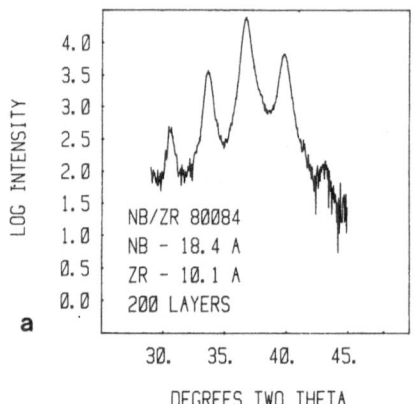

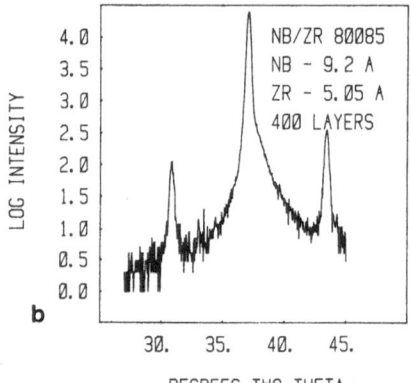

Fig. 7. Powder diffraction scans taken using 1.5418 Å x-rays from niobium-zirconium multilayer samples:

a) 80-084, Theta (110) = 18.33 deg., 200 layer pairs, Zr-10.1 Å, Nb-18.4 Å.

b) 80-085, Theta (110)=18.58 deg., 400 layer pairs, Zr-5.05 Å, Nb-9.2 Å are shown.

Note the side bands as well as the single primary asymmetric (110) Bragg peak which shifts in angular position.

of 10° are shown in Figure 8. The fine single crystal pattern is from the silicon substrate. Note the increased perfection as the multilayer period is decreased--the film having the 14.25 Å period is a mosaic single crystal. This is a unique result as the substrate was amorphous and the substrate temperature less than 50°C. Also, note that reflections from planes

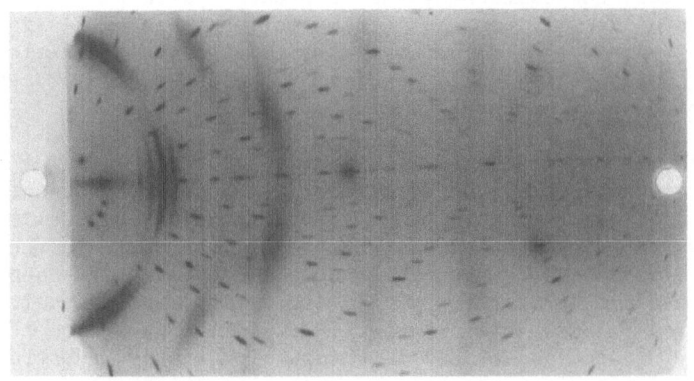

A. 80-084, 200 layer pairs, Zr-10.1 Å, Nb-18.4 Å.

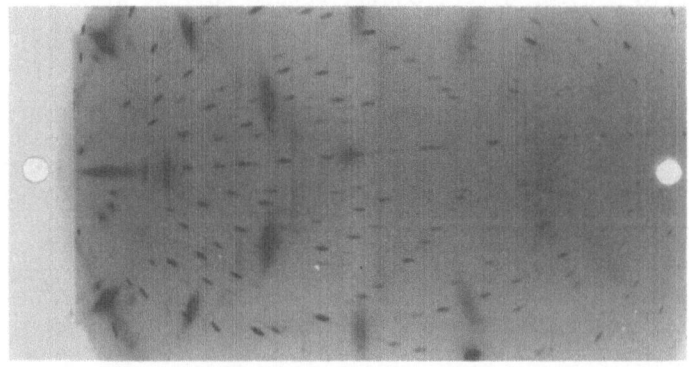

B. 80-085, 400 layer pairs, zr-5.05 Å, Nb-9.2 Å.

Fig. 8. X-ray Read Camera diffraction patterns taken
at an incidence angle of 10 deg. are shown.
Note the significantly increased crystalline
perfection of the smaller period sample (80-
085), both normal to and in the plane of
deposition.

not in the plane of the deposit (200 and 211) also exhibit side bands indicating significant in-plance coherence. Peak broadening analysis for internal strain indicated that the strain in the (110) direction was $\Delta d/d-0.0136$ for the 28.5 Å period sample and $d/d-0.0026$ for the 14.25 Å period sample. No evidence for particle size broadening was found.

It is important to recognize that, in a thermodynamic sense, it is expected that there will be a unique, definable relationship between the deposition process and the deposition product which may be explored by study of the structure-composition-synthesis process parameter relationships. Results of these studies will provide a basis for engineering design of processes for the synthesis of technologically significant materials as well as providing guidance for basic studies of the deposition process and the development of Deposition Phase Diagrams.

Intrinsic process parameters include the equilibration time of an adatom at the deposition surface, the effects of the heat of condensation on local atomic and macroscopic sample bases, the effects of chemical interactions between wanted atomic species at the deposition surface, the allowable free energy differences between metastable and equilibrium deposition products, atomic surface mobilities, atomic deposition rate, deposition surface temperature, bulk substrate temperature, adatom-substrate atomic interaction energy, adatom-adatom atomic interaction energy, substrate orientation and atomic surface morphology, and interaction effects of substrate atomic structure and adatom film atomic structure. These parameters are typically those considered in analysis of thin film growth under ideal conditions in which classical homogeneous nucleation theory is most applicable. They also apply to experimental domains in which non-crystalline deposits are formed for directionally bonded solids, atomic layer growth is observed, and intermediate ordered phases are observed at the substrate-deposit interface.

Extrinsic synthesis process parameters are vapor source-deposition system specific. These include vapor source-deposition surface coupling, the energy distribution of the incident adatoms, the ambient atmosphere in the deposition system, the geometry of the vapor-source substrate configuration and substrate morphological characteristics. It is important to note that these extrinsic parameters are typically assumed to have the greatest effects in producing films of undesirable structure or properties. This is not always the case, though experience indicated that significant deleterious effects are caused by these parameters.

As the fundamental understanding of the bulk properties of condensed materials has deepened, attention has increasingly been focussed on interfacial and surface phenomena. The wide variety of scientifically interesting and technologically important phenomena involved make this an attractive field of research. In particular, thin film synthesis provides a direct challenge to predictive modeling. The basic issues in the experimental approach revolve about the mechanism (kinetics) of vapor quenching and the nature of the adatom/substrate interactions which lead to metastable phases. These issues can be addressed by appropriate surface modeling and computer simulations.

SUMMARY

The unique atom specific building block nature of multilayer synthesis science and technology has been emphasized in this paper. I have conceptually defined the process in terms of the sequential deposition of "atomic" layers of elements A and B to form the compound A_xB_y. The areal densities of atoms in sequential layers will ratio as x/y. The specific areal densities of A and B will be directly related to those expected for particular atomic planes in the crystal structure sought with orientation (h k l). Generally, the specific areal densities of atoms x and y will be submonolayer relative to elemental deposits of A or B, but not necessarily.

The salient experimental features in this approach are the ability to sequentially expose a substrate to fluxes of atoms A, B, C, D....; the ability to control these exposures so that they are specified at a defined level (atoms/cm^2), the ability to control the exposure so that all layers of A are identical (etc.), the ability to control the synthesis environment so that only intended atomic species are incorporated into the films, and the usual requirements for substrate heating, substrate cleaning, and so on. What I have described here is multilayer structure science and technology extended in a very real sense into the domain of Molecular Beam Epitaxy. In doing this we extend the range of elements of interest to the breadth of the Periodic Chart, and the complexity of the materials to compounds of as high an order as of interest--binary, tertiary, quaternary....

Three specific binary systems have been synthesized in this way: copper and chlorine[17], niobium and carbon[17], and palladium and aluminum[17]. In all cases the compounds of interest were formed. In the case of copper/chlorine the desire was to form copperchloride with the metastable zinc blende structure. A rotating substrate was sequentially exposed to copper and chlorine, the copper layer thickness defined by the atom density on (111) planes. Chlorine gas exposures were empirically defined by the stoichiometry of the resulting films. It was possible to synthesize single crystal CuCl in the metastable zinc blende structure on a routine basis. I also note that the equilibrium phases were synthesized with ease.

In the case of niobium/carbon the niobium and carbon layers had areal densities required to form the sodium chloride B2 structure. Highly textured (100) oriented films having superconducting transition temperatures expected for materials formed by high temperature processing (T>1500°C) for the stoichiometries deposited were observed. Note these films were deposited onto ambient temperature substrates.

Palladium/aluminum films were deposited with atom ratios defined by the equilibrium compounds in this binary system: $PdAl_3$, Pd_2Al_3, PdAl, and Pd_2Al. These were all successfully synthesized and have been used in photo-electron spectroscopy studies. It is interesting that PdAl which has an average atomic number Z=29.5 had essentially the same reddish cast as copper Z=29. The excitement here is that compounds of interest were synthesized for experimental investigations designed to study electronic structure using multilayer structure synthesis techniques.

CONCLUSIONS

The threshold I am describing is one placing more of the responsibility of creativity on us as scientists. We are now no longer limited to the combinations of elements allowed (by nature) to form structures dictated by the small differences in forces between atoms and free energies in condensed matter, that define "equilibrium form". This breadth of capability demands that such research be both interdisciplinary and transdisciplinary with theorists, computational experimenters, and physical experimentalists both providing guidance for and being guided by their cohorts. In this way we may develop an understanding of nature which, when obeyed, will allow the design and fabrication of solids at the atomic and electronic levels--ATOMIC ENGINEERING.

ACKNOWLEDGEMENT: Preparation of this manuscript was supported US Dept. of Energy through Lawrence Livermore National Laboratory under Contract W-7405-Eng-48.

REFERENCES

1. T. W. Barbee, Jr., "Synthesis of Multilayer Structures by Physical Vapor Deposition Techniques" in Synthetic Modulated Structures, L.L. Chang and B.C. Geissen, eds., Academic Press, Orlando Florida, USA (1985).

2. T.W. Barbee, Jr., "Multilayers for X-Ray Optics", Opt. Eng. 25:898 (1986).

3. T.W. Barbee, Jr., "Sputtered Layered Synthetic Microstructure (LSM) Dispersion Elements" in Low Energy X-Ray Diagnostics 1981 D.T. Attwood and B.L. Henke, eds., A.I.P. Conf. Proc. No. 75, A.I.P., New York (1981).

4. E. Spiller, "Evaporated Multilayer Dispersion Elements for Soft X-Rays" in Low Energy X-Ray Diagnostics 1981 D.T. Attwood and B.L. Henke, eds., A.I.P. Conf. Proc. No. 75, A.I.P., New York (1981).

5. B. Abeles and T. Tiedje, "Amorphous Semiconductor Superlattices" in Phys. Rev. Lett. 51:2002 (1983).

6. Frans Spaepen, Harvard University, private communication.

7. B.R. Appleton, R.A. Zuhr, T.S. Noggle, N. Herbots, S.J. Pennycook, and G.D. Alton, "Ion Beam Deposition" M.R.S. Bulletin 12:52 (1987).

8. James M.E. Harper "Ion Beam Deposition" in Thin Film Processes, J.L. Vossen and W. Kern, eds., Academic Press, New York (1978).

9. S.V. Gaponov, S.A. Gusev, B.M Luskin, and N. Salashchenko, "Long-wave X-Ray Radiation Mirrors" in Opt. Comm. 38:7 (1981).

10. J.A. Venables and G.L. Price, "Nucleation of Thin Films" in Epitaxial Growth--Part B, J. W. Matthews, ed., Academic Press, New York (1975).

11. B. Lewis and J.C. Anderson, Nucleation and Growth of Thin Films, Academic Press, New York (1978).

12. R.W. Vook, "Nucleation and Growth of Thin Films" in Opt. Eng. 23:343 (1984).

13. F.F. Abraham, "Statistical Surface Physics: A Perspective via Computer Simulation of Microclusters, Interfaces and Simple Films" Rep. Prog. Phys. 45:1113 (1982).

14. Ph. Houdy, V. Bodart, C. Hily, L. Nevot, P. Putnam, M. Arbaoui, N. Alehyine and R. Barchiwitz, "Performances of C-W Multilayers as Soft X-Ray Optics: High Reflectivity in the Range 2d=60 to 100Å" preprint from An International Conference on Soft X-Ray Optics and Technology SPIE, ICC West Berlin (Dec. 1986).

15. T. Sakamoto, H. Funabashi, K. Ohta, T. Nakagawa, N.J. Kawai, T. Kojima and V. Bando, "Well Defined Superlattice Structures Made by Phase-Locked Epitaxy Using RHEED Intensity Oscillations" in Superlattices and Microstructures 1:347 (1985).

16. M.H. Francombe, "Growth of Epitaxial Films by Sputtering" in Epitaxial Growth--Part A J. W. Matthews, ed., Academic Press, New York (1975).

17. T.W. Barbee, Jr., unpublished work.

MOLECULAR BEAM EPITAXY OF ARTIFICIALLY LAYERED SEMICONDUCTOR STRUCTURES -

BASIC CONCEPT AND RECENT ACHIEVEMENTS

Klaus Ploog

Max-Planck-Institut für Festkörperforschung
D-7000 Stuttgart-80
FR-Germany

1. INTRODUCTION

Molecular beam epitaxy has become a widely used technique for growing epitaxial thin films of semiconductors, metals and dielectrics by impinging thermal-energy beams of atoms or molecules onto a heated substrate under ultra-high vacuum (UHV) conditions / 1 /. The method provides atomic abruptness between layers of different lattice-matched and lattice-mismatched crystalline materials at their interfaces or heterojunctions. The interfaces between epitaxial semiconductor layers of different composition or doping are used to confine electrons and holes to two or one-dimensional motion. The challenge for the design and growth of artificially layered materials is to minimize scattering from impurities, alloy clusters or interface irregularities so that carriers can move freely along the interfaces. In this article we first discuss the significant factors to attain high-quality MBE growth of crystalline III-V semiconductors, including growth conditions, in-situ analysis and growth control, and dopant incorporation, with a tutorial emphasis. We then describe a few examples for the control of the interface quality by monitoring the oscillations in the intensity of the reflection high energy electron diffraction (RHEED) and for the investigation of structural interface disorder effects in superlattices (SL) and multi quantum well heterostructures (MQWH) by high-resolution double-crystal X-ray diffraction. In the final chapter we present selected examples for the application of interface formation during MBE to modify the bulk properties of semiconductors through bandgap (or wavefunction) engineering by grading the composition both abruptly (on an atomic scale) or gradually.

2. TECHNOLOGY AND FILM GROWTH PROCESS

2.1 History of MBE and General Considerations

The basic ideas of the MBE growth process were developed by Günther / 2 / in 1958 for the deposition of stoichiometric films of compound semiconductors formed from elements of group III and V of the periodic table (e.g. GaAs, $Al_xGa_{1-x}As$, InAs, InSb etc.). The crucial point for achieving stoichiometry is that the constituent elements of III-V semiconductors have largely different vapour pressures so that these materials exhibit considerable decomposition at their evaporation temperature. In this so-called three-temperature method, Günther used the group-V-element source oven, kept at temperature T_1 (e.g. 300 °C for arsenic), to maintain a steady vapour pressure in his static vacuum chamber. The source oven with the group-III-element, kept at a much higher temperature T_3 (e.g. 950 °C for gallium), was providing a flux of atoms incident on the substrate that was critical for the condensation rate. The choice of the substrate temperature was most crucial. This intermediate temperature T_2 had to be increased to a value that allowed the condensation of the III-V compound but ensured that the excess group-V-component which had not reacted was re-evaporated from the substrate surface. Subsequently, in the middle of 1960's, Arthur / 3, 4 / performed the first fundamental studies of the kinetic behaviour of Ga and As_2 species evaporating from or impinging onto GaAs surfaces using pulsed molecular beam techniques under UHV conditions that led him to the understanding of the growth mechanism. These investigations were soon followed by the application of the same technology, now called molecular beam epitaxy, by Cho / 5, 6 / to the growth of thin films for device fabrication. A detailed bibliography spanning the first three decades of MBE has been compiled for Si by Bean and McAfee / 7 / and for III-V compounds by Ploog and Graf / 8 / .

The characteristic features of MBE are (a) the slow growth rate of one monolayer/s that permits the control of layer thickness within nanometers by compiling atomic layer upon atomic layer, (b) the reduced growth temperature (e.g. 550 °C for GaAs) which causes only negligible bulk diffusion at abrupt material interfaces, (c) the ability to produce extremely abrupt material interfaces by the application of mechanical shutters positioned in the path of each molecular beam, (d) the progressive smoothing of the growing surface for most substrate orientations due to the specific non-equilibrium growth mode, and (e) the facility for in-situ analysis to ensure that the desired surface reaction conditions are reached before growth is started and are maintained while the crystal is growing. The resulting capability of molecular beam epitaxy for deposition control at the monolayer level opened the possibility for the growth of a new generation of complex multilayer structures which include artificial superlattices, semiconductor-metal-semiconductor and semiconductor-dielectric-semiconductor composites, as well as metastable structures.

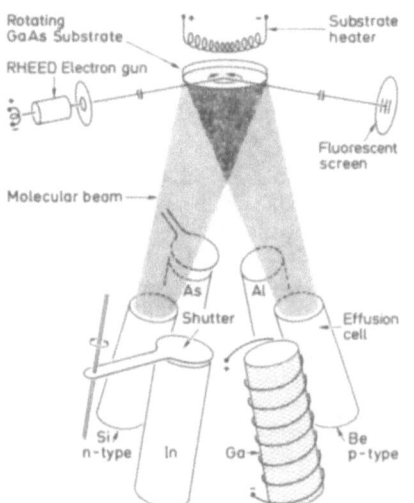

Fig. 1 Schematic illustration of MBE growth process for III-V semiconductors

The schematic illustration in Fig. 1 shows the basic evaporation process for the growth of III-V semiconductors to be carried out under UHV conditions. The molecular beams reacting on the surface of the heated substrate crystal (usually a (OO1)oriented GaAs, InP, or GaSb slice) are generated by thermal evaporation of the constituent elements, contained in temperature-controlled effusion cells. These cells are arranged such that the central portion of their flux distribution intersects the substrate at an orifice-substrate distance ranging from 100 to 180 mm in various MBE systems. Each source is provided with a separate externally controlled mechanical shutter. Operation of these shutters allows the rapid change of the beam species arriving at the substrate in order to alter abruptly the composition of the growing film. At the characteristic MBE growth rates of 0.1 to 1.5 μm/hr the shuttering time is much shorter than the time needed for the growth of a monolayer. Therefore, abrupt material interfaces can be achieved which are not washed out by any bulk diffusion because of the low MBE growth temperatures / 9 / . Uniformity in thickness and composition of the grown film depends on the uniformity of the molecular beams across the substrate and on the uniformity of substrate heating, in particular when growth of III-III-V and III-V-V ternary material is carried out at increased substrate temperature.

The group-III-component for the growth of III-V semiconductors is obtained by evaporation of the respective element which produces only atomic species. The evaporation of group-V-elements, however, produces tetrameric molecules (P_4, As_4, Sb_4), while incongruent evaporation of the III-V compound itself yields dimeric molecules (P_2, As_2) / 10 / . It is not possible to produce monomeric beams of the group-V-elements. The film growth rate and the composition of e.g. $Al_xGa_{1-x}As$ layers is essentially determined by the flux of the group-III-elements, because of their unity sticking coefficient, whereas the stoichiometry is ensured as long as an excess flux of group-V-molecules is supplied (for details see Sect. 2.3). In practice, the III-V compounds are grown with a 2- to 10-fold excess of the group-V-element in order to keep the elemental group-V/III ratio > 1 on the growing surface, also at higher substrate temperatures. The excess arsenic reevaporates from the growth surface. The incorporation of controlled amounts of electrically active impurities into the growing film, for the purpose of doping, is achieved by using separate effusion cells which contain the desired doping elements. The intensities of the molecular beams incident on the growing surface are controlled by the temperature of the respective effusion cells. For typical growth rates of 1 to 2 monolayers per second, the fluxes required at the substrate are approximately 10^{14} to 10^{15} atoms $cm^{-2} s^{-1}$ for the group-III-elements, 10^{15} to 10^{16} atoms $cm^{-2} s^{-1}$ for the group-V-elements, and 10^7 to 10^{11} atoms $cm^{-2} s^{-1}$ for the dopants. These fluxes are produced with cell pressures of 10^{-3} to 10^{-2} Torr for the main constituent elements and correspondingly lower pressures for the dopant elements. Continuous changes in the chemical composition of the growing film are achieved by programmed variation of the cell temperatures. Abrupt changes are obtained by operating the respective mechanical shutters.

2.2 MBE Growth Apparatus

The slow growth rate characteristic for MBE and the obvious film purity requirements necessitate extremely low impurity levels in the growth chamber. The whole operation is thus carried out in a bakeable stainless steel UHV system with a base pressure of 2×10^{-11} Torr which is usually pumped by ion pumps and Ti sublimation pumps. Additional LN_2 cryopanels in the vicinity of the growing crystal keep the partial pressure of gases with high sticking coefficient, e.g. oxygen containing species, even below 10^{-14} Torr. During deposition the total pressure in the growth chamber may rise above 10^{-9} Torr due to scattered beam species and to the excess group-V-element used. However, the UHV condition is always maintained with respect to impurity species. This UHV environment is ideal for surface analytical studies, such as Auger electron spectroscopy (AES), quadrupole mass spectroscopy (QMS) ,and reflection high energy electron diffraction (RHEED), that examine the substrate surface prior to epitaxial growth and provide a high degree of in-situ growth control. In the initial development period, surface analysis performed during deposition played a major role for the understanding of the growth mechanisms of MBE / 11 - 14 / .

The majority of the UHV equipment used for MBE growth until 1977 was home designed and built by commercial UHV manufacturers according to the instruction of the user. It consisted basically of a single deposition-analysis chamber without sample load-lock device. Since 1978 more standardized MBE systems composed of several basic UHV building blocks, such as growth chamber, substrate loading and preparation unit, and optional surface analysis chamber, each with a separate pumping system, have become commercially available. This more expensive but modular design allows the whole instrumentation to be matched easily to more individual requirements. In Fig. 2 we show schematically the top view of a multi-chamber UHV system designed primarily for material growth. The assembly of the effusion cells is positioned such that the center of the beams is directed quasi-horizontally towards the substrate. The system consists of the growth chamber, the sample preparation, and the load-lock chamber, which are separately pumped and interconnected via large-diameter channels and isolation valves. The growth chamber of 450 mm inner diameter (ID) holds a large-diameter removable source flange which provides support to eight effusion cells, their shutters, and the LN_2 shroud. The cells are fixed to individual 38 mm ID flanges for easy refilling and exchanging, and they are spaced and oriented such that their beams converge on the substrate in the growth position. The beam sources are thermally isolated from each other by a LN_2 cooled shroud which prevents chemical cross-contamination. For the growth of reactive Al-containing material ($Al_x Ga_{1-x} As$, $Al_x In_{1-x} As$ etc.) a second cryopanel surrounds the substrate such that the stainless steel bell-jar is internally lined by this cryopanel. This arrangement also reduces contamination due to outgassing of the chamber walls.

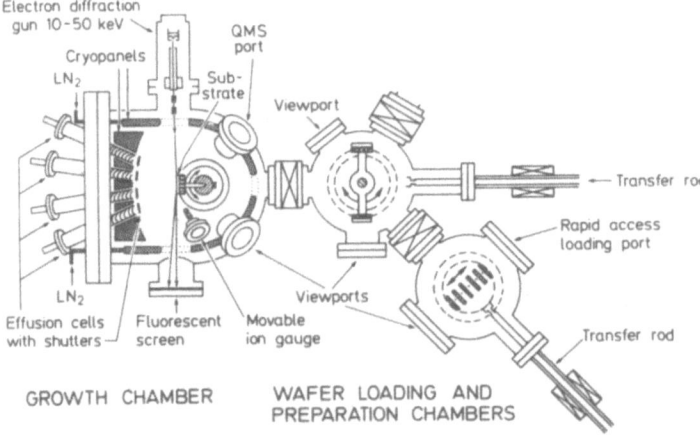

Fig. 2 Schematic diagram (top view) of a multichamber MBE system

In MBE systems used predominantly for continuous film growth, only a few monitoring techniques are required. A RHEED facility with primary energies between 10 and 50 keV, operated in the small glancing angle reflection mode, is used to monitor the structure of the outermost layers of the substrate before and during epitaxial growth. Deviations from established growth patterns can be instantly identified so that adjustments of the operating parameters can be readily made. Originally the RHEED technique was used during MBE growth for the following monitoring purposes: (a) cleaning and annealing process of the substrate surface, (b) initial states of epitaxial growth, and (c) changes of the surface structure during heterostructure growth or when variations of the elemental arrival rate or of the substrate temperature are intentionally performed during growth / 12 / . Recently, temporal intensity oscillations in the features of the RHEED pattern have been employed to study the MBE growth dynamics and the formation of semiconductor heterojunctions / 14 - 17 / . To observe RHEED intensity oscillations, the intensity of particular features (usually the specularly reflected beam) is measured via optical fibre with a well collimated entrance aperture using a photomultiplier. The oscillation period of the specular beam intensity provides a continuous growth rate monitor with atomic layer precision, which is now used for absolute beam flux calibration.

A movable ion gauge is used to measure molecular beam fluxes before commencement of growth by turning it into the substrate position to measure the direct beam density and then out of the beams to measure the background density / 18 / . Although not element specific, the ion gauge beam flux monitor reaches sufficient accuracy for most requirements during MBE growth. In addition to the inexpensive ion gauge, a quadrupole mass spectrometer (QMS) may be used as a true element-specific detector to monitor the species emerging from the molecular beam sources and to control the background-gas composition. The ionizer of the spectrometer must be encircled by a LN_2 cooled shroud, and it should be placed within line of sight of the effusion cells and in close proximity of the substrate position during operation. The QMS is a valuable aid in setting up the initial conditions in a newly installed growth system, because a close correlation of important operating parameters with process results can be readily made.

2.3 Effusion Cells for Molecular Beam Generation

The intensity of the molecular beams is determined by the vapour pressure of the starting materials (elements) and by the evaporation geometry. The design of high-quality molecular beam sources fabricated from non-reactive refractory materials is the most important requirement for successful MBE growth. The evaporation sources must provide stable high-purity molecular beams of the required intensity and uniformity, and they must withstand operating temperatures up to 1400 °C without themselves contributing to the molecular beams.

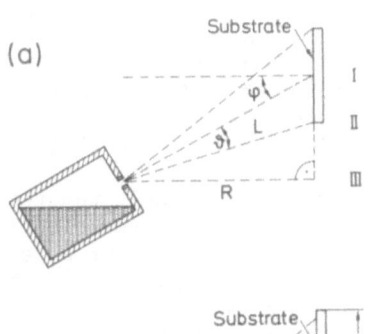

(a)

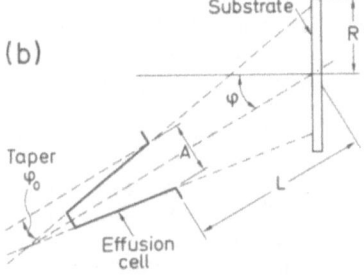

(b)

Fig. 3 Effusion cells for molecular beam epitaxy. (a) Definition of coordinates for the calculation of the flux distribution from a true (equilibrium) Knudsen cell across a non-axially mounted substrate. (b) Nonequilibrium (Langmuir-type) effusion cell used for practical MBE growth

For accurate kinetic studies relevant to the growth process, Knudsen-type effusion cells with small orifice should be used to produce collision-free thermal-energy beams of the constituent elements which are directed to the substrate. When the vapour pressure of the material heated in the cell is such that the mean free path is larger than the orifice diameter, molecular flow of the emerging species is ensured. The beam flux $J(\phi)$ on a substrate at distance L(cm) from the source with angle ϕ between cell axis and the normal of the substrate surface (see Fig. 3a) is then given by

$$J(\phi) = 1.11*10^{22} \ [(AP/L^2)(MT)^{-1/2}] \ \cos \phi \ \text{molecules cm}^{-2} \text{s}^{-1} \qquad (1)$$

where A (cm^2) is the orifice area, P (Torr) the equilibrium vapour pressure in the cell at temperature T (K), and M is the molecular weight. The vapour pressure versus temperature relationship for elemental source materials can be expressed by

$$\log P(T) \underset{\sim}{} \frac{A}{T} + B \log T - C \qquad (2)$$

where A, B, and C are element-specific empirical constants given in Ref. / 19 /. If we take Ga as a typical example, the vapour pressure at T = 1283 K (1010^0C) is 3 x 10^{-3} Torr. Inserting M = 70, A = 1 cm^2 and L = 10 cm, the flux on the substrate is 8 x 10^{14} atoms $cm^{-2} \ s^{-1}$. It is important to note that Eq. (1) is strictly valid only for a true Knudsen effusion cell, i.e. a cell containing condensed phase and vapour at equilibrium and with an ideal orifice of area A in a diaphragm of negligible thickness as effusion aperture. In practice, holes drilled in a lid of finite thickness have to be used as non-ideal orifices with a finite orifice wall. The orifice has thus a somewhat collimating effect on the molecular beam. Herman / 20 / has summarized analytical formulae for the angular intensity distribution of molecular beams from single and multi channel non-ideal orifices. Unfortunately, only very limited experimental data for examination of the accuracy of the calculations are available.

For practical film growth, nonequilibrium (Langmuir-type) effusion cells of sufficient material capacity for reduced filling cycles are used. The cells made of pyrolytic boron nitride (PBN) are of either cylindrical or conical (tapered) shape, and they have a length of about 5 - 10 times greater than their largest diameter. They have a large aperture, i.e. the crucibles have no lid, in order to achieve large epitaxial areas with uniform thickness at the required growth rate of 1 - 2 monolayers/s. Dayton / 21 / has first pointed out that a crucible of straight cylindrical shape acts as a collimator, particularly in the vertical position, so that the flux exhibits a more sharply peaked angular distribution when the charge level falls. A computer model for the effusion process from cylindrically or conically shaped sources was developed by Curless / 22 / and applied to the growth of GaAs. The observed variation of the Ga flux over a 4 cm^2 substrate was in good agreement with the model. The important result was that most of the typical MBE effusion cells yield progressively greater nonuniformity in the films as the charge is depleted. In some MBE systems an effort is now made to balance this variation in uniformity from cell depletion by inclining the rotating substrate with respect to the common central axis of the entire effusion cell assembly. Saito and Shibatomi / 23 / performed a systematic investigation of the dependence of the uniformity of the molecular beams across the substrate on the geometrical relationship between the Langmuir-type effusion cells and the substrate. The authors used conical crucibles having a taper ϕ_0 and a diameter D of the cell aperture, which are inclined at an angle ϕ to the normal of the continuously rotating substrate (see Fig. 3b for illustration). The diameter of the uniform area on the substrate was found to depend primarily on the distance between the substrate and the effusion cell, the taper of the cell wall, and the diameter of the cell aperture. The optimization of these parameters resulted in a reduction of the thickness variation of GaAs and $Al_xGa_{1-x}As$ layers to less than $\pm$ 1% over 3-inch wafers / 23 / .

When using the nonequilibrium (Langmuir-type) effusion cells the formula (1) should be multiplied by a correction factor F which, however, cannot be calculated directly. Therefore, the beam flux emerging from the cell must be monitored intermittantly using the movable ion gauge placed in the substrate growth position. To convert the ion gauge measurements into (approximate) absolute flux densities, Eq. (1) has to be modified by including the ionization efficiency η relative to nitrogen (for which the nominal pressure reading of the ion gauge is calibrated) given by the empirical relation / 24 /

$$\eta \approx 0.6 \ (Z/14) + 0.4 \tag{3}$$

where Z is the number of electrons in the atoms or molecules. The calibration of the ion gauge mounted in the beam path can be made absolute (to $\pm$ 5%) for the group-III-elements by relating the measured ion current to the weight of each element separately deposited on a plate in the beam path. For the group-V-elements, however, only relative calibrations are possible because of non-unity sticking coefficients (an improved calibration procedure for As_4 and P_4 has been described by Foxon et al. / 25 /). However, during practical MBE growth, the ion gauge is mainly used to ensure reproducibility of fluxes, not to measure absolute magnitudes.

Precise temperature stability and reproducibility of the effusion cells are essential. Temperature fluctuations of $\pm$ 1 ^{o}C result in beam flux variations ranging from $\pm$ 2 to 4%. Since the film growth rate is proportional to the arrival rate of the group-III-element, these temperature fluctuations cause a fluctuation in the growth rate of the same order of magnitude. In the case of ternary alloys, such as $Al_xGa_{1-x}As$, $Ga_xIn_{1-x}As$, $Al_xIn_{1-x}As$ etc., these fluctuations produce large variations in alloy composition, e.g. by about $\pm$ 4% if the beam intensity of both group-III-elements fluctuates by $\pm$ 2% each. These fluctuations are deleterious for ternary alloys which require precise control of composition in order to match their lattice to that of the binary substrate. While for the growth of $Al_xGa_{1-x}As$ on GaAs substrates the ternary is closely lattice-matched to the binary substrate for all values of x, growth of $Ga_xIn_{1-x}As$ on InP is only lattice matched if x = 0.47. A variation of 1% in the InAs mole fraction from x = 0.47 causes a lattice mismatch, expressed by the ratio of the difference in lattice constants between the ternary layer and the InP substrate, of $\Delta a/a = 7 \times 10^{-4}$. For high-quality material $\Delta a/a$ should be less than 10^{-3} / 26 / . Therefore, it is necessary to control the alloy composition to better than 1%, which requires an accuracy in temperature control of better than 0.02% for the effusion cells.

In the growth of III-V compounds and alloys, the choice of either dimeric or tetrameric group-V-element species may have a significant effect on the structural, electrical and optical properties of the films, including concentration of majority carrier traps / 27 / , minority carrier lifetime / 28 / , and occupancy of donor and acceptor sites by amphoteric dopants / 29 / . These differences in film properties are related to differences in the surface chemistry of the growth process between dimeric and tetramer species, which are discussed in Sect. 2.6. There are three methods of producing P_2 and As_2. First, incongruent evaporation of the binary III-V compounds yields dimeric molecules / 10 / . Second, the dimeric species are obtained by thermal decomposition of gaseous PH_3 or AsH_3 / 30,31 / which, however, produces a large amount of hydrogen in the UHV system. Third, dimers can be produced from the elements by using a two-zone effusion cell, in which a flux of tetramers is formed conventionally and passed through an optically baffled high-temperature stage where complete conversion to P_2 or As_2 occurs above 900 $^\circ$C. The design of two-zone effusion cells has been described by several authors / 27, 29, 32, 33 / .

Severe problems in the growth of heterojunctions of exact stoichiometry may arise from the existence of transients in the beam flux intensity when the shutter is opened or closed. The temperature of the melt in the crucible is affected by the radiation shielding provided by a closely spaced mechanical beam shutter. Therefore, a flux transient lasting typically 1 - 3 min occurs when the shutter is opened and the cell is establishing a new equilibrium temperature. Several approaches have been proposed to reduce or eliminate these flux transients, including the increase of the distance between cell aperture and shutter to more than 3 cm / 34 / or the application of a two-crucible configuration / 35 / . Elimination of the flux transients is of particular importance for the growth of $Al_xIn_{1-x}As$ /$Ga_xIn_{1-x}As$ heterojunctions lattice-matched to InP substrates.

The PBN effusion cells described here operate at temperatures that range from 250 to 1400 $^\circ$C. In cases where the source material has too low a vapour pressure to be cleanly evaporated by these resistively heated crucibles, electron beam evaporators have to be used. Details of this type of evaporator applied to silicon MBE have been described by Ota / 36 / and by Bean / 37 / .

2.4 Substrate Processing

In general MBE growth of III-V semiconductors is performed on (OO1) oriented ($\pm$ 0.1°) substrate slices 300 - 500 µm thick. The preparation of the growth face of the substrate from the polishing stage to the in-situ cleaning stage in the MBE system is of crucial importance for epitaxial growth of ultrathin layers and heterostructures with high purity and crystal perfection and with accurately controlled interfaces on an atomic scale. We describe in detail the cleaning methods for GaAs and InP, which are the most important substrate materials for deposition of III-V compounds. In any case, the substrate surface should be free of crystallographic defects and clean on an atomic scale, i.e. less than 0.01 monolayer of impurities. The first step always involves chemical etching, which leaves the surface covered with some kind of volatile protective oxide. After insertion in the UHV system this oxide is removed by heating. To avoid surface dissociation of III-V semiconductors, this heating must be carried out in a beam of the group-V-element (P_4/P_2 or As_4/As_2).

Various cleaning methods have been described for (OO1) oriented GaAs, which mainly use chemical etching based on a H_2O_2 / H_2SO_2 / H_2O mixture / 1, 11, 38 / . However, the formation of oval defects on MBE grown GaAs and $Al_xGa_{1-x}As$ surfaces elongated in the < O1$\bar{1}$ > direction and with densities ranging from 10^3 to 10^5 cm^{-2}, has long been a severe problem for practical application. We have recently developed a new method for GaAs substrate preparation which effectively reduces the oval-defect density to less than 10^2 cm^{-2} and allows storage of prepared substrates in air under dust-free conditions for several weeks without any degradation. The reproducible preparation of a contamination-free substrate surface was improved as follows / 39 / : The GaAs wafer is first polished with diamond paste to remove saw-cut damage, followed by etch-polishing on an abrasive-free lens paper soaked with NaOCl solution leaving a mirror-like finish. Then the wafer is simply placed twice in concentrated H_2SO_4 kept at 300 K and stirred ultrasonically. The slice is then carefully rinsed in water to remove all SO_4^{2-} ions (careful ultrasonic stirring may accelerate SO_4^{2-} removal from the surface) and finally blown dry with filtered N_2 gas. The important final step is heating of the wafer to 300 $^\circ$C in air under dust-free conditions for about 3 min. During this heating process a stable Ga-rich surface oxide on the (OO1) GaAs substrate is reproducibly generated to protect the sur-

face from carbon contamination. The passivated wafer is then either soldered with liquid In to a conventional Mo substrate plate or it is fixed to
a special Mo substrate holder designed for direct-radiation substrate heating. After transfer to the MBE growth chamber the surface oxide is thermally
desorbed by heating the substrate wafer to 550 $^\circ$C in a flux of arsenic. When
the desorption process is finished a clear (2 x 4) surface reconstruction
is observed in the RHEED pattern.

The preparation of (001) InP substrates requires extra care during polishing and handling, because InP is much softer than GaAs. In addition,
the thermal cleaning procedure in UHV has to be modified substantially,
since incongruent evaporation of InP starts already at T_s = 365 $^\circ$C while
native oxides on InP are more stable than GaAs oxides / 26, 40 / . After
degreasing the wafer successively in trichloroethylene, acetone, and water,
the saw-cut damage is removed by polishing the substrate on lens paper soaked with 0.5% Br_2/CH_3OH solution to a mirror-like finish. The substrate is
then etched for 10 min in a (3:1:1) solution of $H_2SO_4/H_2O_2/H_2O$ kept at 48°C.
This etch may be followed by an additional 3 min etch in 0.3% Br_2/CH_3OH solution. Finally, the substrate is passivated in water for a short time (about
one minute) and blown dry with filtered nitrogen. The optional oxidizing
Br_2/CH_3OH etch provides only minor carbon contamination while preserving
a protective oxide coverage. In order to remove the surface oxide, the
InP substrate must be heated to above 500 $^\circ$C in UHV, about 140 $^\circ$C its congruent temperature. This would give rise to the formation of In droplets
on the InP surface. Therefore, an impinging flux of arsenic or phosphorus
is required to replace the phosphorus of the InP which is lost by preferential thermal desorption. Heating of the substrate in an As_4 flux of J
$(As_4) = 10^{15} - 10^{16}$ cm^{-2} s^{-1} (or P_4 flux, respectively) for a few minutes
at 505 $^\circ$C is sufficient to completely remove the surface oxide and to provide a clean InP surface / 26, 40 /.

Until recently, most of substrate wafers were soldered with liquid
In (at 160 $^\circ$C) to a Mo mounting plate / 11 / . This practice provides good
temperature uniformity due to the excellent thermal sinking, and it is
advantageous for irregularly shaped substrate slices. However, the increasing demand for production oriented post-growth processing of large-area
GaAs wafers and the availability of large diameter (> 2 in) GaAs substrates
has fostered the development of In-free mounting techniques. Technical details of direct-radiation substrate heaters have been described by several
authors / 41 - 44 / .

Immediately after mounting the etched substrate wafer, the Mo plate
with the freshly prepared substrate is remotely exchanged with a plate holding the processed substrate from the previous growth run using a substrate
exchange load-lock system (Fig. 2). The specimen is transferred between
the chambers by trolley and/or magnetically coupled transfer mechanisms.
In advanced MBE systems, an additional intermediate UHV chamber for storage
of several substrate wafers is added which may contain facilities for further surface treatment prior to epitaxy (ion beam sputtering, preheating
etc.). The Mo mounting plate holding the substrate is fixed by a bayonet
joint to the internally heated Mo heater block which remains always inside
the growth chamber. During sample exchange, the cryopanels in the growth
chamber are held at LN_2 temperature, and the effusion cells are kept at
their usual operating temperature. The sample exchange procedure is completed in a few minutes, and a new growth run can be started without distortion of the growth conditions. The application of interlock systems,
about 7 years ago, has drastically increased film purity and growth reproducibility in MBE technology.

The internally heated Mo block, used for controlled substrate heating
during epitaxial growth, is attached to a special manipulator mounted on

a separate flange which also contains power and thermocouple feedthroughs. This manipulator correctly positions the wafer relative to the sources, heats it to the required temperature, and rotates it azimuthally for good film uniformity. The distance between cell orifices and substrate ranges from 100 to 250 mm in various systems. The temperature of the substrate mounted on the remotely exchanged Mo plate is measured with a W-Re thermocouple which passes through the Mo heater block and is rigidly fixed at a certain distance (e.g. 2 mm) below the surface to which the substrate is fixed. This arrangement does not allow to measure the absolute temperature of the substrate surface accurately. However, it is possible to obtain and maintain a given temperature in a reproducible manner, particularly when the sample exchange device is used, and this is all that is required for successful film growth. A method of calibrating the surface temperature of substrates attached to the heater block uses infrared pyrometry with the correct emissivity setting. For the case of GaAs, e.g., the temperature measured within the heating block and the actual temperature of the substrate surface can be correlated in the range of interest using the well-established temperature of 550 $^{\circ}$C / 11, 39 / , at which the passivating oxide film on GaAs evaporates, and the well-known temperature of 630 $^{\circ}$C / 3, 45 / which indicates the limit of congruent evaporation of GaAs. Both events can be monitored by using the RHEED or - if available - the AES facility.

In modern MBE systems continuously azimuthally rotating substrate holders with a maximum speed of 125 rpm are now used, because even the most sophisticated geometrical arrangement of the effusion cells cannot avoid considerable (> 10%) deviations from layer uniformity across wafers of more than 25 mm diameter / 46 / . When the substrate is rotating about its azimuthal axis during growth with 4 - 10 rpm at 1 μm/hr growth rate, epitaxial films with thickness and compositional uniformity to better than $\pm$ 2% over 50 mm diameter substrates can be achieved routinely. The rotating substrate holder also facilitates growth of high-quality ternary and quaternary III-V compounds which require a close lattice match to the substrate ($Al_xIn_{1-x}As$ or $Ga_xIn_{1-x}As$ on InP, $Ga_xIn_{1-x}P$ on GaAs etc.).

2.5 In-Situ Growth Monitoring by RHEED

In this section we concentrate on the more conventional application of RHEED to monitor the surface symmetry of the various reconstructed surfaces of III-V semiconductors. A more detailed analysis of RHEED patterns provides additional information on surface morphology, disorder and topography. The relation between temporal intensity oscillations in the RHEED pattern and MBE growth dynamics and interface control will be discussed in Sect. 4.1. RHEED provides important information about surface cleaning and proper growth conditions, as its forward scattering geometry makes it fully compatible with the growth process (see Fig. 2 for illustration). It is therefore used in the first few minutes of growth for every run. The availability of this facility in the growth chamber is essential.

A semiconductor surface of the same fundamental crystallographic orientation can have different structures because of the phenomenon of surface reconstruction. The surface would have the (1 x 1) structure if the bulk atom position would be maintained at the surface. However, in general the symmetry at the surface is reduced by reconstruction which results from the rehybridization of bound orbitals of surface atoms in order to lower the free energy of the surface. This reordering of the outermost layers often results in a surface symmetry which is modified with respect to that of the bulk lattice, i.e. the size of the unit cell has a larger periodicity. The notation of Wood / 47 / describes the new primitive cell in terms of the dimensions of the unreconstructed unit cell. In the case of the (001) surface of GaAs there are several reconstructions. The most

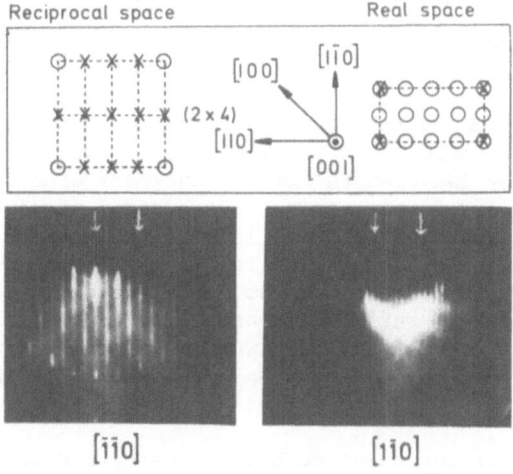

Fig. 4 (a) Real-Space and reciprocal space representation for the (2 x 4)
surface reconstruction on (001) GaAs. (b) RHEED patterns of
the (2 x 4) surface structure taken at different azimuths

stable is the (2 x 4) structure (Fig. 4) where a twofold and fourfold
increase in the periodicity occurs along two orthogonal [110] directions.
The existence of surface reconstruction leads to additional features in
the RHEED pattern at fractional intervals between the bulk diffraction
streaks. Therefore, information contained in the RHEED patterns comprises
(a) the symmetry and periodicity of ordered layers near the surface and
(b) the position of atoms within the unit mesh. The form of the reconstruction of the (001) surface of III-V semiconductors can qualitatively be
correlated to the surface stoichiometry, which is an important growth parameter / 48 - 50 / .

An important aspect of RHEED patterns obtained from atomically flat
surfaces is the appearance of streaks normal to the shadow edge instead of
spots, as shown in Fig. 4. For explanation the thermal diffuse scattering
mechanism was proposed by Holloway and Beeby / 51 / , but so far a true
satisfactory interpretation does not exist. The simplest treatment of this
phenomenon is the assumption that the penetration of the electron beam under conditions of grazing incidence is restricted to the outermost layer
of the crystal surface. In this case only this first layer contributes to
the diffracted intensities, i.e. the usual picture of the reciprocal lattice
point is drawn out into a one-dimensional rod perpendicular to the surface
(in practice, however, the penetration depth of the electron beam is not
totally restricted to the outermost layer, and a model using a modulated
reciprocal lattice rod is probably more appropriate). Using this relaxation
of the third Laue condition, the diffraction pattern can then be visualized
as an intersection of the Ewald sphere, whose radius is large at electron
energies > 5 keV, with a set of reciprocal lattice rods, as shown in Fig. 5.
In this way the spacing between reciprocal lattice rods can be related to
the spacing between rows of atoms in the surface layer. For practical MBE
growth it is further important that, as the topography changes from a flat
to a rough surface, the streaks change from intense, short, and well-defined
to longer, diffused, and spotty, and finally to a pure spotted pattern
arising from transmission of electrons through surface asperities. We have
to realize, however, that the length and width of streaks are not necessarily related to the topography / 50 / .

Even when the kinematic treatment based on the reciprocal lattice rod concept of the Ewald sphere is used, the detailed interpretation of RHEED data from reconstructed III-V semiconductor surfaces during MBE growth has remained controversial. In general, the surface periodicity normal to the beam azimuth can be evaluated from the Ewald sphere construction taking the distance between the streaks on the fluorescent screen, the camera constant, and the energy (wavelength) of the incident electrons. Measurements along two or three different azimuths are often required to determine the surface symmetry unambiguously. For illustration we show in Fig. 4 the real and reciprocal space representation together with the corresponding zero-order Laue zone RHEED patterns at two different azimuths for the most important (001) GaAs-(2 x 4) surface reconstruction. Multiazimuthal measurements become particularly important, if two-dimensional surface disorder or domains exist which are separated by one-dimensional or antiphase boundaries, respectively.

The analysis also of intermediate Laue zones of a RHEED pattern from measurements at a single azimuth is possible in the absence of a pronounced surface disorder. An elegant method given by Hernandez-Calderon and Höchst / 52, 53 / will be described in some detail by means of Fig. 5. The continuous rods of the reciprocal lattice are normal to the surface plane and possess a two-dimensional translational symmetry in that plane. The Ewald sphere intersects all rods contained in the projection of this sphere onto the surface. Diffracted beams are thus observed in the angular directions α and β. The electron wave vector k_O in a typical RHEED experiment is about 10^2 A^{-1}. Using a reciprocal surface vector g, we obtain a characteristic value of g/k_O of about 10^{-2}. Thus, the Ewald sphere touches only a few rods on both sides of the (00) rod. The angle between these reflections is given by / 52 /

$$\beta_O = \tan^{-1} [g_{\perp}/(k_O^2 - g_{\perp}^2)^{1/2}] = \tan^{-1}(t/L) \tag{4}$$

where t is the distance between streaks and L is the sample-to-screen distance. From Eq. (4) we obtain the relation

$$d_{||} = t = L\omega_O \tag{5}$$

as $g_{\perp} \ll k_O$ and $g_{\perp} = 2\pi/d_{||}$, where $d_{||}$ is the distance between equivalent rows of atoms parallel to the incident beam, and where λ_O is the electron wavelength according to

$$\lambda_O = 12.27/[V_O(1 + 0.978*10^{-6}V_O)]^{1/2} \tag{6}$$

where V_O is the electron accelerating voltage. With a fixed energy and geometry we can thus determine surface lattice constants with a precision of 10^{-3}.

The intersections of each row of rods normal to the incident beam are projected on a screen over arcs of circumference. The distance between these arcs is determined by the periodicity in the $g_{||}$ direction. The separation between spots on the same arc provides information on the periodicity in the $g_{||}$ direction [Eq. (4)]. Hernandez-Calderon and Höchst / 52 / have described the determination of the magnitude of $g_{||}$ which allows a detailed analysis of intermediate Laue zones of a RHEED pattern (Fig. 5b). From this analysis the authors were able to determine the surface reconstruction of (001) α-Sn. A more empirical method to obtain the reconstruction of the surface, which is also used in practical MBE growth, is via the measurement of the streak separation under different azimuths, as those shown in Fig. 4.

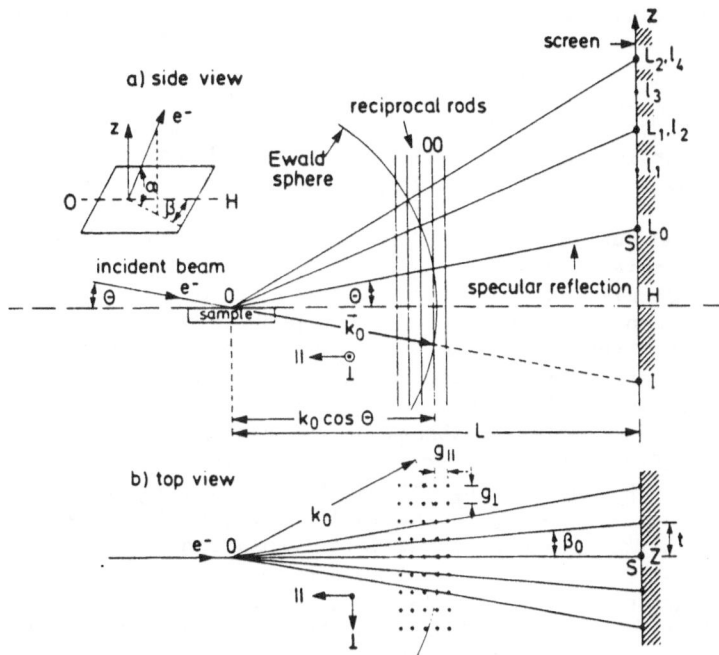

Fig. 5 Ewald construction to depict the origin of a typical RHEED pattern during MBE growth (not to scale). (a) Side view where L_n corresponds to the Laue zones of the ideal surface. The intermediate spots l_n that are present in the case of $g_{||} = \frac{1}{2}g_{||}$ are shown for illustration (b) Top view with a projection of the Ewald sphere on a plane parallel to the sample surface (according to / 52 /)

We have already mentioned that in some cases the length and width of streaks in the RHEED pattern may be related to the topography of the growth surface. The simplest case is surface roughness or aspherities. They usually appear during improper heat treatment of the chemically cleaned substrate on such a scale that the glancing incidence beam produces a transmission diffraction pattern consisting of diffraction spots instead of streaks. At typical electron energies of 20 - 30 keV, the asperities must have a thickness of less than 100 nm in beam direction to allow the formation of transmission patterns. If GaAs is grown on a slightly C-contaminated (001) GaAs substrate, {511} facets are formed during the onset of growth / 54 / . The presence of these facets produces additional streaks in the RHEED pattern which are not normal to the shadow edge due to diffraction from the facet planes. As a result, arrowhead-like features show up in the RHEED pattern with an angle of about 19° between these streaks and the surface normal, which are characteristic for {511} facet planes / 54 / .

In addition to the information on topography, the streak shape of RHEED patterns can be used to evaluate the effects of two-dimensional (2D) surface disorder / 53 / . The average size of ordered surface regions may be restricted due to the lack of perfect ordering in a particular direction (azimuth). In this case the ideal one-dimensional (1D) reciprocal lattice rod becomes two-dimensional, i.e. it forms a solid ellipsoidal cylinder in the Ewald sphere construction. The existence of domains having a strong ordering direction on the surface is indicated by the lengthening or broadening of the fractional and integral order beams in that direction for two orthogonal [110] azimuths, where the domain extension is restricted While the streaks are thus long and broad with the beam parallel to the short domain side, short and narrow streaks are produced when the beam is parallel to the long domain side. Joyce et al. / 53 / pointed out that the domains on the (2x4) reconstructed (001) GaAs surface are separated by one-dimensional (or antiphase) boundaries which give rise to curved streaks in the [110] azimuth. A detailed analysis of the curved streaks in intermediate azimuths allowed the authors to relate the (2x4) and c(2x8) reconstruction occurring on many (001) oriented III-V semiconductor surfaces simply by the surface disorder effect.

We have pointed out in Sect. 2.2 that most III-V semiconductors are grown with a substantial excess of the more volatile group-V-element. For GaAs these As-stabilized growth conditions yield a (2x4) [or c (2x8)] surface reconstruction as shown in the RHEED pattern of Fig. 4. The As-stabilized (2x4) structure is stable over a wide range of substrate temperatures / 55 / . Smooth (001) surfaces with steps down to atomic dimensions can be routinely achieved with these conditions, provided the starting substrate surface is sufficiently clean. If MBE growth is performed with an elemental III/V flux ratio of 1 or at substrate temperatures close to the limit of congruent evaporation, the Ga-stabilized (4x2) surface reconstruction is obtained. Depending on the incident arsenic and gallium fluxes and on the substrate temperature, reversible transitions between the two principal surface structures on (001) GaAs are possible; these involve a change of the surface composition. Several intermediate structures, e.g. (3x1), (1x6), (4x6), (3x6) and mixtures can be observed within very narrow ranges of growth conditions. Whereas the transitions between these structures are not sharp, the final change to the Ga-stabilized (4x2) structure is very abrupt. However, only a very small increase in the Ga flux above the minimum value required to produce the Ga-stabilized structure results in the formation of free gallium on the surface. This means that the practical use of conditions with a Ga (4x2) structure is limited, because it yields samples with dull surfaces and in some cases with evidence of a build up of Ga droplets. The close relationship between reconstruction effects and surface chemistry as well as the correlation with gain or loss of arsenic from the surface has been determined by several authors / 11, 48 - 50 / .

Wood / 56 /, Neave et al. / 14 / , van Hove / 15 / , and Sakamoto / 16 / have discovered pronounced periodic intensity oscillations in the specularly reflected and diffracted beams in the RHEED pattern during growth of GaAs, AlAs, $Al_x Ga_{1-x} As$ etc. (Fig. 6). The period of the oscillations corresponds exactly to the time required to grow a monolayer of GaAs (or AlAs or $Al_x Ga_{1-x} As$) on the (001) substrate surface. A monolayer GaAs is defined as one complete layer of Ga plus one complete layer of As having a thickness equivalent of $a_o/2$. These RHEED intensity oscillations provide direct evidence that MBE growth occurs predominantly in a 2D layer-by-layer growth mode. The changing intensity reflects variations in the step density of each layer when growth proceeds. To a first approximation we can assume that the oscillation amplitude reaches its maximum when the monolayer is completed (maximum reflection). Although the fundamental principles underlying the occurrence of these oscillations and the damping of their amplitude are not completely understood, the method is now widely used to monitor and to calibrate absolute growth rates in real time with monolayer resolution (see Sect.4.1).

2.6 Kinetic and Mechanism of Growth Processes

The growth of binary and ternary III-V semiconductors, especially GaAs, by MBE has been studied by two methods. Modulated molecular beam spectroscopy has been used to investigate the surface chemical processes / 57 / and dynamic RHEED measurements have been used to determine the involved growth mechanisms / 58 /. Additional Monte Carlo simulations of growth / 59 / have improved our understanding of the kinetic processes. From these results detailed kinetic models were established for the growth of GaAs from beams of Ga and As_2 and Ga and As_4 (Fig. 7) / 10, 60 - 62 /. Gallium has a unity sticking coefficient on (001) GaAs at 500 °C. Condensation of As_2 and As_4 occurs only when Ga adatoms are present on the GaAs surface. Dissociation of adsorbed As_2 in a simple first-order process occurs as the molecules encounter single Ga lattice atoms while migrating on the surface. In the absence of free Ga adatoms on the surface, no condensation of As_2 occurs although the species exhibit a measurable surface lifetime. The sticking coefficient of As_2 is therefore a function of the arrival rate of Ga. It is unity only for a monolayer coverage of Ga atoms. Consequently, stoichiometric GaAs can be grown from As_2 with a relative arsenic-to-gallium flux ratio larger than or equal to unity.

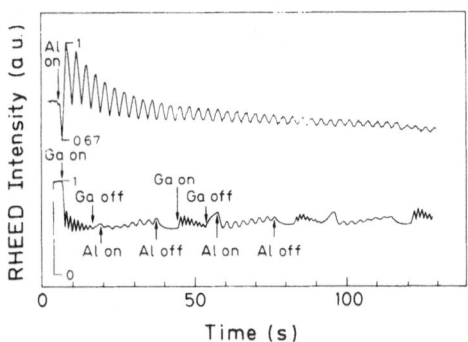

Fig. 6   RHEED intensity oscillation of the specular beam obtained in [100] on (001) GaAs substrate during continuous growth of GaAs / $Al_xGa_{1-x}As$ / AlAs heterostructures

A similar model is valid for the interaction of Ga and As_4 molecular beams with a heated (001) GaAs surface (Fig. 7b). The As_4 molecules, generated from elemental arsenic, are first adsorbed into a mobile precursor state, while the Ga adatom population controls the condensation and reaction of As_4. The sticking coefficient of As_4, however, can never exceed 0.5 and become unity, even when a monolayer of Ga is present on the surface. This crucial behaviour is caused by the complex pairwise interaction process of As_4 molecules chemisorbed on adjacent Ga lattice atoms (second-order process), as indicated in Fig. 7b. Consequently, because of the smaller sticking coefficient, a larger amount of excess As_4 as compared to As_2 is required for growth of stoichiometric GaAs with an As-stabilized (2x4) surface reconstruction. The differences in growth mechanism between the two arsenic species have pronounced effects on the non-equilibrium concentration of native defects incorporated during MBE growth.

The growth models depicted in Fig. 7 imply that almost all Ga atoms incident on the surface are incorporated into the growing epitaxial layer and the growth rate is thus controlled by the Ga flux / 11 / . These models are also valid for other binary III-V semiconductors and to a good approximation also for ternary III-III-V alloys / 63 / . A good compositional control of the growing III-III-V alloy films can be achieved by supplying

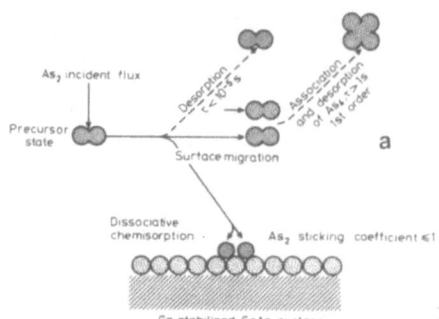

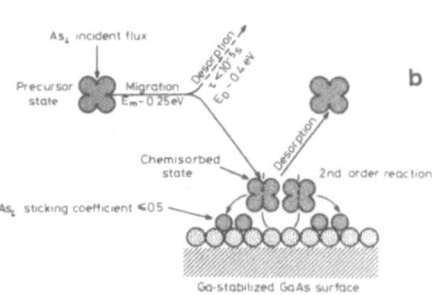

Fig. 7 Models for MBE growth of GaAs (a) from Ga and As_2 species and (b) from Ga and As_4 species, developed by Foxon and Joyce / 61, 62 /

excess group-V-species and adjusting the flux densities of the impinging group-III-beams, as long as the substrate temperature is kept below the congruent evaporation limit of the less stable of the constituent binary III-V compounds (e.g. GaAs in the case of $Al_xGa_{1-x}As$). At higher growth temperatures, however, preferential desorption of the more volatile group-III-element (i.e. Ga from $Al_xGa_{1-x}As$) occurs so that the final film composition is not only determined by the added flux ratios but also by the differences in the desorption rates. To a first approximation we can estimate the loss rate of the group-III-elements from their vapour pressure data / 19 / . This assumption is reasonable because the vapour pressure of these elements over their compounds, i.e. Ga over GaAs, is similar to the vapour pressure of the element over itself / 45 / . The results are summarized in Table 1. The surface of alloys grown at high temperatures will thus be enriched in the less volatile group-III-element. As a consequence, we expect a significant loss of In in $Ga_xIn_{1-x}As$ films grown above 550 $^\circ$C and a loss of Ga in $Al_xGa_{1-x}As$ films grown above 650 $^\circ$C. An intermittant detailed calibration based on measured film composition is thus recommended for accurate re-adjustments of the effusion cell temperatures.

Table 1 Approximate loss rate of group-III-elements in monolayer per second estimated from vapour pressure data

Temperature ($^\circ$C)	Al	Ga	In
550	–	–	0.03
600	–	–	0.3
650	–	0.06	1.4
700	–	0.4	8
750	0.05	2	30

The growth of ternary III-V-V alloy films (e.g. GaP_yAs_{1-y}) by MBE is more complicated, because even at moderate substrate temperatures the relative amounts of the group-V-element incorporated into the growing film are not simply proportional to their relative arrival rate / 25 / . The factors controlling this incorporation behaviour are at present not well understood. It is therefore extremely difficult to obtain a reproducible compositional control during MBE growth of III-V-V alloys when solid phosphorus and arsenic sources are employed.

2.7 Dopant Incorporation

Application of MBE grown films in device structures requires the control of the electrical and optical material properties by the incorporation of small amounts of impurity (dopant) elements from additional effusion cells. The electronic properties of the grown films depend on the flux of the dopant atoms, on their sticking coefficient and on their incorporation behaviour and electrical activity. The required dopant concentration is typically in the range 10^{16} - 10^{18} atoms cm^{-3} or 10^{-6} - 10^{-4} atomic fraction. The most common dopants used during MBE growth have unity sticking coefficients over a large range of growth conditions / 10 / . This behaviour is indicated by a linear behaviour of carrier concentration versus reciprocal dopant cell temperature (Clausius-Clapeyron type plots), which are shown in Fig. 8 for four representative dopants of MBE GaAs. In Table 2 we summarize a few important properties of the most promising dopant elements used for GaAs grown by MBE. Many of these dopants are also suitable for other III-V compounds. Hereafter we will discuss in some detail the characteristics of the four dopants displayed in Fig. 8.

Be is used for p-type doping, while for n-type it is either Si, Ge, or Sn. At doping levels below 1×10^{19} cm^{-3} Be behaves as an almost ideal shallow acceptor in MBE grown III-V compounds / 64 / . Each incident Be atom produces one ionized impurity species providing, in GaAs, an acceptor level 29 meV above the valence band edge. The observed doping level is thus simply proportional to the arrival rate. The low diffusion coefficient of Be in MBE GaAs ensures excellent control of the depth distribution of this p-type

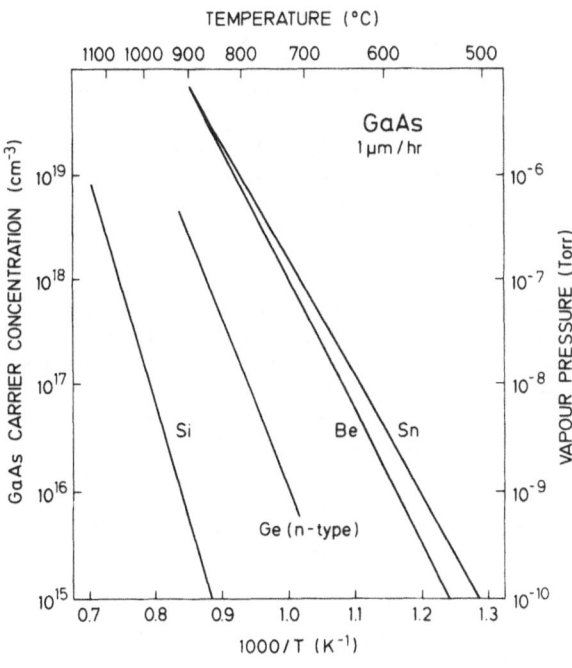

Fig. 8   Room-temperature carrier concentration in MBE grown GaAs as a function of four dopant effusion cell temperatures for constant growth rate, substrate temperature and As_4-to-Ga flux ratio. The data were obtained from Hall effect and capacitance-voltage measurements They are also valid for other III-V semiconductors.

Table 2 Properties of widely used dopant elements in GaAs grown by MBE

Element	Incorporation behaviour	Degree of compensation	Maximum achievable carrier concentration (cm^{-3})	Remarks
Be	acceptor only	low	6×10^{19}	vapour is highly toxic
Si	predominantly donor	fairly low with As-stabilized conditions	1×10^{19}	high source temperature required
Ge	depends strongly on growth conditions	can be high	4×10^{18}	acceptor with Ga-stabilized and donor with with As-stabilized conditions
Sn	donor only	low	6×10^{19}	tends to accumulate at the growth surface
S Se Te	donors only	low	3×10^{19}	molecular sources like PbS, PbSe, PbTe or electrochemical cells for controllable incorporation required

dopant, and abrupt doping profiles can be achieved / 65 / . At doping levels above 5×10^{19} cm^{-3}, however, the surface morphology and the luminescence properties degrade / 66 / and the diffusion of Be is enhanced / 67, 68 / if the samples are grown at a substrate temperature above 550 $^{\circ}$C. Lowering of the substrate temperature to 500 $^{\circ}$C makes feasible Be acceptor levels up to 2×10^{20} cm^{-3} with perfect surface morphology and reduced Be diffusion / 69 / . It is important to note that Be is highly toxic. In addition, the commercially available Be source material is at best 99.99% pure and contains a certain amount of metallic impurities. Unfortunately, no other suitable p-type dopant is available for MBE growth of GaAs and other III-V semiconductors.

The group-IV-element Si is primarily incorporated on Ga sites during MBE growth under As-stabilized conditions, yielding n-type material of fairly low compensation / 65 / . The observed doping level is simply proportional to the dopant arrival rate provided care is taken to reduce the water and carbon monoxide level during growth. The upper limit of $n = 1 \times 10^{19}$ cm^{-3} for the free-electron concentration in GaAs was originally attributed to the enhanced autocompensation, i.e. the incorporation of Si on As sites and on interstitials / 70, 71 / . Recent investigations / 72, 73 / indicate, however, that more probably nitrogen evolving from the PBN crucible containing Si and heated to about 1300 $^{\circ}$C causes the compensating effect. This effec can be overcome by the use of a very low growth rate of about 0.1 μm/hr / 72 / . The possibility of Si migration during MBE growth of $Al_xGa_{1-x}As$ films at high substrate temperatures and/or with high donor concentration has been the subject of controverse discussions, because of its deleterious effects on the properties of selectively doped $Al_xGa_{1-x}As$ / GaAs hetero-structures (see / 74 / and references therein). Gonzales et al. / 74 / provided some evidence that only at high doping concentrations (> 2×10^{18} cm^{-3}) Si migration might occur in $Al_xGa_{1-x}As$ films as the result of a concentration-dependent diffusion process which is enhanced at high substrate temperatures. Additionally, it is important to note that the incorporation of Si atoms on either Ga or As sites during MBE growth depends on the orientation of the GaAs substrate. Wang et al. / 75 / found that in GaAs grown on (111)A, (211)A and (311)A orientations the Si atoms predominantly occupy As sites and act as acceptors, while they occupy Ga sites and act as donors on (OO1), (111)B, (211)B, (311)B, (511)A, (511)B and higher-index orientations. Based on these results, Miller / 76 / and Nobuhara et al. / 77 / could grow a series of lateral p-n junctions on graded steps of a (OO1) GaAs substrate surface.

The group-IV-element Ge is an amphoteric dopant in MBE GaAs, and it can be used to prepare either p- or n-type films depending on the growth conditions / 78 - 80 / . On As sites Ge acts as an acceptor, while on Ga sites Ge acts as a donor. The site incorporation depends thus critically on the arsenic-to-gallium flux ratio and on the substrate temperature. When the surface of the growing GaAs film exhibits a Ga-stabilized (4x2) reconstruction, the Ga atom surface population is increased. The codeposited Ge atoms are incorporated predominantly on As sites, and thus p-doped films result. The major problem with MBE growth of p-type GaAs doped with Ge is the small stability range of the required Ga-stabilized growth conditions. With As-stabilized growth conditions Ge is predominantly incorporated on Ga sites and acts as donor in GaAs. The observed free-carrier concentration versus reciprocal Ge effusion cell temperature plot (see Fig. 8) implies that the degree of auto-compensation of Ge-doped n-GaAs does not depend on the doping level. Owing to the amphoteric nature of the dopant, compensation of highly Ge-doped GaAs films is considerable and the maximum achievable free-carrier concentration is lowered to 4×10^{18} cm^{-3}.

A widely used n-type dopant for many III-V semiconductors in any epitaxial growth technique is Sn which is not amphoteric and acts as a shallow donor only. The measured free-electron concentration is directly proportional to the Sn flux in the molecular beams, and high concentrations up to 5×10^{19} cm^{-3} can be achieved. The major disadvantage of using Sn is its tendency to accumulate at the GaAs surface during growth / 81, 82 / . Wood and Joyce / 83 / observed that the Sn incorporation is surface rate limited. Before a steady-state donor concentration in the growing GaAs film can be achieved, it is necessary to build up a steady-state surface population of Sn, which may be as large as O.1 monolayer. This produces dips in the carrier-concentration profile at the film-substrate interface, which can be overcome only by predeposition of Sn prior to GaAs growth. Consequently, no sharp doping profiles can be realized with Sn doping.

The group-VI-elements S, Se, Te cannot be incorporated during MBE growth by simply evaporating the elements in separate dopant cells. A method to use these elements for MBE GaAs involves synergic reactions using the lead chalcogenides PbS, PbSe, or PbTe / 84, 85 / . A second method for the incorporation of S and Se into GaAs uses the electrolysis of a solid state silver electrolyte / 86 / . At elevated growth temperature (> 6OO $^{\circ}$C), however, the incorporation efficiency of the group-VI-elements in GaAs as well as in $Al_xGa_{1-x}As$ is reduced. To a certain extent this loss of the dopant may be suppressed by increasing the As_4 overpressure during growth.

The discussion of this section has shown that as yet no one element of Table 2 can be considered truly as the best choice for MBE growth. The nature of the required doping profile and the application of the wafer determine to a large extent which dopant element is used.

3. INITIATION OF GROWTH

For III-V compounds, the oxide passivation layer serves as a protection for the chemically etched substrate from atmospheric contamination before epitaxial growth. After mounting the Mo transfer plate with the fixed substrate, the introduction chamber is pumped down to 10^{-8} Torr, and the substrate is heated (350 °C for GaAs) to desorb the water from the substrate mounting plate. In the meantime, the LN_2 shrouds of the growth chamber are cooled and the effusion cells are brought up to the desired temperatures. When the pressure in the introduction (or intermediate preparation) chamber has reached the low 10^{-9} Torr range, the substrate is transferred to the growth chamber. There it is heated to a higher temperature (< 630 °C for GaAs) under a flux of the group-V-element to evaporate the oxide from its surface (see Sect. 2.4). At this stage the substrate, provided proper preparation procedures were followed, is nearly atomically clean and ready for epitaxial growth. This thermal cleaning of a chemically etched surface produces a surface which is sometimes rough on a microscopic scale as indicated by a spotty RHEED pattern. After the deposition of several tens of nm (of GaAs, e.g.) by MBE the surface has become atomically flat and produces a streaked diffraction pattern with additional features due to reconstruction (see Fig. 4).

Using the ability to produce abrupt interfaces and doping variations, multilayered single-crystal structures with dimensions of only a few atomic layers can be obtained by actuating the shutters in front of the effusion cells in a predetermined manner. This potential for excellent dimensional control of MBE is indicated in the transmission electron micrographs (TEM) of Fig. 9. The growth rate and the layer thickness are calibrated by the following procedure. First, estimated layer thicknesses of 1 to 5 µm are grown on the substrate using different temperatures of the group-III-element effusion cells in sequential growth runs. The cleaved cross-sections of the layers are then stained or etched to delineate the interface and then examined with a scanning electron microscope. The actual growth rate is then determined by the measured layer thickness divided by the growth time. To account for possible flux transients in periodic structures, the period

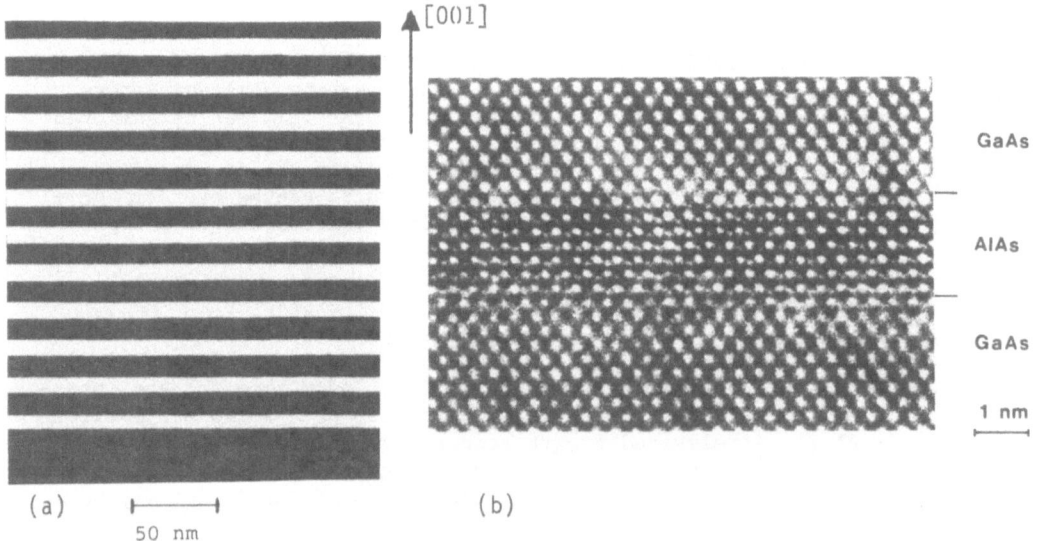

Fig. 9   (a) (110)cross-sectional TEM of a large-period AlAs/GaAs superlattice; (b) High-resolution lattice image of a 2-nm AlAs/
2.5-nm GaAs superlattice using (110)electron beam incidence

and the layer thickness in properly designed superlattice configurations are determined by high-resolution double-crystal X-ray diffraction (see Sect. 4.2 for details).

The unique capability of MBE to create a variety of mathematically complex compositional and doping profiles in semiconductors arises from the conceptual simplicity of this process. Accurately controlled temperatures have a direct, calculable effect upon the growth process. This simplicity allows composition control from $x = 0$ to $x = 1$ in $Al_xGa_{1-x}As$ with a precision of ± 0.001 and doping control, both n- and p-type, from 10^{14} cm^{-3} to the 10^{19} cm^{-3} range with a precision of $\pm 1\%$. The accuracy is largely determined by the care with which the growth rate and doping level were previously calibrated in test layers.

4. EXAMINATION OF INTERFACE QUALITY

The heterointerfaces between epitaxial layers of different composition ($Al_xGa_{1-x}As/GaAs$, $Al_xIn_{1-x}As/Ga_xIn_{1-x}As$, $Al_xGa_{1-x}Sb/GaSb$, Si/Si_xGe_{1-x}, etc.) are used to confine electrons or holes to two-dimensional (2D) motion. Recent developments in materials characterization have helped significantly to elucidate complex processes and phenomena connected with the microstructure of materials and interactions at abrupt heterointerfaces. However, the experimentally realisable interface perfection is now so high that conventional methods of profiling, which involve sputtering to section the material followed by some method of composition determination, such as AES or secondary ion mass spectroscopy (SIMS), do not have the required resolution. The available methods to obtain structural and compositional information with the needed level of spatial resolution are limited to transmission electron microscopy (TEM) of cross sections, double-crystal X-ray diffraction, and to a certain extent photoluminescence combined with absorption measurements. In this chapter we present a few examples for monitoring the interface quality by RHEED intensity oscillations and for studying interface disorder effects by X-ray diffraction.

4.1 RHEED Intensity Oscillations

For closely lattice-matched systems it is possible to prepare interfaces by MBE so that compositional changes occur over no more than one monolayer, because MBE growth occurs predominantly in a 2D layer-by-layer growth mode. Direct evidence for this 2D growth mode is provided by the observation of oscillations in the intensity of the RHEED pattern / 14 - 16 / , as shown in Fig. 6. Inspection of this figure reveals that damped oscillations occur immediately after the initiation of growth. We assume that the equilibrium surface is smooth and strongly reflective to electrons so that the specular beam intensity is high. As growth commences, 2D clusters form randomly on the surface, leading to a decrease in specular beam intensity. The minimum occurs at half-layer coverage. The damping of the oscillations is probably due to deviations from this ideal model in that the surface becomes statistically distributed over several incomplete layers. When growth is suspended by closing the group-III-element shutters and the layer is maintained at its growth temperature in a beam of arsenic, the reflectivity recovers to its original value, i.e. the surface becomes smoother. Briones et al. / 89 / have recently shown that damping of the RHEED oscillations during MBE growth of GaAs can be totally eliminated by a periodic phase-locked perturbation of the growth front by short synchronized interruptions of the arsenic flux. This cessation of the arsenic flux greatly enhances the surface migration of Ga atoms and thus promotes the formation of single-phase surface domains.

The oscillatory nature of the RHEED intensities provides direct real-time evidence of compositional effects and growth modes during the formation of heterointerfaces. As for the widely used $Al_xGa_{1-x}As$/GaAs heterojunctions, the sequence of layer growth is critical for compositional gradients and crystal perfection, which in turn is important for optimizing 2D transport properties. When the Al flux is switched at the maximum of the intensity oscillations, the first period for the growth sequence from ternary alloy to binary compound corresponds neither to the $Al_xGa_{1-x}As$ growth rate nor to the steady-state GaAs rate, but shows some intermediate value. For the growth sequence from binary compound to ternary alloy or between the two binaries an intermediate period does not exist. A possible explanation for this phenomenon can be found in the relative surface diffusion lengths of the group-III-elements Al and Ga, which were estimated to be $\lambda_{Al} \cong 3.5$ nm and $\lambda_{Ga} \cong 20$ nm on (100) surfaces under typical MBE growth conditions /17/. These differences in cation diffusion rates have striking consequences on the nature of the interface. While a GaAs layer should be covered by smooth terraces of 20 nm mean length between monolayer steps, those on an $Al_xGa_{1-x}As$ layer would only be 3.5 nm apart. The important result of this qualitative estimate is that the GaAs/$Al_xGa_{1-x}As$ interface is much smoother on an atomic scale than the inverted structure. Direct experimental evidence for this distinct difference in binary-to-ternary layer growth sequence is obtained from the high-resolution TEM investigations of Suzuki and Okamoto / 88 / . Their lattice image of an $Al_{0.2}Ga_{0.8}As$/AlAs superlattice shows clearly that the heterointerface is abrupt to within one atomic layer only when the ternary alloy is grown on the binary compound but not for the inverse growth sequence.

Since the nature of the heterointerface is critical for optimizing excitonic as well as transport properties in quantum wells, various attempts have been made to minimize the interface roughness (or disorder) by modified MBE growth conditions. The most successful modification is probably the method of growth interruption at each interface. Growth interruption allows the small terraces to relax into larger terraces via diffusion of the surface atoms. This reduces the step density and thus simultaneously enhances the RHEED specular beam intensity which can be used for real-time monitoring. The time of closing both the Al and the Ga shutter (while the As shutter is left open) apparently depends on the actual growth condition. Values ranging from a few seconds to several minutes have been reported by different authors / 90, 91 / , in particular for the $Al_xGa_{1-x}As$/GaAs interface and if the full recovery of the specular beam intensity has been allowed for. We have found that in most cases (i.e. growth rate < 1 μm/hr and growth temperature < 650 °C) growth interruption of less than 100 s is sufficient to minimize the interface roughness of the Al-Ga-As system.

In Fig. 10 we show the RHEED intensity sequence for the growth of ultrathin-layer $(GaAs)_m(AlAs)_m$ superlattices with m = 1, 2, and 3. The growth rates of GaAs and AlAs monolayers were accurately controlled by the period of the intensity oscillations of the specularly reflected beam in the RHEED pattern, and actuation of the effusion cell shutters was synchronized to these oscillations. The synthesis of each superlattice period requires four steps. At first a monolayer (or bi- or tri-layer) of GaAs is deposited. In the second step the crystal growth is stopped for 5 s and the surface is only exposed to the arsenic flux. The third step involves the deposition of a monolayer (or bi- or trilayer) of AlAs. Finally, the fourth step is the same as the second one. The shutters in front of the Ga and Al effusion cells, respectively, are opened at the time when the intensity of the specularly reflected beam has reached its maximum value and closed after one (or two or three) period(s). The maximum intensity indicates that the growth surface has become extremely smooth. A low growth rate of 1 monolayer per 5 s ensures that the crystal growth occurs indeed in a 2D layer-by-layer mode. The purpose of the second and the fourth step of this growth sequence is that each layer is completed before the next one starts. The formation of high-quality layered crystals has been demon-

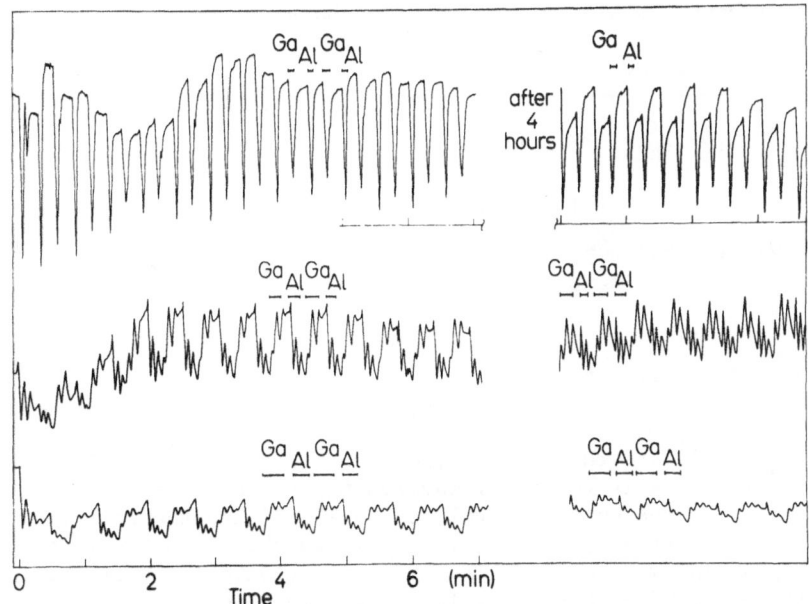

Fig. 10 RHEED intensity oscillations of specularly reflected electron
beam from (001) surface in [100] azimuth during growth of
$(GaAs)_m (AlAs)_m$ superlattices with m = 1, 2 and 3 (from top to
bottom).

strated by the appearance of distinct satellite peaks around the Bragg
reflections in the X-ray diffraction patterns (see next Sect. 4.2). Our
detailed luminescence investigations / 92 / revealed that the ultrathin-
layer $(GaAs)_m (AlAs)_m$ superlattices with m < 4 indeed represent a new artificial semiconductor material with novel electronic properties.

4.2 Double Crystal X-Ray Diffraction

High-angle X-ray diffraction is a powerful non-destructive technique
for investigation of interface disorder effects in superlattices and multi
quantum well heterostructures, if a detailed analysis of the diffraction
curves is performed. We have recently developed a semi-kinematical approach
of the dynamical theory of X-ray diffraction to determine the strain profile,
the composition (periodicity), and the interface quality of $Al_x Ga_{1-x} As/GaAs$
and of $Al_{0.48} In_{0.5} As/Ga_{0.47} In_{0.53} As$ heterostructures and superlattices
/ 93 / . In the following we briefly outline those parts of the theory that
are relevant to the experiments.

The scattering of X-rays from strained single crystals is described by
Taupin's formalism of the dynamical theory of X-rays / 94 / . The semikine-
matical approximation of Petrashen / 95 / uses the first iteration of
Taupin's equation for the amplitude ratio of the diffracted and incident
waves and is valid if the thickness of the deformed layer is small compared
with the X-ray extinction length. The reflectivity is then given in integral
form which reduces the calculation time for a diffraction curve. For our
calculation the epitaxial layer is divided into n lamellae of equal thickness $\Delta = z_0/n$, where z_0 is the total thickness of the layer measured in
units of the extinction length divided by π. The reflectivity R(y) is given
by the product of the diffraction curve for an undeformed crystal $R_p(y) = |g - y|^2$ and a deformation dependent factor

$$R(y) = R_p(y) \left| 1 - 2iy \sum_{j=1}^{n} \exp(i\phi_j) \frac{\sin(y-s_j)\Delta}{y-s_j} \right| . \qquad (6)$$

For a crystal without a centre of inversion we have

$$y = \frac{-2b \sin(2\Theta_B)\Delta\Theta - (1-b)(\chi_o^r + i\chi_o^i)}{2C \left(|b|(\chi_h^r\chi_{\bar{h}}^r - \chi_h^i\chi_{\bar{h}}^i)\right)^{1/2}} \qquad (7)$$

$$g = \text{sign}(y)\ (y^2-1-2i\ \frac{\chi_{\bar{h}}^r\chi_h^i + \chi_h^r\chi_{\bar{h}}^i}{\chi_h^r\chi_{\bar{h}}^r + \chi_h^i\chi_{\bar{h}}^i})^{1/2} \qquad (8)$$

and

$$\phi_j = -y\Delta\ (2(n-j)+1) + \Delta s_j + 2\Delta + \sum_{j=j+1}^{n} s_j \qquad (9)$$

where $\chi_h^{r,i} = r_e \frac{\lambda^2}{\pi V} F_h^{r,i}$ are the h-th Fourier coefficient of the polarizability multiplied by 4π, λ is the X-ray wavelength, r_e is the classical electron radius, V is the unit cell volume and F^r, F^i are the real and imaginary part of the structure factor. $b = \gamma_o/\gamma_h$ is the asymmetry factor, and $\gamma_o = \sin(\theta_B-\alpha)$ and $\gamma_h = \sin(\theta+\alpha)$ are the direction cosines of the incident and diffracted waves respectively, θ_B is the kinematical Bragg angle and α the angle between crystal surface and reflecting lattice plane. Δ is the deviation from the Bragg angle. C is the polarization factor, which is equal to unity for σ-polarization and $\cos 2\theta_B$ for π-polarization. s_j is the normalized strain for the j-th lamella. This constant value within each lamella is given by the equation

$$s = \frac{\lambda\ |b|^{1/2}}{2\pi C\ (\chi_h^r\chi_{\bar{h}}^r - \chi_h^i\chi_{\bar{h}}^i)^{1/2}}\ \frac{\partial}{\partial s_h}\ (\vec{h}\vec{u}) \qquad (10)$$

where $\bar{h}$ is the diffraction vector, $\bar{u}$ is the displacement field, and s_h is the direction of the diffracted wave. If there is no shear strain, as in epitaxial layers grown in (001) and (111) orientation, and if the crystal is bound by the xy plane and the diffraction plane is the xz plane, Eq. (10) takes the form

$$(11)$$

$$s(z) = \frac{2|b|^{1/2}\sin^2\theta_B}{(\chi_h^r\chi_{\bar{h}}^r - \chi_h^i\chi_{\bar{h}}^i)^{1/2}}\ (\varepsilon_{zz}(z)\cos^2\alpha+\varepsilon_{xx}(z)\sin^2\alpha+(\varepsilon_{zz}(z)-\varepsilon_{xx}(z))\frac{\sin 2\alpha}{2\tan\theta_B}).$$

Here the depth dependence of the normalized strain is taken into account, and $\varepsilon_{zz}(z)$ and $\varepsilon_{xx}(z)$ are the strains at depth z perpendicular and parallel to the crystal surface, respectively.

The sinusoidal term in Eq. (6) describes the amplitude of a single lamella, whereas ϕ_j in the exponential term describes the phase of a single lamella. One of the interesting consequences of Eq. (6) is the possibility to observe oscillations on the X-ray diffraction curve, i.e. Pendellösung fringes, as a result of interferences of waves scattered at different depths in the epitaxial layer (see $\Delta\omega$ in Fig. 11). If we assume that s(z) is constant in the whole epilayer of the thickness D, the sinusoidal term of the amplitude in Eq. (6) oscillates with the period π. For the argument of the sinusoidal term we then obtain the relation

$$yD = \pi. \qquad (12)$$

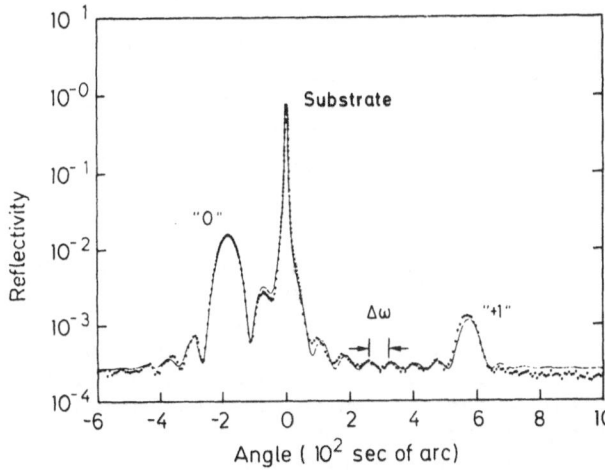

Fig. 11 Experimental (dotted curve) and theoretical (solid line) X-ray diffraction curve of an $(AlAs)_{44}(GaAs)_{46}$ superlattice in the vicinity of the (004) reflection using $CuK\alpha_1$ radiation

Using Eq. (7) and considering the fact that D must be multiplied by the extinction length l_{ex}, we find the known relation between the angular spacing of the Pendellösung fringes $\Delta\omega$ and the thickness D of the strained surface layer, i.e.

$$D = \frac{\lambda \mid \gamma_h \mid}{\Delta\omega \sin(2\theta_B)} \quad . \tag{13}$$

If we apply the same evaluation to a superlattice, i.e. a one-dimensional periodic structure with a period length T, we find the identical relation as given in Eq. (13). The whole epilayer thickness is replaced by the superlattice period T, and the oscillation spacing is replaced by the angular distance between the satellite peaks which appear in X-ray diffraction curves from superlattices. The amplitude of the deformation factor in Eq. (6) which describes the scattering of the deformed layer, has a maximum if $y = \bar{s}$, where $\bar{s}$ is the average value of s(z) in the strained layer. This consideration leads to the relation between (i) the angular spacing of the main diffraction peaks of the strained layer and of the unstrained crystal, $\Delta\theta^0$, and (ii) the average strain value $\bar{\varepsilon}_{zz}$, which exists in the strained layer.

Semiconductor superlattices are one-dimensional periodic structures consisting of thin layers of alternating composition. Therefore, the depth variation of the strain in Eq. (11) and the depth variation of the structure factor in the deformation dependent term of Eq. (6) must be included. The depth distribution of the strain and of the chemical composition are determined by a comparison of the calculated X-ray diffraction curve with the experimental data. A certain strain and an approximate composition distribution is initially presumed from the employed growth conditions. The best coincidence of the theoretical and experimental X-ray diffraction curve is then found by varying successively the values for the structural parameters, i.e. the strain and composition profiles as well as the epilayer thickness.

Our X-ray diffraction measurements are performed with a computer-controlled double-crystal X-ray diffractometer in non-dispersive (+, -) Bragg geometry. An asymmetrically cut (100) Ge crystal is used for monochromizing and collimating the X-rays / 93 / . As the lattice parameter and the scattering factors of superlattices are subject to a one-dimensional modulation in growth direction, the diffraction patterns consist

57

of satellite reflections located symmetrically around the Bragg reflections, as shown in Fig. 11 for an AlAs/GaAs superlattice. From the position of the satellite peaks the superlattice periodicity can be deduced. Detailed information about thickness fluctuations of the constituent layers, inhomogeneity of composition, and interface quality can be extracted from the halfwidths and intensities of the satellite peaks. The excellent agreement between the experimental and the theoretical diffraction curve in Fig. 11 indicates extremely abrupt AlAs/GaAs interfaces to within one monolayer.

Since MBE growth occurs predominantly in a two-dimensional layer-by-layer growth mode, the compositional changes at heterointerfaces of closely lattice-matched materials, like AlAs/GaAs, should occur over no more than one monolayer. However, for the widely used $Al_xGa_{1-x}As$/GaAs heterojunction it is well established that the sequence of layer growth is critical for compositional gradients and crystal perfection, which in turn strongly affect the excitonic and the transport properties of quantum wells. While the $GaAs/Al_xGa_{1-x}As$ heterointerface is abrupt to within one monolayer when the ternary alloy is grown on the binary compound, this is not the case for the inverse growth sequence under typical MBE growth conditions. In X-ray diffraction the existence of interface disorder manifests itself in an increase of the halfwidths and a decrease of the intensities of the satellite peaks, as shown in Fig. 12. During MBE growth of these two $(AlAs)_{42}$ $(GaAs)_{34}$ superlattices, the adjustment of the shutter motion at the transition from AlAs to GaAs and vice versa was changed in the two growth runs. Sample A was grown with growth interruption at each AlAs/GaAs and GaAs/AlAs interface, whereas sample B was grown continuously. While the positions of all the diffraction peaks of sample A coincide with those of sample B, the halfwidths of the satellite peaks from sample A are narrower and their reflected intensities are higher. A growth interruption of 10 s was sufficient to smooth the growing surface which then provides sharp heterointerfaces. When the heterojunctions are grown continuously, the monolayer roughness

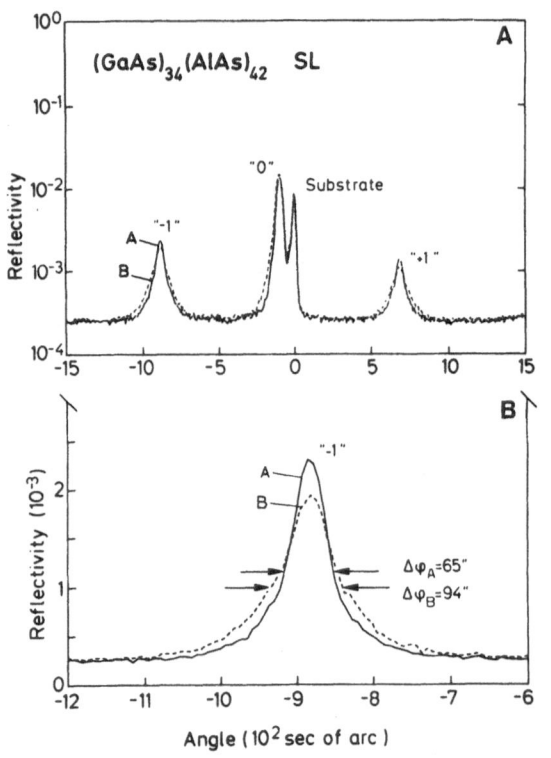

Fig. 12 X-ray diffraction curves of two AlAs/GaAs super-lattices grown (A) with and (B) without growth interruption at the heterointerfaces, recorded with CuKα_1 radiation in the vicinity of the quasi-forbidden (002) reflection

of the growth surface leads to a disorder and thus broadening of the interface. In X-ray diffraction this broadening manifests itself as a random variation of the superlattice period of about one lattice constant (~ 5.6 Å) for sample B.

The quantitative evaluation of the interface quality by X-ray diffraction becomes even more important if the lattice parameters of the epilayer have to be matched to those of the substrate by appropriate choice of the layer composition, as for $Al_{0.48}In_{0.52}As/Ga_{0.47}In_{0.53}As$ superlattices lattice-matched to InP substrates. In Fig. 13 we show the X-ray diffraction pattern of such a superlattice with a periodicity of 20.4 nm. For the investigation of this all-ternary material system the X-ray diffraction patterns were recorded in the vicinity of the symmetric (002) and (004) reflections and of the asymmetric (224) and (044) reflections using $CuK\alpha_1$ radiation. Both symmetric and asymmetric diffraction data yield the average lattice strain perpendicular, $\bar{\varepsilon}_{zz}$, and parallel, $\bar{\varepsilon}_{xx}$, to the (001) substrate surface. The lattice strains $\bar{\varepsilon}_{zz}$ and $\bar{\varepsilon}_{xx}$ are correlated with the angular distance $\Delta\theta_I$ and $\Delta\theta_{II}$ between the substrate diffraction maximum and the main epitaxial layer peak ("0"-peak) by the equation

$$\begin{pmatrix} \bar{\varepsilon}_{zz} \\ \bar{\varepsilon}_{xx} \end{pmatrix} = \begin{pmatrix} A_I & B_I \\ A_{II} & B_{II} \end{pmatrix}^{-1} \begin{pmatrix} \Delta\theta_I \\ \Delta\theta_{II} \end{pmatrix} \tag{14}$$

with

$$A_{I,II} = \cos\alpha_{I,II} * [\cos\alpha_{I,II} * \tan\theta_{I,II} + \sin\alpha_{I,II}]$$

$$\tag{15}$$

$$B_{I,II} = \sin\alpha_{I,II} * [\sin\alpha_{I,II} * \tan\theta_{I,II} - \cos\alpha_{I,II}]$$

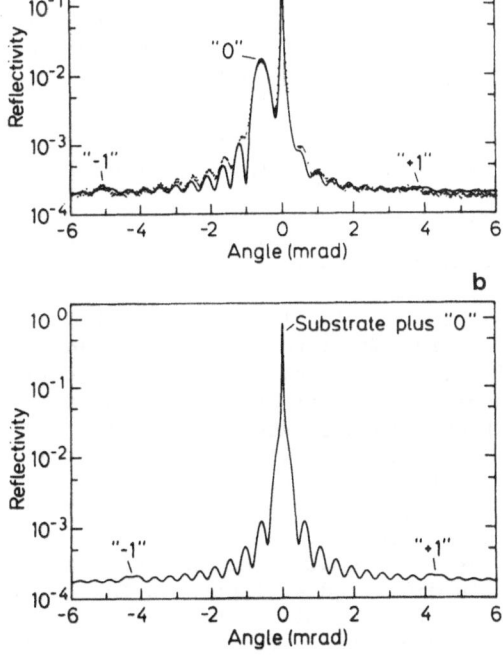

Fig. 13 (a) $CuK\alpha_1$ (004) diffraction pattern of $Al_xIn_{1-x}As/Ga_xIn_{1-x}As$ superlattice on (001) InP with $L_z = L_B = 10.2$ nm (.... experiment, —— theory), (b) Calculated diffraction pattern of a perfectly lattice-matched $Ga_xIn_{1-x}As/Al_xIn_{1-x}As$ superlattice on (001) InP [$CuK\alpha_1$ (004) reflection].

and

$$\bar{\epsilon}_{zz} = (\bar{d}_e^{\perp} - d_s) / d_s$$

$$\bar{\epsilon}_{xx} = (\bar{d}_e^{\parallel} - d_s) / d_s \;\;.$$

(16)

The indices I and II hold for symmetric and asymmetric reflections, resp. Here, $\bar{d}_e^{\perp}$ and $\bar{d}_e^{\parallel}$ are the average interplanar spacings of the epitaxial layer perpendicular and parallel to the crystal surface, while d_s is the interplanar spacing of the substrate crystal, respectively. θ_I and θ_{II} are the kinematic Bragg angles, and α_I, α_{II} the angles between crystal surface and reflection planes I and II. The evaluation of our experimental diffraction data reveals that the superlattice is not misoriented with respect to the substrate crystal and the $\bar{\epsilon}_{xx} = 0$ for all samples. i.e. the lattice spacing parallel to the crystal surface in the constituent $Ga_xIn_{1-x}As$ and $Al_xIn_{1-x}As$ layers and in the InP substrate crystal are the same. Hence it follows that the lattice strain in the $Ga_xIn_{1-x}As$ and $Al_xIn_{1-x}As$ epilayers perpendicular to the crystal surface is a measure of their mole fraction x. The relation between elastic strain and chemical composition in the $Ga_xIn_{1-x}As$ layers is given by

$$x = \frac{1}{a_{GaAs} - a_{InAs}} * [a_{InP}(1 + \frac{c_{11}}{c_{11} + 2c_{12}} \epsilon_{zz}) - a_{InAs}]$$

(17)

where c_{11} and c_{12} are the elastic stiffness constants of the epilayer material. For the $Al_xIn_{1-x}As$ layers the lattice constant a_{GaAs} in Eq. (17) must be replaced by that of a_{AlAs}.

Table 3 Structural parameters of four 10-period $Al_xIn_{1-x}As/Ga_xIn_{1-x}As$ superlattices grown by MBE with different temperature settings of the group-III-element effusion cells, as determined by X-ray diffraction.

Sample No.	Thickness of $Ga_xIn_{1-x}As$ layers (nm)	Thickness of $Al_xIn_{1-x}As$ layers (nm)	Average lattice mismatch of superlattice $\bar{\epsilon}_{zz}$ (x 10^{-4})	Lattice strain in $Ga_xIn_{1-x}As$ layers $\bar{\epsilon}_{zz}$ (x 10^{-4})	Lattice strain in $Al_xIn_{1-x}As$ layers $\bar{\epsilon}_{zz}$ (x 10^{-4})	Mole fraction x of $Ga_xIn_{1-x}As$ layers	Mole fraction x of $Al_xIn_{1-x}As$ layers
5649	10.6	10.6	- 4.5	+ 6.0	- 15.0	0.464	0.487
5653	10.2	10.2	+ 9.4	9.5	9.3	0.462	0.470
5654	10.2	10.2	- 4.0	0.0	- 8.0	0.468	0.482
5655	10.2	10.2	- 3.5	7.0	- 14.0	0.464	0.486

In Fig. 13a we show the experimental (dotted line) and the theoretically fitted (solid line) diffraction patterns obtained from a representative $Al_xIn_{1-x}As/Ga_xIn_{1-x}As$ superlattice in the vicinity of the symmetrical (004) $CuK\alpha_1$ reflection. From the theoretical diffraction pattern we obtain the thickness as well as the lattice strain of the individual layers. The chemical compositions are determined by using Eq. (17). In Table 3 we summarize the measured and the calculated structural data from four $Al_xIn_{1-x}As/Ga_xIn_{1-x}As$ superlattices grown under different conditions. The growth conditions were adjusted such that for all samples the barrier width L_B equals the well width L_z. The lattice mismatch perpendicular to the (001) growth face for the tetragonally distorted epilayers on InP is less than 2.8×10^{-3}, i.e. less than the lattice mismatch in the AlAs/GaAs system. The Pendellösung fringes observed between the main diffraction peaks and the satellite peaks "-1" and "+1" demonstrate the excellent thickness and composition homogeneity perpendicular and parallel to the crystal surface. In Fig. 13b we show the theoretical diffraction pattern for a perfectly latticematched $Ga_{0.468}In_{0.532}As/Al_{0.477}In_{0.523}As$ superlattice with $L_B = L_z = 10.6$ nm. It should be noted that the satellite peaks "-1" and "+1" have almost disappeared. This finding is in contrast to that observed in strained layer superlattices, where a strain periodicity produces strong satellite peaks. The low intensity and the Pendellösung fringes are caused only by the periodicity of the structure factors in the superlattice. A weak intensity of the satellite peaks is also observed if the lattice strains of the $Ga_xIn_{1-x}As$ layers are of the same magnitude, which occurs in the sample displayed in Fig. 13a (see also Table 3).

These few examples indicate that important details of the interface structure can be extracted from high-angle X-ray diffraction if a semi-kinematical treatment of the dynamical theory is applied to analyse the experimental data.

5. CURRENT RESEARCH ACTIVITIES

In this section we will briefly define some selected areas of current research activities on the application of interface formation during MBE growth to modify the bulk properties of semiconductors through bandgap engineering. An extensive survey on these research activities can be found in Refs. / 96 - 98 / . In the following we classify the various heterojunctions and superlattice systems and give references for the most active areas of research.

5.1 $GaAs/Al_xGa_{1-x}As$

The $GaAs/Al_xGa_{1-x}As$ system has a lattice mismatch of only 0.16% for $x = 1$ and represents the prototype materials system for research on artificially layered semiconductor structures.

5.1.1 Dynamical Properties of Two-Dimensional Excitons (excitons in an electric field, optical dephasing of excitons)

E.O. Göbel et al., J. Lumin 30, 541 (1985)
Y. Horikoshi et al. Jpn. J. Appl. Phys. 24, 955 (1985)
H.J. Polland et al., Phys. Rev. Lett. 55, 2610 (1985)
R.T. Collins et al., Phys. Rev. B33, 4378 (1986); Appl. Phys. Lett. 49, 406 (1986)
L. Vina et al., Phys. Rev. B 33, 5939 (1986)
L. Schultheis et al., Phys. Rev. B 34, 9027 (1986); Superlattices and Microstructures 2, 441 (1986)

5.1.2 Quantum Hall Effect

K. von Klitzing, Rev. Mod. Phys. 58, 519 (1986)
R.E. Prange and S.M. Girvin, The Quantum Hall Effect, Springer Ser. Contemp. Phys. 2 (1987)
H.L. Störmer et al., Phys. Rev. Lett. 56, 85 (1986)
Z. Schlesinger et al., Phys. Rev. Lett. 58, 73 (1987)

5.1.3 (Resonant) Tunneling and Ballistic Transport

T.C.L.G. Sollner et al., Appl. Phys. Lett. 43, 588 (1983)
M. Tsuchiya et al., Jpn. J. Appl. Phys. 24, L 466 (1985)
N. Yokoyama et al., Jpn. J. Appl. Phys. 24, L 853 (1985)

T. Nakagawa et al., Appl. Phys. Lett. $\underline{49}$, 73 (1986)
R.A. Davies et al., Phys. Rev. Lett. $\underline{55}$, 1114 (1985)
E.E. Mendez et al., Phys. Rev. B $\underline{34}$, 6026 (1986)
T.W. Hickmott, Phys. Rev. B $\underline{32}$, 6531 (1985)
M. Heiblum et al., Phys. Rev. Lett. $\underline{55}$, 2200 (1985)
F. Capasso et al., IEEE J. Quantum Electron. QE-22, 1853 (1986)
M. Heiblum and L.F. Eastman, Scientific American, No. 2,65 (1987)

5.1.4 Folded Phonons and Interface Modes in Superlattices

A.K. Sood et al., Phys. Rev. Lett. $\underline{54}$, 2111 (1985); Phys. Rev. Lett. $\underline{54}$, 2115 (1985); Phys. Rev. Lett. $\underline{56}$, 1751 (1986)

5.1.5 Plasmon Resonances and Single Particle Excitations

G. Fasol et al., Phys. Rev. Lett. $\underline{56}$, 2517 (1986)

5.1.6 Electron-Phonon Interactions

M.A. Brummel et al., Phys. Rev. Lett. $\underline{58}$, 77 (1987)

5.1.7 GaAs on Si Substrates (lattice mismatch is about 4%)

The following problems have to be solved: (i) antiphase disorder, (ii) misfit dislocations, and (iii) interface charge and cross doping.
R. Fischer et al., J. Appl. Phys. $\underline{60}$, 1640 (1986)
H. Kroemer, Mater. Res. Soc. Symp. Proc. $\underline{67}$, 3 (1986)
H. Morkoc, Mater, Res. Soc. Symp. Proc. $\underline{67}$, 149 (1986).

5.2 $GaSb/Al_xGa_{1-x}Sb$

The lattice mismatch is 0.66% for x = 1. The accomodation of the lattice mismatch by elastic deformations (strain) may affect the subband levels.
Y. Ohmori et al., Jpn. J. Appl. Phys. $\underline{24}$, L 657 (1985); J. Appl. Phys. $\underline{59}$, 3760 (1986)
K. Ploog et al., Appl. Phys. Lett. $\underline{47}$, 384 (1985)
A. Forchel et al., Phys. Rev. Lett. $\underline{57}$, 3217 (1986);Appl. Phys. Lett. $\underline{50}$, 182 (1987)

5.3 InAs/GaSb

The lattice mismatch is 0.61% for this system. The unique broken-gap band line-up at the interface leads to a spatial separation of carriers with electrons in the InAs layers and holes in the GaSb layers. In particular, the two-dimensional gas with coexisting electrons and holes has been studied.
E.E. Mendez et al., Phys. Rev. Lett. $\underline{55}$, 2216 (1985)
S. Washburn et al., Phys. Rev. B $\underline{33}$, 8848 (1986)
L.M. Claessen et al., Phys. Rev. Lett. $\underline{57}$, 2556 (1986).

5.4 $Ga_xIn_{1-x}As/InP$ and $Ga_xIn_{1-x}As/Al_xIn_{1-x}As$ Lattice Matched to InP

These two materials systems are of significant importance for advanced photonic devices.
Y. Kawamura et al., Electron Lett. $\underline{21}$, 1168 (1985); Jpn. J. Appl. Phys. $\underline{25}$, L928
J. Wagner et al., Solid State Commun. $\underline{57}$, 781 (1986)
W.T. Tsang and E.F. Schubert, Appl. Phys. Lett. $\underline{49}$, 220 (1986)
T. Fujii et al., Jpn. J. Appl. Phys. $\underline{25}$, L598 (1986)

5.5 GaAs/$Ga_{0.8}In_{0.2}As$ on GaAs Substrate

The lattice mismatch of $Ga_{0.8}In_{0.2}As$ is 1.43% with respect to GaAs, and below a certain critical thickness the $Ga_{0.8}In_{0.2}As$ layers are heavily strained ("strained layer superlattices"). The built-in strain in $Ga_{0.8}In_{0.2}As$/GaAs superlattices was used to remove the degeneracy of the valence band at the Brillouin zone center so that the light-hole band can be preferentially populated.
J.E. Schirber et al., Appl. Phys. Lett. <u>46</u>, 187 (1985)

5.6 Si/Si_xGe_{1-x} Strained-Layer Superlattices

The lattice mismatch between Si and Ge is about 4%. Modification of the band line-up by strain distribution was used to form 2D electron or hole gas. Observation of folded phonons.
G. Abstreiter et al., Phys. Rev. Lett. <u>54</u>, 2441 (1985)
H. Brugger et al., Phys. Rev. <u>B 33</u>, 5928 (1986)
D.J. Lockwood et al., Phys. Rev. <u>B 35</u>, 2243 (1987)

5.7 CdTe/HgTe

Transition from finite to zero-gap semiconductor.
J.P. Faurie, IEEE J. Quantum Electron. <u>QE-22</u>, 1656 (1986)

5.8 ZnSe/$Zn_xMn_{1-x}Se$

2D magnetic semiconductor superlattices
L.A. Kolodzieski et al., IEEE J. Quantum Electron. <u>QE-22,</u> 1666 (1986).

5.9 PbTe/$Pb_{1-x}SnTe$ and PbTe/$Pb_xEu_{1-x}Se_yTe_{1-y}$

Transport and magneto-optical properties
G. Bauer, Surf. Sci. <u>168</u>, 462 (1986)
D.L. Partin et al., Superlattices and Microstructures <u>2</u>, 459 (1986).

5.10 Doping Superlattices in GaAs

Space-charge induced modulation of real-space energy bands
K. Ploog and G.H. Döhler, Adv. Phys. <u>32</u>, 285 (1983)
K. Ploog et al., Surf. Sci. <u>174,</u> 120 (1986)

ACKNOWLEDGEMENT

This work was sponsored by the Bundesministerium für Forschung und Technologie of the Federal Republic of Germany.

REFERENCES

1. For a recent extensive review on MBE see: <u>The Technology and Physics of Molecular Beam Epitaxy</u>, ed. by E.H.C. Parker (Plenum, New York 1985)
2. K.G. Günther: Z. Naturforschg. <u>A 13</u>, 1081 (1958)
3. J.R. Arthur: J. Phys. Chem. Solids <u>28</u>, 2257 (1967)
4. J.R. Arthur: J. Appl. Phys. <u>39</u>, 4032 (1968)
5. A.Y. Cho: J. Appl. Phys. <u>41</u>, 782 (1970)
6. A.Y. Cho: J. Vac. Sci. Technol. <u>8</u>, S 31 (1971)
7. J.C. Bean and S.R. McAfee: J. Physique <u>43</u>, Colloque C5, 153 (1982)
8. K. Ploog and K. Graf: <u>Molecular Beam Epitaxy of III-V Compounds</u> (Springer, Berlin, Heidelberg 1984)
9. A.C. Gossard: Thin Solid Films <u>57</u>, 3 (1979)

10. C.T. Foxon and B.A. Joyce: In Current Physics in Materials Science, Vol. 7, ed. by E. Kaldis (North-Holland, Amsterdam 1981) 1
11. A.Y. Cho and J.R. Arthur: Progr. Solid State Chem. $\underline{10}$, 157 (1975)
12. K. Ploog: J. Vac. Sci. Technol. $\underline{16}$, 838 (1979)
13. R. Ludeke: J. Vac. Sci. Technol. $\underline{17}$, 1241 (1980)
14. J.H. Neave, B.A. Joyce, P.J. Dobson, and N. Norton: Appl. Phys. $\underline{A\ 31}$ 1 (1983)
15. J.M. van Hove, C.S. Lent, P.R. Pukite, and P.F. Cohen: J. Vac. Sci. Technol. $\underline{B\ 1}$, 741 (1983)
16. T. Sakamoto, H. Funabashi, K. Ohta, T. Nakagawa, N.J. Kawai, T. Kojima, and K. Bando: Superlattices and Microstructures $\underline{1}$, 347 (1985)
17. B.A. Joyce, P.J. Dobson, J.H. Neave, K. Woodbridge, J. Zhang, P.K. Larsen, and B. Bölger: Surf. Sci. $\underline{168}$, 423 (1986)
18. C.E.C. Wood: In Physics of Thin Films, Vol. 11, ed. by W.R. Hunter and G. Haas (Academic, New York 1980) 35
19. R.E. Honig and D.A. Kramer: RCA Review $\underline{30}$, 285 (1969)
20. M.A. Herman: Vacuum $\underline{32}$, 555 (1982)
21. B.B. Dayton: Trans. 2nd American Vacuum Soc. Symp. $\underline{5}$, 5 (1961)
22. J.A. Curless: J. Vac. Sci. Technol. $\underline{B\ 3}$, 531 (1985)
23. J. Saito and A. Shibatomi: Fujitsu Tech. J. $\underline{21}$, 190 (1985)
24. T.A. Flaim and P.D. Ownby: J. Vac. Sci. Technol. $\underline{8}$, 661 (1971)
25. C.T. Foxon, B.A. Joyce, and M.T. Norris: J. Cryst. Growth $\underline{49}$, 132 (1980)
26. K.Y. Chen, A.Y. Cho, W.R. Wagner, and W.A. Bonner: J. Appl. Phys. $\underline{52}$, 1015 (1981)
27. J.H. Neave, P. Blood, and B.A. Joyce: Appl. Phys. Lett. $\underline{36}$, 311 (1980)
28. G. Duggan, P. Dawson, C.T. Foxon, and G.W.t'Hooft: J. Physique $\underline{43}$, Colloque C5, 129 (1982)
29. H. Künzel, J. Knecht, H. Jung, K. Wünstel, and K. Ploog: Appl. Phys. $\underline{A\ 28}$, 167 (1982)
30. M.B. Panish: J. Electrochem. Soc. $\underline{127}$, 2729 (1980)
31. A.R. Calawa: Appl. Phys. Lett. $\underline{38}$, 701 (1981)
32. T. Henderson, W. Kopp. R. Fischer, J. Klem, H. Morkoc, L.P. Erickson, and P.W. Palmberg: Rev. Sci. Instrum. $\underline{55}$, 1763 (1984)
33. D. Huet, M. Lambert, D. Bonnevie, and D. Dufresne: J. Vac. Sci. Technol. $\underline{B\ 3}$, 823 (1985)
34. T. Mizutani and K. Hirose: Jpn. J. Appl. Phys. $\underline{24}$, L119 (1985)
35. P.A. Maki, S.C. Palmateer, A.R. Calawa, and B.R. Lee: J. Electrochem. Soc. $\underline{132}$, 2813 (1985)
36. Y. Ota: Thin Solid Films $\underline{106}$, 3 (1983)
37. J.C. Bean: J. Vac. Sci. Technol. A1, 540 (1983)
38. L.L. Chang and K. Ploog: Molecular Beam Epitaxy and Heterostructures (Martinus Nijhoff, Dordrecht 1985) NATO Adv. Sci. Inst. Ser. E 87
39. H.J. Fronius. A. Fischer, and K. Ploog: Jpn. J. Appl. Phys. $\underline{25}$, L137 (1986)
40. G.J. Davies, R. Heckingbottom, H. Ohno, C.E.C. Wood, and A.R. Calawa: Appl. Phys. Lett. $\underline{37}$, 290 (1980)
41. K. Oe and Y. Imamura: Jpn. J. Appl. Phys. $\underline{24}$, 779 (1985)
42. L.P. Erickson, G.L. Carpenter, D.D. Seibel, P.W. Palmberg, P. Pearah, W. Kopp, and H. Morkoc: J. Vac. Sci. Technol. $\underline{B\ 3}$, 536 (1985)
43. D.E. Mars and J.N. Miller: J. Vac. Sci. Technol. $\underline{B\ 4}$, 571 (1986)
44. A.J. Springthorpe and P. Mandeville: J. Vac. Sci. Technol. $\underline{B\ 4}$, 853 (1986)
45. C.T. Foxon, J.A. Harvey, and B.A. Joyce: J. Phys. Chem. Solids $\underline{34}$, 1693 (1973)
46. A.Y. Cho and K.Y. Cheng: Appl. Phys. Lett. $\underline{38}$, 360 (1981)
47. E.A. Wood: J. Appl. Phys. $\underline{35}$, 1306 (1964)
48. A.Y. Cho: J. Appl. Phys. $\underline{47}$, 2841 (1976)
49. K. Ploog and A. Fischer: Appl. Phys. 13, 111 (1977)
50. J.H. Neave and B.A. Joyce: J. Cryst. Growth $\underline{44}$, 387 (1978)
51. S. Holloway and J.L. Beeby: J. Phys. $\underline{C\ 11}$, L247 (1978)
52. I. Hernandez-Calderon and H. Höchst: Phys. Rev. $\underline{B\ 27}$, 4961 (1983)

53. B.A. Joyce, J.H. Neave, P.J. Dobson, and P.K. Larsen: Phys. Rev. B 29, 814 (1984)
54. G. Laurence, F. Simondet, and P. Saget: Appl. Phys. 19, 63 (1979)
55. A.Y. Cho: J. Appl. Phys. 42, 2074 (1971)
56. C.E.C. Wood: Surf. Sci. 108, L441 (1981)
57. C.T. Foxon, M.R. Boundry, and B.A. Joyce: Surf. Sci. 44, 69 (1974)
58. B.A. Joyce, P.J. Dobson, J.H. Neave, and J. Zhang: Surf. Sci. 174 (1986)
59. A. Madhukar: Surf. Sci. 132, 344 (1983)
60. J.R. Arthur: Surf. Sci. 43, 449 (1974)
61. C.T. Foxon and B.A. Joyce: Surf. Sci. 50, 434 (1975)
62. C.T. Foxon and B.A. Joyce: Surf. Sci. 64, 293 (1977)
63. C.T. Foxon and B.A. Joyce: J. Cryst. Growth 44, 75 (1978)
64. M. Ilegems: J. Appl. Phys. 48, 1278 (1977)
65. K. Ploog, A. Fischer, and H. Künzel: J. Electrochem. Soc. 128, 400 (1981)
66. N. Duhamel, P. Henoc, F. Alexandre, and E.V.K. Rao: Appl. Phys. Lett. 39, 49 (1981)
67. D.L. Miller and P.M. Asbeck: J. Appl. Phys. 57, 1816 (1985)
68. Y.C. Pao, T. Hierl, and T. Cooper: J. Appl. Phys. 60, 201 (1986)
69. J.L. Lievin and F. Alexandre: Electron Lett. 21, 413 (1985)
70. E. Nottenburg, H.J. Bühlmann, M. Frei, and M. Ilegems: Appl. Phys. Lett. 44, 71 (1984)
71. J.M. Ballingall, B.J. Morris, D.J. Leopold, and D.L. Rode: J. Appl. Phys. 59, 3571 (1986)
72. M. Heiblum, W.I. Wang, L.E. Osterling, and V. Deline: J. Appl. Phys. 54, 6751 (1983)
73. R. Sacks and H. Shen: Appl. Phys. Lett. 47, 374 (1985)
74. L. Gonzales, J.B. Clegg, D. Hilton, J.P. Gowers, C.T. Foxon, and B.A. Joyce: Appl. Phys. A41, 237 (1986)
75. W.I. Wang, E.E. Mendez, T.S. Kuan, and L. Esaki: Appl. Phys. Lett. 47, 826 (1985)
76. D.L. Miller: Appl. Phys. Lett. 47, 1309 (1985)
77. H. Nobuhara, O. Wada, and T. Fujii: Electron. Lett. 23, 35 (1987)
78. A.Y. Cho and I. Hayashi: J. Appl. Phys. 42, 4422 (1971)
79. C.E.C. Wood, J. Woodcock, and J.J. Harris: Inst. Phys. Conf. Ser. 45, 28 (1979)
80. K. Ploog, A. Fischer, and H. Künzel: Appl. Phys. 22, 23 (1980)
81. A.Y. Cho: J. Appl. Phys. 46, 1733 (1975)
82. K. Ploog and A. Fischer: J. Vac. Sci. Technol. 16, 838 (1978)
83. C.E.C. Wood and B.A. Joyce: J. Appl. Phys. 49, 4854 (1978)
84. C.E.C. Wood: Appl. Phys. Lett. 33, 770 (1978)
85. D.S. Jiang, Y. Makita, K. Ploog, and H.J. Queisser: J. Appl. Phys. 53, 999 (1982)
86. D.A. Andrews, R. Heckingbottom, and G.J. Davies: J. Appl. Phys. 54 4421 (1983)
87. D.A. Andrews, R. Heckingbottom, and G.J. Davies: J. Appl. Phys. 60, 1009 (1986)
88. Y. Suzuki and H. Okamoto: J. Appl. Phys. 58, 3456 (1985)
89. F. Briones, D. Golmayo, L. Gonzales, and A. Ruiz: J. Cryst. Growth 81, 287 (1987)
90. M. Tanaka, H. Sakaki, and J. Yoshino: Jpn. J. Appl. Phys. 25, L155 (1986)
91. F. Voillot, A. Madhukar, J.Y. Kim, P. Chen, N.M. Cho, W.C. Tang, and P.G. Newman: Appl. Phys. Lett. 48, 1009 (1986)
92. T. Isu, D.S. Jiang, and K. Ploog: Appl. Phys. A 43, 75 (1987)
93. L. Tapfer and K. Ploog: Phys. Rev. B 33, 5565 (1986)
94. D. Taupin: Bull. Soc. Fr. Minéral. Cristallogr. 87, 496 (1964)
95. P.V. Petrashen: Fiz. Tverd. Tela (Leningrad) 16, 2168 (1974); ibd. 17, 2814 (1975)
96. Special issue of IEEE J. Quantum Electron. QE-22, No. 9 (1986)
97. Proc. 6th Int. Conf. Electron. Prop. Two-Dimens. Systems (EP2DS-VI), Surf. Sci. 170 (1986)
98. Proc. 2nd Int. Conf. Modulated Semicond. Struct. (MSS-2), Surf. Sci. 174 (1986).

THE STRUCTURE OF MULTILAYERS:

X-RAY STUDIES

D. B. McWhan

AT&T Bell Laboratories

Murray Hill, New Jersey 07974 U.S.A.

INTRODUCTION

The quality and complexity of multilayer structures continues to increase, and so does the need for non-destructive accurate structural information. The purpose of these lectures is threefold. First, the importance of using x-ray scattering as the primary routine probe of sample quality is stressed. Second, model calculations of the evolution from disordered one-dimensional to ordered three-dimensional structures are presented. Finally, the different types of x-ray scattering geometries which are being used are compared. These lectures are meant to supplement the review of the structure of chemically modulated films prepared in 1982,[1] a discussion of coherence in multilayers with components of differing structure,[2] and a survey of the interfaces in multilayers.[3]

The early x-ray studies concentrated on the determination of the modulation wavelength of the multilayer and the coherency strain imposed by the epitaxial growth on a substrate with a different lattice parameter. In addition to this information, which is obtained from the position of reflections in reciprocal space, the limited intensity information was used to obtain approximate models for the modulations of both the chemical composition and the interplanar spacing. As more detailed engineering of the electronic band structure or the magnetic spin structure is done, more accurate knowledge of the actual composition and the position of each layer in an average modulation wavelength is needed rather than an approximate waveform. This requires measuring the intensities of a large set of reflections so as to do a complete crystal structure determination. Using the high flux and tunability of synchrotron sources and the anomalous dispersion effect it is now possible to contemplate obtaining such a data set for a few representative multilayers. As a multilayer is a one-dimensional system with various types of random errors which are cumulative, the structures do not have true long-range order, and it is important to determine the type and magnitude of the fluctuations in the structure. Are the interfaces atomically sharp or diffused, or are they rough with uncorrelated steps? This information is contained in the size and shape of the reflections in reciprocal space and the diffuse scattering between reflections.

If one looks at the problems involved in growing a multilayer, the quality of the materials which have been prepared is truly remarkable. Typically the layers are deposited at rates of the order of an Angstrom per second in MBE growth or 10 Angstroms per second in sputtering processes. This translates into 10^{15} to 10^{16} atoms/sec arriving on a substrate which is 1 cm × 1 cm. If the parameters to favor layer-by-layer growth have been found, then in order to deposit exactly a monolayer with an accuracy of 1% one must have a shutter system which is capable of turning on and off the beam of atoms arriving at any point on the substrate with an accuracy of the order of 0.01 second. This of course assumes that the deposition rate can be calibrated and maintained with a similar accuracy. Next one has to consider the surface of the substrate. If a monolayer has a typical height of 2 to 3Å then in order not to have any steps on the surface of a 1 cm × 1 cm substrate, the surface must be cut parallel to the crystallographic planes with an accuracy of better than 10^{-2} seconds of arc (10^{-6} degrees). More typically the substrate is cut with an accuracy of a few tenths of a degree, and this translates into a surface which has steps every few hundred Angstroms. These simple considerations show that any multilayer will not be a perfect single crystal with interfaces that are spaced so as to form a superlattice with true long range order or that are smooth on an atomic scale throughout. The problem then is to assess the type and magnitude of the deviations from this ideal picture.

There are a number of probes which are used routinely to measure the structural perfection of a multilayer. To begin with RHEED photographs are used to establish that the substrate surface is fairly smooth and then oscillations in the RHEED intensity during deposition are used to show that the growth is in a layer-by-layer mode. The difficulty of this technique is one of calibration. How does one know what fraction of the growth is in this mode? After deposition the structure can be studied by x-ray diffraction, electron diffraction or Rutherford back scattering. X-ray scattering contains a wealth of information, but it does present an average picture of the structure over the area of the sample illuminated by the x-ray beam. Transmission electron microscopy provides beautiful lattice image pictures on an atomic scale, but it too is an average through the sample and is therefore a projection of all the planes in the thin direction. Also the effect of the thinning process is unclear. Rutherford backscattering provides another measure of the average perfection of the channels in a crystal over the area irradiated by the beam. It has the added advantage of being sensitive to composition as a function of depth with a resolution of 100 Angstroms or so. After preparation, different physical properties of the material also provide indirect measures of structural perfection. In the following sections we will concentrate on X-ray scattering as a probe. We begin with a discussion of the scattering function, S(Q), for different types of disorder. In the second section the three-dimensional structure of multilayers is considered. In the final section the experimental aspects of x-ray scattering are presented with a special emphasis on contrasting the two and three crystal scattering geometries which are commonly used.

THE SCATTERING FUNCTION FOR 1-D MODEL MULTILAYER STRUCTURES

In this section the scattering function is calculated for a number of approximate models which qualitatively demonstrate some of the types of imperfections that can occur in all multilayer structures. Throughout this paper the notation will be that of the x-ray or neutron scattering community. The scattering will be described in terms of positions in reciprocal space rather than various diffractometer angles. In this notation Bragg's law has the form

$$|\vec{Q}| = 2k_i \sin\theta$$

$$\vec{Q} = h\vec{a}* + k\vec{b}* + l\vec{c}* \qquad (1)$$

$$k_i = 2\pi/\lambda \quad .$$

Throughout h, k, and l will be considered as continuous variables. For most of the paper the discussion will center on one-dimensional problems and unless otherwise noted, Q will be taken to be parallel to $c*$ with the vector notation eliminated. In general the growth process of a multilayer is such that the modulation wavelength will not be exactly an integral number of layers, and the modulation direction may not be parallel to one of the major crystallographic directions. In this case, a modulation wavevector is defined, and a reciprocal lattice is defined using a four-index notation. For systems with an underlying lattice of orthorhombic or higher symmetry the reciprocal lattice is defined as

$$\vec{Q} = H\vec{a}* + K\vec{b}* + L\vec{c}* + M\vec{k}_m$$

$$|\vec{a}*| = \frac{2\pi}{a} \;,\; |\vec{b}*| = \frac{2\pi}{b} \;,\; |\vec{c}*| = \frac{2\pi}{c} \;,\; |\vec{k}_m| = \frac{2\pi}{\lambda_m} \;, \qquad (2)$$

where H, K, L, M are integers; a, b, and c are the lattice parameters in real space and λ_m is the modulation wave length. In reciprocal space this gives a set of average reciprocal lattice points given by HKL with each point surrounded by a set of satellites which are equally spaced on each side of the average reciprocal lattice point at multiples of the modulation wave vector. In practice for simple modeling it is convenient to round off the modulation wave vector to the nearest integral number. In the following the scattering expected for more and more complex structures is developed starting with a set of N parallel planes which are repeated N_3 times and progressing to multilayers with N_1 layers of one material alternating with N_2 layers of a second material with a different structure.

In order to demonstrate the effect of fluctuations in a one-dimensional system, consider a multilayer of the type used for monochromators in the soft x-ray region.[4] Typically these consist of regions of a heavy metal such as tungsten or rhenium alternating with regions of a light material such as carbon in order to provide as large an x-ray contrast as possible. The tungsten is textured when deposited by sputtering techniques and has the (110) planes parallel to the film. The polycrystalline grains are random in the other two directions. By contrast the carbon regions are amorphous. The multilayer consists of a sequence of N_1 layers of tungsten with a spacing between layers of d_1 followed by regions of carbon with a spacing of $c_o + c_1$, where c_o is the average thickness and c_j is the deviation of the jth layer from the average. This sequence is then repeated N_3 times. For simplicity it is assumed that the scattering from the carbon regions is negligible relative to that of the tungsten. The scattering amplitude for this model is given by

$$F(Q) = f(Q) \sum_{n=o}^{N_1-1} e^{iQnd_1}\left[1 + e^{iQ[N_1 d_1 + c_o + c_1]} + \ldots \right.$$

$$\left. \ldots + e^{iQ[N_3(N_1 d_1 + c_o) + \sum_{j=1}^{N_3-1} c_j]} \right] \qquad (3)$$

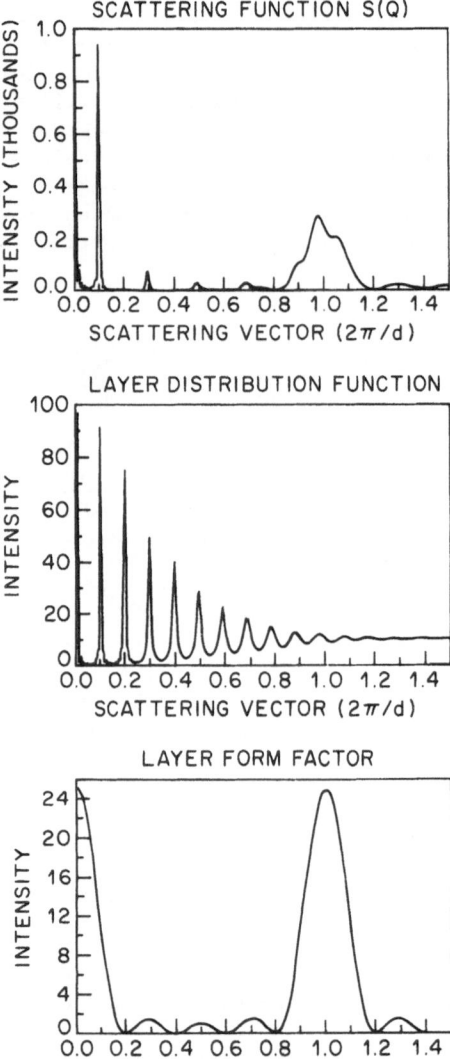

Figure 1. (Bottom) Form Factor for five equally spaced layers of atoms. (Middle) The layer distribution function for 10 repeats with a random fluctuation in distance of 0.5 of a layer spacing. (Top) The scattering function which is the product of the middle and bottom curves.

where $f(Q)$ is the scattering power per layer of tungsten. As pointed out by Sevenhans et al.[5] in order to calculate the total scattering cross section, $S(Q)$, one must average $\langle F(Q)F*(Q)\rangle$ in order to include both the Bragg and the diffuse scattering instead of calculating $\langle F(Q)\rangle\langle F*(Q)\rangle$ which is just the Bragg component (as the author had done previously).[6] Expanding and collecting terms leads to the expression

$$S(Q) = f^2(Q) \frac{\sin^2(N_1 Q d_1/2)}{\sin^2(Q d_1/2)} \left[N_3 + 2 \sum_{j=1}^{N_3-1} (N_3-j)\cos Qj(N_1 d_1 + C_o) \right.$$

$$\left. x \langle e^{iQc_j} \rangle^j \right] .$$

(4)

This expression assumes that the individual c_js are statistically independent variables. If the c_j have a continuous Gaussian distribution then the errors are cumulative. This is the model for a paracrystal in which the most probable spacing between tungsten regions is c_o, $2c_o$, $3c_o$, ..., Mc_o, but unlike a crystal the corresponding widths of the probability distribution for successive neighbors increase as σ, 2σ, 3σ, ..., $M\sigma$. For the distribution the average value is

$$\langle e^{iQc_j} \rangle = \frac{1}{\sigma\sqrt{\pi}} \int_{-\infty}^{+\infty} e^{iQc} e^{-c^2/\sigma^2} dc = e^{-Q^2\sigma^2/4} .$$

(5)

Substituting the average value from Eq. 5 into Eq. 4 gives $S(Q)$ for this model of a tungsten-carbon multilayer. If the scattering vector is written in the form $Q = 2\pi\ell/d$ where d is the spacing between tungsten layers and ℓ is a continuous variable, then we can follow the evolution of the scattering as a function of the width of the distribution, σ. The first term in Eq. 4 is the scattering from each individual block of N_1 layers, and the second is the Fourier transform of the layer-layer correlation function. These two terms are plotted in Fig. 1 along with the total intensity for a distribution with a width $\sigma = 0.5d_1$. The calculation was done for a sequence of 5 layers of tungsten followed by a nonscattering region with $c_o = 5.22d_1$ with the sequence repeated 10 times. The layer distribution function approaches 10^2 in the limit of $\ell = 0$ and the peaks become broader and weaker with increasing ℓ. The layer form factor has a maximum of 5^2 and the product of the two terms shows sharp reflections for $\ell \to 1$ and broad unresolved reflections for $\ell \approx 1$. In Fig. 2 the scattering near $\ell = 1$ is compared for two values of σ. Reducing the width of the distribution from $0.5d_1$ to $0.2d_1$ leads to an increase in the intensity and a narrowing of each reflection. These calculated curves are similar to those reported in the literature for multilayer structures of this type and this suggests that the variations in thickness during growth are of the order of 2% to 5%.

Recently Clemens and Gay have shown that the results are quite different if instead of an amorphous region with a continuous distribution of layer thickness, the second region is composed of layers with fluctuations in the integral number of layers in each region.[7] In this case if the layer spacing for the second component is d_2, then the average is given by

$$\langle e^{iQnd_2} \rangle = \frac{1 + 2 \sum_{n=1}^{\infty} \cos(Qnd_2) e^{-n^2 d_2^2/\sigma^2}}{1 + 2 \sum_{n=1}^{\infty} e^{-n^2 d_2^2/\sigma^2}} .$$

(6)

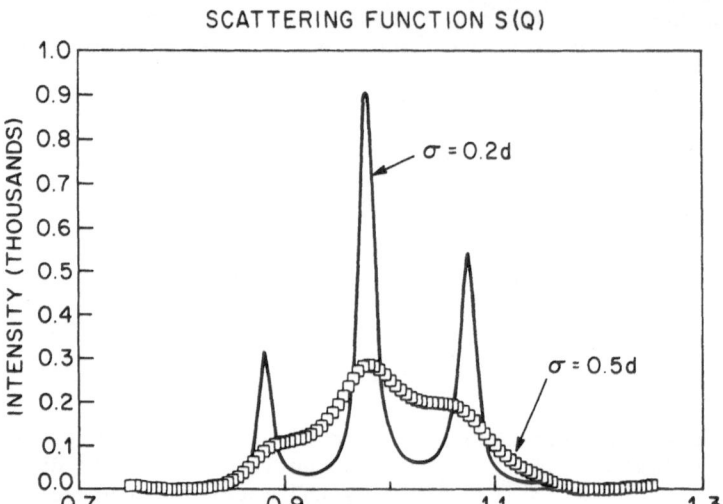

SCATTERING FUNCTION S(Q)

Figure 2. Comparison of S(Q) for values of σ.

The average value calculated using the continuous distribution (Eq. 5) and the discrete distribution (Eq. 6) are markedly different. Both start with an average value of one at Q = 0. With increasing Q the average value of Eq. 5 drops smoothly, but the average value of Eq. 6 is periodic and has a value of one for all integral multiples of $Q = 2\pi/d_2$ because all the cosine terms equal one and then the numerator and the denominator are identical. The average value of Eq. 6 is a superposition of a series of curves of the form of Eq. 5 on each side of every integral value of ℓ. The resulting layer distribution function will have sharp peaks at integral values of $2\pi\ell/d_2$ with satellites on each side. The intensities of the satellites decrease and the widths of the satellites increase with increasing order. There is a marked difference in the resulting diffraction patterns. The high angle lines are all broad for the continuous distribution but the lines are sharper for the discrete distribution. Clemens and Gay[7] compare their calculations with measurements on Ti-Ni and Mo-Ni multilayers. The former shows a continuous distribution while the latter is more compatible with a layered model.

The models presented above are for scattering from N_1 layers of one material with different types of fluctuations in the thickness of the second component. We now progress to multilayers in which both components are layered but with different layer spacings and scattering powers. What would one expect for such multilayers if the growth conditions are such that layer-by-layer growth is achieved for both components? Even if it is assumed that the substrate has been prepared so that it is atomically smooth with no steps, there will be errors in the opening and closing of shutters such that a layer will only be partially complete when one stops depositing component A and begins to deposit component B. This will lead to steps in the composition and presumably there will be steps in composition at each interface throughout the multilayer structure. As an extreme example, assume that the shutters close 5% early at the changeover and that nucleation takes place at one random place for each new layer and that the nominal expected composition was to be $N_1 = 5$ and $N_2 = 5$. A structure generated with this prescription is shown schematically in Fig. 3. Instead of having 5 layers of A followed by 5 layers of B, the average modulation wavelength would be $4.75 + 4.75 = 9.5$ layers. The composition modulation

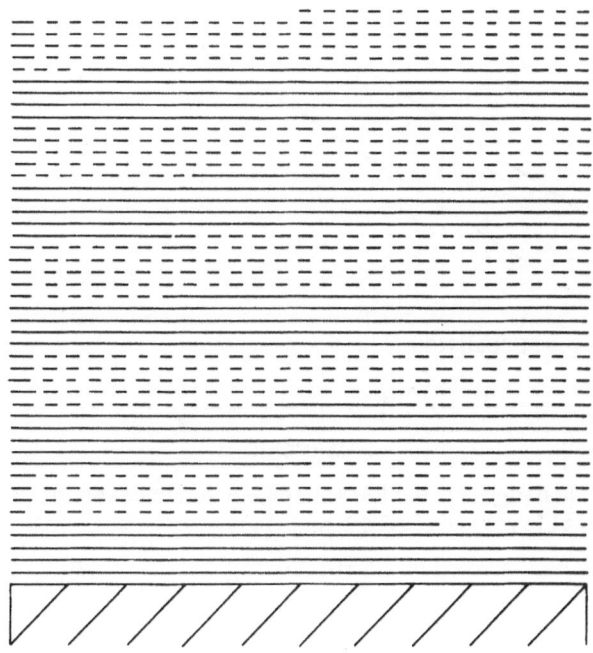

Figure 3. A nominal 5/5 multilayer grown with a 5% error in deposition rates.

fluctuates from place to place in the film. For example on the left side the sequence is 5, 4, 5, 5, 4, 5, 5, 5, 4, 5 whereas on the right it is 4, 6, 4, 5, 5, 4, 5, 5, 4, 6. If the lateral coherence of each sequence is large compared with the coherence normal to the film then a reasonable model might be to assume a random distribution of repeat sequences such as (5,5), (5,4), (4,5), (6,5), (5,6) which had probabilities given by a Gaussian distribution about the average composition of (4.75, 4.75). $S(Q)$ for this type of model has been developed in the context of clays with differing water content in different layers. The original calculation is due to Hendricks and Teller,[8] and the problem has been formulated for a finite number of layers in matrix form by Kakinoki and Komura.[9] The general result of the Hendricks-Teller model is that the widths of the diffraction lines are increased because locally there are repeat sequences of 9, 10, and 11 layers each with a different average spacing and modulation wavelength. This is to be contrasted with models in which the interfaces are assumed to be flat but which have interdiffusion of one component into another. In the case of diffused interfaces there are no local deviations in the periodicity so that no broadening of the reflections occurs. In practice most multilayer structures show linewidths which increase with the order of the harmonic (see, for example, Fig. 4), but as there are many possible causes of line broadening, it is difficult to unambiguously conclude that the structure

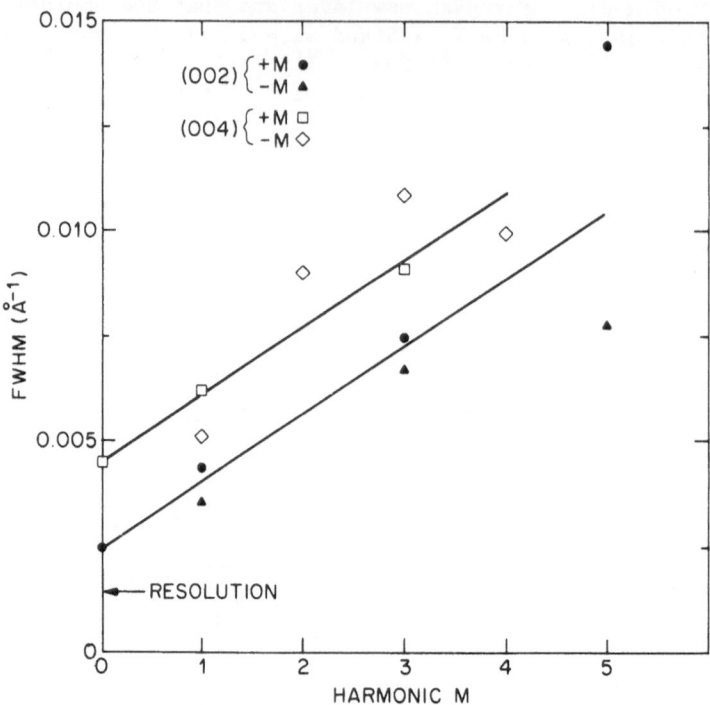

Figure 4. Broadening or harmonics as a function of M in a $Gd_{21}Y_{21}$ Multilayer (Refs. 12 and 14).

has stepped interfaces rather than diffused interfaces on the basis of the measured linewidths. For example, if the ovens are not normal to the substrate surface, then there may be a chemical gradient across the sample, and this would lead to a spread in modulation wavelength across the sample. This is a more macroscopic wavelength spread than the local fluctuations shown in Fig. 3. The effect on S(Q) will be qualitatively similar, namely a linear increase in the widths of the harmonics with order.

The discussion above assumes that the planes are all parallel in which case the widths of the reflections in the directions transverse to the surface normal would be infinitely sharp. Measurements of the transverse widths of real multilayers show that they are seldom limited by the resolution of the diffractometer. Usually the widths of the satellite reflections increase with satellite order around each average Bragg reflection. Often the transverse widths are oscillatory functions of the azimuthal angle, i.e. as one rotates the transverse component of the scattering vector around in the plane of the film. The transverse width is a measure of the coherence lengths in the plane of the film. There are a number of obvious ways in which the coherence can be limited. If the surface of the substrate is rough then the average distance over which there are no errors in the structure in the transverse direction is limited. Furthermore as the satellites are related to the Fourier components of both the chemical modulation and the interplanar spacing modulation, the higher order satellites are more sensitive to the interface. In effect the transverse coherence of the higher order satellites will be less for rough interfaces and therefore the transverse

widths of the satellites will increase with order. An extreme example of
this is a multilayer grown on a substrate which is miscut. In this case
the average surface will not be parallel to the crystallographic planes
and therefore the modulation wavevector will not be parallel with c*. As
illustrated in Fig. 5 the satellites, which lie along k_m, will be shifted
progressively further and further away from c*. There is some evidence
that the average coherence length is approximately equal to the average
step spacing in the direction perpendicular to the steps and longer in
the direction parallel to the steps.[1],[10] One could envision a rough
surface as a collection of miscut regions and therefore each satellite
would be an average over all such regions leading to increased
linewidths.

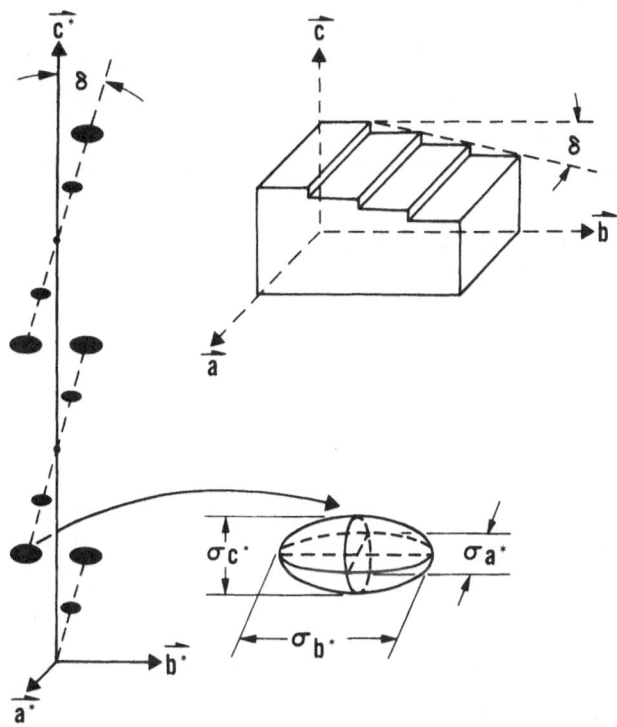

Figure 5. Reciprocal lattice for a multilayer grown on a
stepped surface.

The models discussed above demonstrate that some disorder will be
expected in the interfaces of multilayer structures. Furthermore this
disorder is reflected in the broadening of the satellite reflections.
This broadening is of two types, namely along the surface normal and
transverse to it. The former will be related to the height of the
interfacial roughness. In our example there are deviations from the
average modulation wavelength of plus or minus 1, 2, or more planes. The
latter will give a measure of the length scale parallel to the film for
the oscillation in the interface position perpendicular to the film.
Next we turn our attention to the composition of the multilayer. In the
example shown in Fig. 3, a scattering experiment would give average Bragg
reflections flanked by satellites at positions corresponding to an

average repeat distance of 9.5 layers. A priori there is no reason
.without resorting to intensity arguments to choose between compositions
given by (4.5, 5), (4,75, 4.75), (5, 4.5) or any ratio in between which
sums to 9.5. An approximate model for the average structure can
determine whether one component has a larger average composition.

The average structure can be determined at different levels of
detail depending on the amount of intensity data available. For example
for the hypothetical structure shown in Fig. 3, one might start with the
approximation that there are 5 layers of each component. This would lead
to a large supercell containing 10 layers. One would need both an
average composition and a positional parameter along c for each layer
plus a scale factor and at least an average temperature factor for each
of the components. This amounts to 23 parameters. In order to have more
measurements than parameters one would need to measure the intensities of
5 orders of harmonics on both sides of at least three orders of average
reflections. This is substantially more than are usually available, and
one must resort to various approximate solutions. The first order
approximation is to assume that the interfaces are sharp, that the
interplanar spacing is constant for each component and that at the
interfaces the spacing is the average of that of the two components.
This is the "step" model introduced by Segmuller and Blakeslee,[11] and it
is given by

$$
S(Q) = \left[\frac{\sin(N_3 \lambda_m Q/2)}{\sin(\lambda_m Q/2)} \right]^2 \left[f_1 \left[\frac{\sin(N_1 d_1 Q/2)}{\sin(d_1 Q/2)} \right]^2 \right.
$$

$$
+ f_2 \left[\frac{\sin(N_2 d_2 Q/2)}{\sin(d_2 Q/2)} \right]^2 \tag{7}
$$

$$
\left. + 2 f_1 f_2 \, \cos(\lambda_m Q/2) \, \frac{\sin(N_1 d_1 Q/2)}{\sin(d_1 Q/2)} \, \frac{\sin(N_2 d_2 Q/2)}{\sin(d_2 Q/2)} \right],
$$

where

$$
\lambda_m = N_1 d_1 + N_2 d_2 \quad .
$$

This model is illustrated in Fig. 6 which is based on work in the
Nb-Al multilayer system.[6] It illustrates how significant information
about a multilayer structure can be obtained just from visual inspection
of the data. In this case the Nb is expected to form with the (110)
planes parallel to the substrate and the Al(111) planes. The d spacings
for these two planes are almost the same in the pure materials. In the
absence of a difference in interplanar spacing the intensities of the
harmonics vary as the square of the difference of the atomic scattering
power times the Fourier component of the composition modulation. The
intensities of the plus and minus harmonics should be the same. This is
illustrated in the top boxes where the dotted lines are the function in
the brackets and the solid shaded curves are S(Q). However, the measured
intensities showed that the minus harmonics were more intense. This in
turn indicates that the d spacing in the Nb region is larger than the
corresponding d spacing in the Al region, and therefore, that the
multilayer must have some amount of coherency strain at the interface to
cause this difference. This, in fact, was confirmed by measurements of
the in plane lattice parameters of the Nb as a function of modulation
wavelength. The step model gives information about the relative
scattering powers and corresponding d spacings. If the heavier component

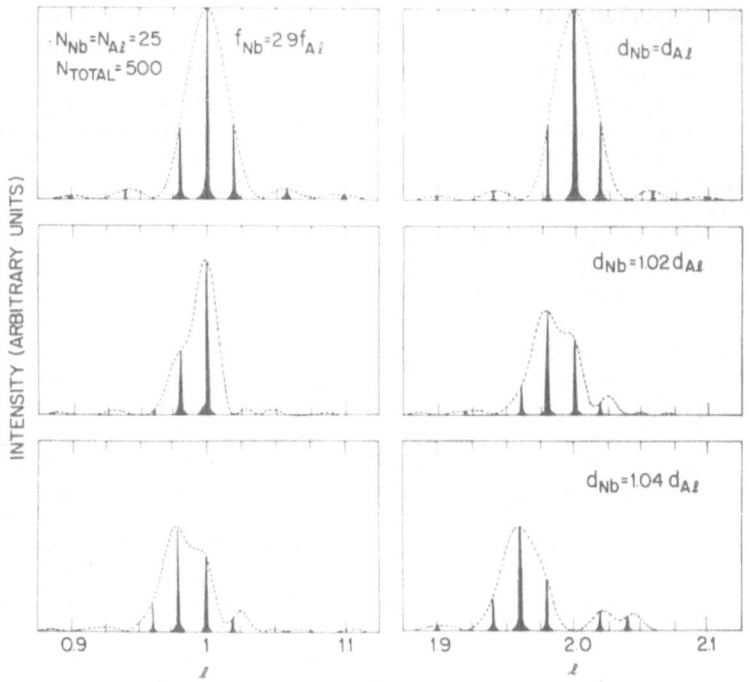

Figure 6. Scattering function and layer form factor (dotted curves) for a step model (Eq. 7) with different interplanar spacings.

is larger then the minus harmonics are stronger and vice versa. As the difference in d spacing increases the asymmetry becomes more pronounced. The intensity of the central average reflection decreases and the intense harmonics are centered around the d spacing of the individual components. This is commonly seen in strained layer superlattices.

The second approximation to the average structure is to allow the average interface to have a finite width. This can be approximated by a trapezoidal wave or if one wants an approximation that gives more rounded interfaces then a sum of error functions can be used. In both cases there are independent waves both for the composition and for the interplanar spacing which for the error function model are given by

$$C(N) = \frac{C_o}{2} \left[ERF \left(\frac{N_1 + 2N}{N_I^c} \right) + ERF \left(\frac{N_1 - 2N}{N_I^c} \right) \right]$$

$$\Delta d(N) = \frac{e}{2} \left[ERF \left(\frac{N_1 + 2N}{N_I^e} \right) + ERF \left(\frac{N_1 - 2N}{N_I^e} \right) \right] \quad .$$

(8)

This model has eight parameters which include the average composition, the asymmetry in the form of the number of layers of the first component, the amplitude of the interplanar spacing modulation, two interface width parameters, two temperature factors and a scale factor. With a typical data set of 20 to 30 reflections the parameters are reasonably well determined. Typical weighted quadratic R factors are in the range of 0.1 to 0.2 which is rather poor by crystallographic standards.[12] It must be remembered, however, that the dynamic range represented by a typical data set is over 5 orders of magnitude, and this is larger than most crystallographic data. These approximate models are sufficiently accurate to provide information on changes in either interplanar spacing or in composition which result from annealing, chemical treatment or changes due to magnetostriction. For example, when hydrogen is absorbed in a transition metal the lattice parameter expands. Miceli et al. studied the hydrogen uptake in multilayers of Nb/Ta.[13] The hydrogen goes preferentially into the Nb and the resulting changes in the interplanar spacing in both the Nb and the Ta can be obtained from the intensities of the harmonics using approximate models similar to those discussed above. A second example involves multilayers composed of alternating magnetic and non-magnetic components. In a Gd/Y multilayer the interplanar spacing in the Gd regions increases with decreasing temperature below the Curie temperature as a result of magnetostriction. At the same time the spacing in the Y regions decreases as a result of normal thermal contraction. Analysis of the x-ray data taken as a function of temperature using the approximate model given by Eq. 8 yielded values for the magnetostriction in the ferromagnetic regions which compared favorably with the magnetostriction observed in bulk Gd.[14] In the cases where the interplanar spacing is similar for both components a number of workers have successfully Fourier transformed the satellite intensities to determine the composition modulation. Typically the interfaces are of the order of one to two atomic layers thick. This type of measurement has been used to study interdiffusion and a number of studies have been made as a function of annealing. These results will be discussed by Spaepen.[15]

In order to do substantially better in the determination of the average structure, it will be necessary to obtain larger data sets. The obvious way to do this is to use anomalous dispersion. Synchrotron sources provide a tunable source and because of the higher flux one can observe the weaker reflections. The atomic scattering factor has dispersion corrections near absorption edges which can substantially change the scattering power. The structure factor is given by

$$F = \sum_{j} (f_j^{\circ} + f_j' + if_j'')e^{iQz\lambda_m} , \qquad (9)$$

where f' and f" are the dispersion corrections and z is the positional parameter as a fraction of the repeat distance. If data sets are collected at several energies the correlations between the position and scattering power parameters in a least squares refinement is reduced. The composition component of the intensity of the harmonics depends on the difference between the atomic scattering factors of the two components. For a multilayer composed of Gd and Y the difference is (64-39) at Q = 0 for energies far away from an absorption edge, but the difference actually changes sign near the edge because f' for Gd approaches -30 electrons.[16] The size of this effect is illustrated in Fig. 7 for the intensity of the (004) and the (004-1) near the Gd L_3 edge.[17] It is hoped that with synchrotron data of this type much more accurate average structures will be determined.

Up to this point in the development of S(Q) for multilayers it has been tacitly assumed that the thickness of the film was small enough that the kinematic theory of x-ray scattering was a reasonable approximation. This in fact has been true for almost all the past work, but there are two areas where the samples fall somewhere in between the limits where kinematic and dynamical theory are valid. The first is in the limit of small scattering vectors and the second is in some of semiconductor multilayers. In the latter case the strongest reflections suffer from extinction effects, and the measured intensities of these reflections can only be used with caution. The first term in Eq. 7 is the scattering for N_3 repeats of the modulation wavelength. The full width at half maximum, FWHM, of the main reflection is given by $\delta Q/Q = 0.89/N_3$. If the layers continue to add coherently, then the width continues to get smaller. This model neglects the fact that at each plane part of the incident beam is reflected so that the incident beam arriving at the next plane is less. When this process is repeated a large number of times the incident beam is reduced to a negligible intensity, and the distance at which this occurs is referred to as the extinction depth. This can be a distance that is much smaller than the depth given by normal photoelectric

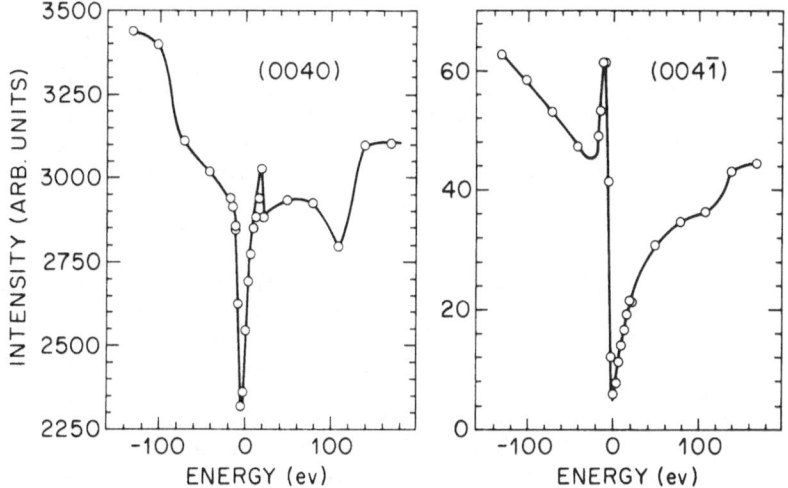

Figure 7. Intensity vs. energy around L_3 absorption edge of Gd in a $Gd_{21}Y_{10}$ multilayer (Ref. 17).

absorption. In essence only a thin layer near the top of the crystal does the diffracting. This, in fact, is the origin of the width of a reflection within dynamical theory. Expressing the extinction depth as a function of Q and substituting it into the expression for the width of a reflection from a small crystal yields an expression that is functionally identical to that for the Darwin width given by dynamical theory with only minor differences in the numerical factors as shown in Eqs. 10 and 11. These equations have been simplified by assuming that the incident beam is linearly polarized and that the imaginary part of the structure factor can be neglected.

$$t = \frac{VQ}{4\pi r_e |F|} \quad \text{extinction depth}$$

$$\Delta Q_{crystal} = \frac{70.3 r_e |F|}{VQ}$$

(10)

$$\Delta Q_{darwin} = \frac{50.2 r_e |F|}{VQ}$$

(11)

where $r_e = \dfrac{e^2}{4\sigma mc^2} = 2.82 \times 10^{-5} \text{Å}$

V is unit cell volume .

Dynamical effects become important when the observed FHWM approaches the Darwin Limit from above. For the majority of multilayers prepared to date this limit is seldom reached except at small Q or for reflections with large values of $|F|$. As discussed above and as shown in Fig. 4, there is a large variation in the relative linewidths as a result of disorder, so some lines are effected more than others. A table of extinction depths and Darwin widths for many relevant semiconductor multilayer materials is given by Tapfer and Ploog.[18] A complete discussion of this problem is given by Bartels et al. who use the criterion that the observed peak reflectivity must be less than 10% before one is safely in the kinematic regime.[19] Unfortunately many studies are not done on an absolute scale so that the true reflectivity is not determined. As pointed out by Macrander et al.[20] and by Speriosu and Vreeland[21] for structures with graded strain layers the kinematic model works well far from the substrate peak but dynamical theory is needed as the bulk substrate peak is approached.

In addition to dynamical effects becoming important in the limit of Q = 0, there are also optical effects which result from the fact that the index of refraction is less than one. This leads to the phenomena of total external reflection and the shifting of the position of the low order satellites in a multilayer. As will be discussed by Spiller,[21] reflectivity measurements near total reflection provide detailed information about the surface layers. Here we only refer to two recent uses of the refractive index shift. The refractive index is given by Eq. 12 which when combined with Snell's law gives the shift in Q:

$$n = 1 - \sum_j \frac{c_j \lambda^2 r_e n_j}{2\pi} (f_j{}^\circ + f_j{}' + i f_j{}'')$$

(12)

where n_i is atomic number density.

$$Q_{obs} \approx Q_{bragg} + \frac{1}{Q_{bragg}} \sum_j 8 c_j \pi r_e n_j (f_j{}^\circ + f_j{}') .$$

(13)

Miceli et al.[23] have measured the shift for a series of satellites around the origin in several different multilayers. If the dispersion corrections are known for each of the components then the only unknowns in the equation are the concentrations, c_j. The authors demonstrate that the composition of the multilayer can be determined with reasonable accuracy by this technique. This then provides an independent x-ray measurement of the composition from the analysis of the intensity data discussed above. Again the size of the shift can be tuned by varying the energy and consequently f' for each of the components. On the other hand if the composition is known then tuning the energy through the absorption edge and measuring the shift gives directly f' as a function of energy (see Barbee et al.[24]).

In this section we have tried to build up a picture of the structure of typical multilayers and to show how this structure is reflected in x-ray scattering measurements. The broadening of the reflections both normal and parallel to the surface is a measure of the interface roughness. The intensities of the reflections are used to develop approximate models for the average composition and interplanar spacing modulations. Comparison of these models with x-ray measurements for a range of multilayer systems suggests that the typical interface thickness is of the order of one to two atomic layers, and the models are accurate enough to give reliable measurements of the composition. This is supplemented by results based on the reflective index shift of low angle reflections. Reasonable models for the interplanar spacing modulation are also derived from the intensity measurements and they can be used to study changes resulting from magnetostriction, diffusion, or to study changes in coherency strain with external parameters such as thermal treatment.

THE THREE-DIMENSIONAL STRUCTURE OF MULTILAYERS

We now turn to the structure in the plane of the multilayer. The models given above are one dimensional and are based on measurements made with the scattering vector normal to the multilayer. A multilayer structure often involves a series of different regions that are deposited sequentially. As illustrated in Fig. 8, there may be a buffer layer deposited on the substrate. This may provide a chemical barrier as in the case of rare earth multilayers grown on sapphire, where a niobium buffer prevents the reaction of the rare earth with the sapphire.[12]

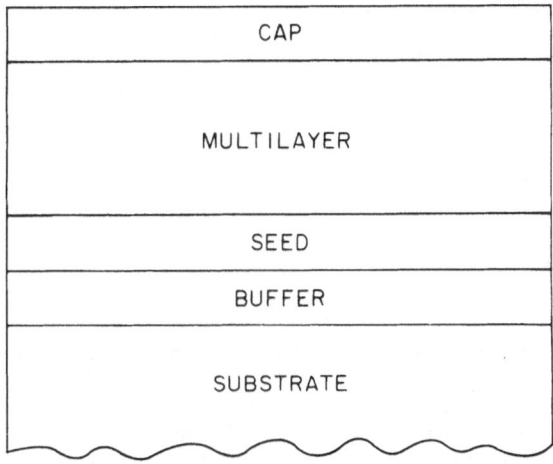

Figure 8. Typical multilayer structure.

In the case of strained layer superlattices the buffer may be a layer of graded composition which brings the average lattice parameter parallel to the film to a value close to that of the desired superlattice. Next there is usually a seed layer of material with the appropriate average composition, and this layer is grown until RHEED measurements indicate a smooth surface has been obtained. Only then is the superlattice grown. After the superlattice is grown a protective cap of some type is likely to be deposited. One would like to determine the epitaxial relations between the substrate and the buffer layer and then the relation between the buffer layer and the seed layer. In addition to the relative crystallographic orientations, one or both of the layers may have strains induced by the epitaxial growth. In order to determine these effects x-ray measurements must be made over a much wider region of reciprocal space than just along the normal or near to the normal to the film as discussed in the last section.

The most important problem is that of the coherence of the multilayer itself. If the lattice parameters of the two components of a multilayer are different, then one or both of the components will be strained so as to give a single lattice parameter in each direction parallel to the film. As the multilayer is grown on a seed layer, the in-plane lattice parameter may be tied to that of the seed layer in which case one or both of the components may be under a biaxial strain. At the other extreme the multilayer may be floating and only have perhaps an orientational relationship with the seed layer. These different cases are shown schematically in reciprocal space in Fig. 9. (The satellites have been excluded for clarity.) The figure exaggerates the example of a multilayer composed of an equal number of layers of Gadolinium and Yttrium. Both elements are hexagonal close packed and the multilayer grows with the close-packed (001) planes parallel to the film. The in-plane lattice parameters of the pure elements differ by 0.3% with a (Gd) = 3.6360A and a(Y) = 3.6474A. The average strained lattice parameter, a_s, of the multilayer is calculated using the appropriate combination of elastic constants for hexagonal symmetry. The inplane strain in each material leads to an expansion or contraction along the c axis given by ϵ_c. The resulting expressions are:[24]

$$a_s = a_Y - (a_Y - a_{Gd}) \frac{c_{Gd}}{c_{Gd} + \frac{a_{Gd}}{a_Y} c_Y}$$

$$\text{with} \quad c = c_{11} + c_{12} - 2 \frac{c_{13}^2}{c_{33}} \tag{14}$$

$$\epsilon_c = -2 \frac{c_{13}}{c_{33}} \left(\frac{a_s - a}{a} \right) \quad .$$

Using the elastic constants for Gd and Y and equation 14 gives a_s = 3.6422A and c axis strains of ϵ_c = -9.8×10^{-4} (Gd) and ϵ_c = $+7.5 \times 10^{-4}$ (Y). Adding these strains to the lattice parameters for the pure elements gives strained parameters of c(Gd) = 5.7769 and c(Y) = 5.7349. The diffraction pattern for the multilayer will have an average c parameter of c_s = 5.7559. Because the elastic constants of Gd and Y are similar in magnitude the relative strains in each material are of the same order of magnitude and the average lattice parameter is quite close to the average of the two components, namely 5.7566. However the interplanar spacing in the Gd and Y regions will be different from the pure elements with the spacing in the Gd regions being 2.8885 instead of 2.8913 and in the Y regions being 2.8675 instead of 2.8653. These are the values that one would observe if the multilayer does not have a strain imposed by the substrate. As illustrated in Fig. 9, the diffraction pattern in this case will be composed of the reflections from the Y substrate and then of an average reciprocal lattice for the multilayer in which the component of each reciprocal lattice vector normal to the film will be displaced to smaller values than the corresponding lattice for the yttrium. At the same time the component parallel to the film will be displaced to larger values because the average strained a lattice parameter is smaller in real space and therefore larger in reciprocal space. If the multilayer grows epitaxially on the yttrium seed layer then the in-plane lattice parameter of the gadolinium will be strained to match that of yttrium. It follows from Eq. 14 that the average c axis in each gadolinium region will be reduced even further with the strain being $e(Gd) = -18.1 \times 10^{-4}$ The Gd c lattice parameter will be $c(Gd) = 5.7722$ and the average of the Gd and the Y will be $c_s = 5.7514$. This is a value for the average c lattice parameter which is in between the value for the yttrium substrate and that for a floating multilayer. Since in this limit the lattice parameter parallel to the film is clamped to that of the seed layer there is no displacement of the reciprocal lattice points parallel to the film. Careful measurements of the reciprocal lattice parameters for the

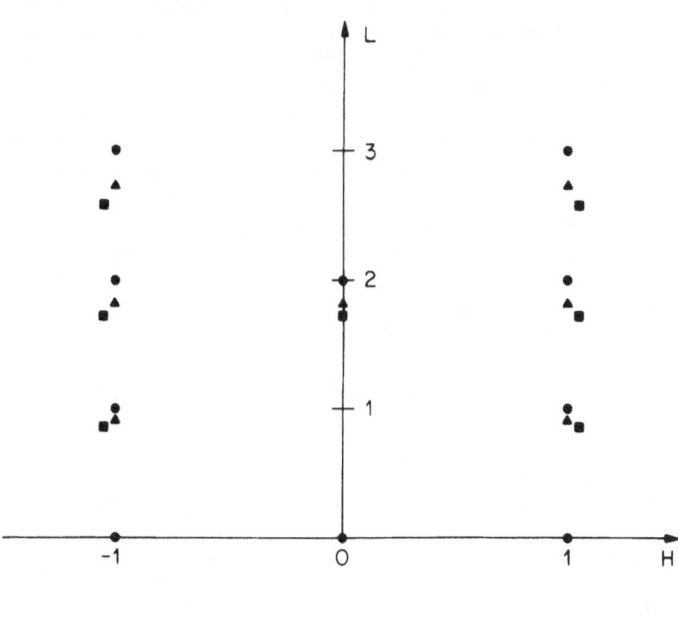

Figure 9. Reciprocal lattice showing exaggerated shift in peak position from Y lattice for a clamped and a floating Gd/Y superlattice.

multilayer relative to the seed layer provide a measure of the average coherence of the multilayer. In our example of Gd/Y multilayers, the in-plane component was found to be much closer to the value expected for multilayers which are floating. The a lattice parameters expected for clamped and floating conditions are 3.6474A and 3.6422A respectively, whereas the measured values for samples with repeat sequences of 10/10 and 21/21 were 3.6427 and 3.6416 respectively. Properties which depend on the axial ratio such as magnetocrystalline anisotropy may change significantly depending on the degree of coherence. In the case of multilayers composed of yttrium and the heavier rare earth metals, the normal sequence of magnetic phase transitions is altered. For example, dysprosium metal orders magnetically into a spiral structure with decreasing temperature. At a lower temperature there is a transition in the pure metal from the spiral, antiferromagnetic structure to a ferromagnetic structure. In multilayers the uniaxial strain imposed on the Dy leads to a suppression of the second transition.[25,26]

In addition to measuring the average reciprocal lattice parameters, the widths of the reflections can be measured and some information about the correlation lengths can be obtained. In previous sections the widths of the harmonics around the average reflections were shown to be related to fluctuations in the modulation wavelength and also to the roughness of the interfaces. By measuring the widths of the average reflections, information about the perfection of the average lattice can be obtained. For example, is the lattice coherent all the way through the film? Is the film coherent in the plane of the film for large distances or are there many misfit dislocations which relieve the strains? All models of film growth predict that for very thin films there will be a high degree of coherency but at some critical thickness the coherency will start to decrease. This will first be seen in transmission electron microscopy, but the loss of coherence will also be observable as an increase in the in-plane width of the average reflections. The system in which the most detailed measurements have been made is that of bilayers of Si and Si_xGe_{1-x}. By varying x the degree of epitaxial strain varies from 4% at x = 0 to 0% at x = 1. The critical thickness was measured as a function of strain using a combination of x-ray, RBS, and TEM measurements.[27] The results were not in agreement with the early models such as those due to Franck and Van der Merwe. These are equilibrium theories and it is apparent that there are barriers to the nucleation of misfit dislocations and these have to be taken into account when calculating the critical thickness which is measured in the laboratory.[28]

Another example is the changes in local order, coherence, or composition which may occur during annealing. It has been reported that a multilayer of alternating regions of 225Å of Si and 75Å of $Si_{0.4}Ge_{0.6}$ which is repeated 20 times undergoes an ordering of the Si and Ge in the mixed layer during annealing.[29] X-ray diffraction studies of this multilayer before and after annealing showed that the sample had broken up into two distinct regions after annealing because all the reflections (both average reflections and harmonics) split on annealing.[30] The modulation wavelength of each region was the same with one set of reflections moving closer to the substrate peak and the other set moving away. From the discussion above, it is evident that there will be three contributions to the position of the average peak of the multilayer with respect to the substrate peak. First the lattice parameter of an alloy with a given value of x will be different from that of Si so that in the multilayer the average of the Si layers and the alloy layers will be different from that of Si. This is a contribution from the composition and any local changes in composition during annealing might result in regions which are richer in or and poorer in Ge. This would lead to the average peak broadening or splitting. The latter would occur if there were two definite compositions that were thermodynamically stable. The second contribution to the position of the average peak is that an ordered phase could have a different lattice parameter than the disordered alloy. The splitting of the reflections would result from regions of ordered and regions of disordered Si-Ge. The third contribution to the position of the reflections is the coherency strain resulting from the epitaxial growth of the film on a Si substrate. The alloy will be compressed in the plane and, as a result, the lattice parameter of the alloy normal to the film will increase. If, during the annealing, parts of the sample lose some of the coherency strain then the average alloy peak would broaden or split depending on whether there was a continuous or discrete distribution of coherency strains. The relative importance of the three contributions is not known at this time.

Up to this point we have been discussing various contributions to the coherence of a multilayer structure and how coherency strains lead to changes in both the in-plane lattice parameters and to the interplanar spacings of the layers in the different components. Some information can be inferred from x-ray measurements with the scattering vector normal to the film but the in-plane lattice parameter should be measured directly by determining the positions of off-axis reflections in reciprocal space (as illustrated in Fig. 9). A large fraction of the multilayers which have been studied are composed of materials with similar structures such as close-packed metal structures, the diamond or zinc blend structure or the NaCl structure. These structures may grow with different orientations or have stacking faults so that the resulting multilayer may have growth twins or different stacking sequences depending on the modulation wavelength or the growth conditions. These aspects of the structure are determined from measurements along rows of reciprocal space which are normal to the film but with constant in-plane components. As an example consider metals which grow with the hexagonal close-packed planes parallel with the film. The reciprocal lattices for the two orientations of the face-centered cubic structure and the hexagonal close-packed structure are sketched in Fig. 10. If a fcc structure is

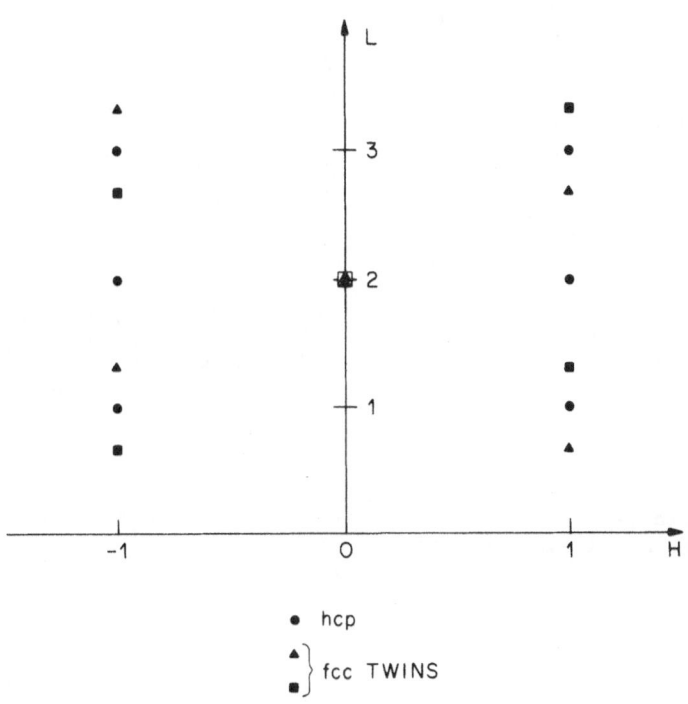

Figure 10. Reciprocal lattice for superimposed hcp and fcc structures.

growing on a substrate with hexagonal symmetry then there is no mechanism to favor one orientation over the other and the multilayer will be composed of regions of each orientation. An example of this type of twin growth is the Cu-Ni system grown on mica.[31] A more complicated example is the growth of a multilayer composed of materials which have the fcc and the hcp structure. A recent example is the Ru-Ir system.[32]

Ruthenium is hcp and iridium is fcc. For short modulation wavelengths
the hcp structure is stabilized in the Ir regions so that the multilayer
is hcp throughout, but only at long modulation wavelengths does the Ir
adopt the fcc structure which is found in the bulk material. X-ray
measurements made along the [10.ℓ] show reflections only at hcp positions
at short wavelengths and a mixture of fcc and hcp at long wavelengths.
In hcp metals the hexagonal symmetry precludes growth twins, and stacking
faults result in regions which have the fcc structure. The existence of
stacking faults leads to broadening of the off axis reflections and the
broadening has a well-known characteristic pattern. In the multilayers
composed of the hcp rare earth metals the stacking fault density is not
measurable as illustrated in Fig. 11 which shows a scan along the [10.ℓ]

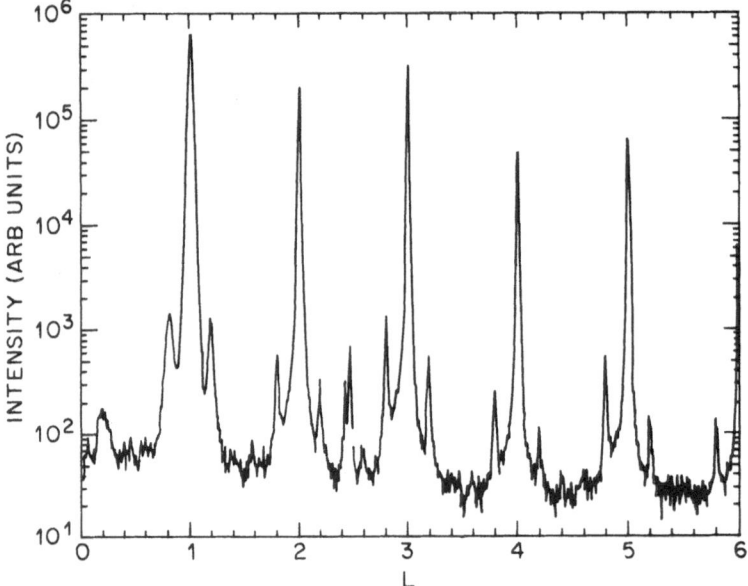

Figure 11. Intensity vs. ℓ along [10ℓ] direction in a
Gd_5Y_5 multilayer showing hcp structure.

direction in a multilayer of 5 layers of Gd alternating with 5 layers of
Y. All the main average Bragg peaks are sharp and the satellites
resulting from the modulation are also sharp. (The reflection at $\ell = 2.4$
is from the sapphire substrate.) Much less work has been done on trying
to grow multilayers which have different structures such as alternating
body-centered cubic and face-centered cubic materials. Usually the
grains of each material have a random orientation in the plane and a
strong texture with the close-packed planes of each structure being
parallel to the film. Some information can be obtained from x-ray
measurements of the positions of the powder rings which occur at well
defined angles with respect to the texture direction. This is especially

important if the components of the multilayer may form an intermetallic compound.

EXPERIMENTAL METHODS

The field of multilayers draws people from a number of different disciplines and therefore the types of x-ray scattering geometries which are employed are also diverse. There are basically three different groups and in this section the different techniques will be compared. First is the area of intermetallic compounds. Usually these are polycrystalline and therefore people from this field use powder diffractometers. Second is the area of perfect crystals, and the standard technique is the double crystal diffractometer. The third area involves people coming from the neutron and x-ray scattering community where the instrument of choice is the triple axis spectrometer. Often these different groups use the same terms to mean entirely different things. As discussed below the optimum use of x-rays in the determination of the structure of multilayers involves the use of several different geometries. The three different scattering geometries are shown in Fig. 12. In the powder diffractometer the incident beam and the

SCATTERING GEOMETRIES

POWDER DIFFRACTOMETER

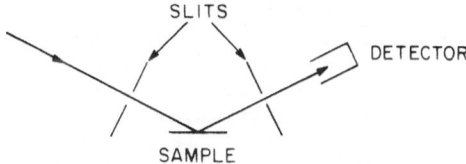

DOUBLE CRYSTAL DIFFRACTOMETER

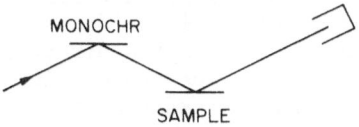

TRIPLE AXIS DIFFRACTOMETER

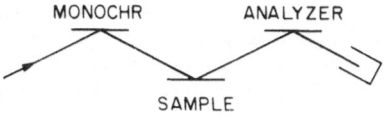

Figure 12. Comparison of three commonly used scattering geometries.

diffracted beam are defined by slits, and the angle that the scattered beam makes with the incident beam is increased at twice the rate that the sample is rotated. The theta-two-theta scan measures the scattering along a line in reciprocal space which is normal to the surface of the flat sample. For many metallic multilayers the sample has a texture in which the close packed planes are oriented parallel to the substrate. There is usually a distribution of grains in which the orientation of the grains varies over an angular range of anywhere from 0.1 degrees to many degrees. When a sample of this type is put on a powder diffractometer the texture direction is usually aligned close enough to the normal to the film that a theta-two-theta scan accurately records the position and the intensity of the average Bragg reflections and the surrounding harmonics. If the substrate is miscut or the growth direction is not normal to the surface then it may be necessary to have a sample holder capable of adjusting the two eulerian angles so as to align the texture direction parallel to the scattering vector measured by the diffractometer. The diffraction angle is determined by slits and therefore the measured angles are sensitive to the position of the sample.

The double crystal diffractometer is used extensively in the study of semiconductor multilayers, and it is a natural extension of the study of defects in perfect crystals such as Si, Ge, and GaAs. In this geometry a crystal of the substrate is used as the first crystal and then the second crystal is the sample which is the matching substrate with the multilayer on top. In this way the system is dispersionless which means that the two crystals are parallel when Bragg's law is satisfied. The incoming beam will have a range of energies and some angular spread. At the Bragg condition any x-ray which is diffracted by the first crystal will hit the second crystal at the same angle and therefore be diffracted. There is no broadening of the reflection due to a spread in energies in the incident beam because the resulting spread in Bragg angles all diffract together. The reflections from the multilayer in the vicinity of the same average Bragg reflection are only displaced by a small angle from the Bragg angle for the substrate, and the broadening is negligible. Implicit in this geometry is the assumption that the crystal is perfect in the sense that all the planes are parallel to the film. A scan is made by rotating the second crystal through the harmonics and the average Bragg reflections for the multilayer and the substrate, and hence it is referred to as a "rocking curve." Because of the assumption of parallel planes this is equivalent to a theta-two-theta scan in a powder diffractometer, i.e. it is a scan with the scattering vector normal to the surface. The scattered x-rays are detected with an open detector, and therefore the scattering angle, which is the angle between the incident beam and the scattered beam, is not determined independently. If the planes are not parallel or if the interfaces are rough so that the reflections in reciprocal space broaden in the direction transverse to the scattering vector, then this technique cannot distinguish between this and broadening resulting from, for example, fluctuations in modulation wavelength or finite size effects. This effect is clearly seen in studies of InP grown on GaAs by MOCVD.[34] In a powder diffractometer scan the reflection from the InP was very slightly broader than the reflection from the GaAs substrate and the latter was limited by the resolution of the instrument. In a rocking curve on the same sample, the InP reflection was 4 to 5 times broader than the GaAs reflection. This showed that there was a broadening of the InP reflection in the transverse direction which suggests that there is a range of orientations of the planes in the film. A big advantage of this geometry is that with an open detector the full integrated intensity of each reflection is recorded.

The third geometry is the triple axis spectrometer shown at the bottom of Fig. 12. By adding an analyzer crystal both theta and the scattering angle, two-theta, are measured independently. If the monochromator and analyzer crystals are matched then the volume of reciprocal space which is sampled for a given setting of all the different angles is symmetric with respect to the scattering vector. This is illustrated in Fig. 13 where the Ewald construction to demonstrate Bragg's law and the scattering triangle are on the left. On the right both k_i and k_f are allowed to accept a range of directions given approximately by the Darwin widths of the monochromator and the analyzer. These ranges trace out a diamond in reciprocal space whose area is given (in this idealized case) by the equation. In practice the volume of the resolution function is determined by measuring the volume for a series of reflections from a perfect crystal, and then the measured resolution function is used to deconvolute measurements on an unknown sample. With this geometry it is possible to map out the reflections illustrated in Fig. 9 and to determine both the in-plane lattice parameters and the parameter normal to the film directly. At the same time the widths of the reflection can be measured in all three directions and various types of broadening discussed in the first section and illustrated in Fig. 5 can be measured.

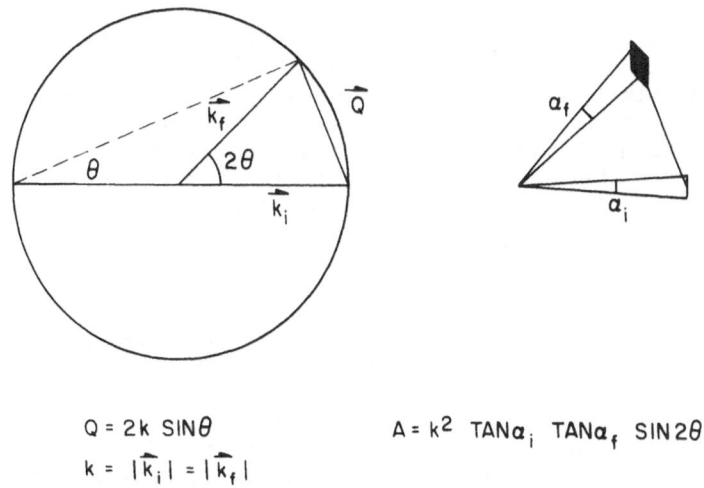

$$Q = 2k \, \text{SIN} \theta \qquad\qquad A = k^2 \, \text{TAN} \alpha_i \, \text{TAN} \alpha_f \, \text{SIN} 2\theta$$
$$k = |\hat{k}_i| = |\hat{k}_f|$$

Figure 13. Ewald sphere construction of scattering triangle and Bragg's Law (left) and area, A, of resolution function resulting from variations in $\vec{k}_i$ and $\vec{k}_f$.

The transverse width is often approximated by setting the spectrometer so as to record the maximum intensity from a given reflection and then rocking theta. This "rocking curve" is obtained in the same way as the "rocking curve" from a double crystal diffractometer but in this case it is approximately the transverse width of the reflection in a given scattering plane. In the double crystal geometry, for reflections normal to the film, the rocking curve is assumed to be the radial or longitudinal width of the reflection, i.e. the "rocking curves" referred to in the literature may be at right angles to each other in reciprocal space depending on which geometry the authors are using to make the measurements. As mentioned above, the lattice parameters both normal and parallel to the film can be determined directly using a triple axis geometry because the positions of off-axis reflections in reciprocal space can be measured. In the double crystal geometry it is necessary to measure the Bragg angle for two off-axis reflections and to solve for the two lattice parameters. We have discussed how in a perfect crystal the reflections are broadened in reciprocal space in the direction normal to the film because the diffraction is essentially from a thin layer at the top of the crystal. This particle size broadening is the same for all reflections in reciprocal lattice units, and a rocking curve in a double crystal diffractometer is a scan normal to the surface. In a double axis geometry an off-axis reflection is brought into the Bragg condition in two steps. First the crystal is rotated so as to bring the reciprocal lattice vector parallel to the scattering vector and then the crystal is rotated in order to increase the length of the scattering vector until the reflection is reached. Consider a (001) plate of silicon. For the (404) reflection the two angular components add so that the "rocking curve" is broader than for the (004) where the first angular component is zero. Conversely for the (-404) the two angular components oppose each other and the "rocking curve" is narrower. The differences in "rocking curve" widths result from the geometric factors involved in scanning identical volumes of reciprocal space and this is exactly the principle of asymmetrically cut crystals. These differences are of great importance in considerations of x-ray optics but are not relevant to the understanding of the structure of a film or multilayer.

As indicated throughout this article, there are several different types of information that need to be collected in order to understand the structure of a multilayer. First the positions of the reflections and their respective widths have to be measured. The modulation wavelength, coherency strains, and various types of fluctuations in the structure can be determined from these measurements. The triple axis geometry gives the greatest flexibility in obtaining this information. This geometry can be further improved by utilizing a four bounce monochromator so as to reduce the spread in energy of the incident beam.[35] It is difficult to use the triple axis geometry to collect integrated intensities because the widths of the reflections and the harmonics often cover a large range. In order to deconvolute the data easily it is advantageous to be in one of two regimes. Either the reflections from the sample are sharper than the resolution function or vice versa. Often one finds that the average reflections are narrower than the resolution width but the higher harmonics are broader. In principle, one can deconvolute the data, but a more reliable way is to use a double crystal geometry with an open detector to insure that the integrated intensity is being measured. Finally in order to have a large enough data set to refine the structure the data should be collected at a synchrotron in order to take advantage of the tunability for anomalous dispersion and of the flux for observing more weak harmonics. In all the calculations above it has been assumed that the measured intensities have been corrected for Lorentz and polarization factors and for absorption using standard techniques.

We thank J. D. Axe, B. W. Batterman, R. M. Fleming, and C. Vettier for many helpful discussions.

REFERENCES

1. D. B. McWhan, Structure of Chemically Modulated Films in Synthetic Modulated Structures, L. L. Chang and B. C. Giessen, Eds., Academic Press, Inc., New York (1985).
2. D. B. McWhan, Structure and Coherence of Metallic Superlattices in Layered Structures, Epitaxy and Interfaces, Gibson and Dawson, Eds., Mat. Res. Soc. Proc., Vol. 37, 493 (1985).

3. D. B. McWhan and C. Vettier, Interfaces in Metallic Superlattices in Frontiers in Electronic Materials and Processing, L. J. Brillson, Ed., AIP Conf. Proc. No. 138, New York (1986).

4. T. W. Barbee, Jr., Layered Synthetic Microstructures (LSM): Reflecting Media for X-ray Optical Elements and Diffracting Structures for the Study of Condensed Matter; Superlattices and Microstructures $\underline{1}$, 311 (1985).

5. W. Sevenhans, M. Gijs, Y. Bruynseraede, H. Homma, and I. K. Schuller, Cumulative Disorder and X-ray Line Broadening in Multilayers, Phys. Rev. B $\underline{34}$, 5955 (1986).

6. D. B. McWhan, M. Gurvitch, J. M. Rowell, and L. R. Walker, Structure and Coherence of NbAℓ Multilayer Films, J. Appl. Phys. $\underline{54}$, 3886 (1983).

7. B. M. Clemens and J. G. Gay, The Effect of Layer Thickness Fluctuations on Superlattice Diffraction, Phys. Rev. B $\underline{35}$, 9337 (1987).

8. S. Hendricks and E. Teller, X-ray Interference in Partially Ordered Layer Lattices, J. Chem. Phys. $\underline{10}$, 147 (1942).

9. J. Kakinokl and Y. Komura, Intensity of X-ray Diffraction by a One-dimensionally Disordered Crystal, J. Phys. Soc. Japan $\underline{7}$, 30 (1952).

10. D. A. Neumann, H. Zabel, and H. Monkoi, X-ray Evidence for a Terraced GaAs/ALAS Superlattice, Appl. Phys. Letts. $\underline{43}$, 59 (1983).

11. A. Segmuller and A. E. Blakeslee, X-ray Diffraction from One-Dimensional Superlattices in $GaAs_{1-x}P_x$ Crystals, J. Appl. Cryst. $\underline{6}$, 19 (1973).

12. J. Kwo, E. M. Gyorgy, D. B. McWhan, M. Hong, F. J. DiSalvo, C. Vettier, and J. E. Bower, Magnetic and Structural Properties of Single-Crystal Rare-Earth Gd-Y Superlattices, Phys. Rev. Letts. $\underline{55}$, 1402 (1985).

13. P. F. Miceli, H. Zabel, and J. E. Cunningham, Hydrogen-Induced Strain Modulation in Nb-Ta Superlattices, Phys. Rev. Letts. $\underline{54}$, 917 (1985).

14. C. Vettier, D. B. McWhan, E. M. Gyorgy, J. Kwo, B. M. Buntschuh, and B. W. Batterman, Magnetic X-ray Scattering Study of Interfacial Magnetism in a Gd-Y Superlattice, Phys. Rev. Letts. $\underline{56}$, 757 (1986).

15. F. Spaepen (this volume).

16. L. K. Templeton, D. H. Templeton, R. P. Phizackene, and K. O. Hodgson, L_3-Edge Anomalous Scattering by Gadolinium and Samarium Measured at High Resolution with Synchrotron Radiation, Acta Cryst. A $\underline{38}$, 74 (1982).

17. R. M. Fleming, D. B. McWhan, J. Kwo, and P. Marsh, to be published.

18. L. Tapper and K. Ploog, Improved Assessment of Structural Properties of $AL_xGa_{1-x}As$/GaAs Heterostructures and Superlattices by Double-Crystal X-ray Diffraction, Phys. Rev. B $\underline{33}$, 5565 (1986).

19. W. J. Bartels, J. Hornstra, and D. J. W. Lobeek, X-ray Diffraction of Multilayers and Superlattices, Acta Cryst. A $\underline{42}$, 539 (1986).

20. A. T. Macrander, E. R. Minami, and D. W. Berreman, Dynamical X-ray Rocking Curve Simulations of Nonuniform InGaAs and InGaAsP using Abeles' Matrix Method, J. Appl. Phys. $\underline{60}$, 1364 (1986).

21. V. S. Speriosu and T. Vreeland, Jr., X-ray Rocking Curve Analysis of Superlattices, J. Appl. Phys. $\underline{56}$, 1591 (1984).

22. E. Spiller (This volume).

23. P. F. Miceli, D. A. Neumann, and H. Zabel, X-ray Refractive Index: A Tool to Determine the Average Composition in Multilayer Structures, Appl. Phys. Letts. $\underline{48}$, 24 (1986).

24. T. W. Barbee, Jr., W. K. Warburton, and J. H. Underwood, Determination of the X-ray Anomalous Dispersion of Titanium Made with a Titanium-Argon Layered Synthetic, J. Opt. Soc. Am. B $\underline{1}$, 691 (1984).

25. C. Vettier, private communication (1986).

26. M. B. Salamon, S. Sinha, J. J. Rhyne, J. E. Cunningham, E. Ross, J. Borchens, and C. P. Flynn, Long Range Incommensurance Magnetic Order in Dy-Y Multilayer, Phys. Rev. Lett. $\underline{56}$, 259 (1986).

27. M. Hong, R. M. Fleming, J. Kwo, L. F. Schneemeyer, J. V. Waszczak, J. P. Mannaerts, C. F. Majkrzak, Doon Gibbs, and J. Bohr, Synthetic Magnetic Rare-Earth Dy-Y Superlattices, J. Appl. Phys. $\underline{61}$, 4057 (1987).

28. J. C. Bean, L. L. Feldman, A. T. Fiory, S. Nakahara, and I. K. Robinson, Ge_xSi_{1-x}/Si Strained-Layer Superlattice Grown by Molecular Beam Epitaxy, J. Vac. Sci. Tech. A $\underline{2}$, 436 (1984).

29. R. People, Physics and Applications of Ge_xSi_{1-x}/Si Strained-Layer Heterostructures, J. Quant. Elect. $\underline{QE-22}$, 1696 (1986).

30. A. Ourmazd and J. C. Bean, Observation of Order-Disorder Transitions in Strained-Semiconductor Systems, Phys. Rev. Lett. $\underline{55}$, 765 (1985).

31. W. P. Lowe, J. C. Bean, and A. Ourmazd, High Resolution X-ray Studies of GeSi/Si Strained Layer Superlattices, National Synchrotron Light Source Annual Report 1986, Brookhaven National Laboratory, Upton, NY, pg. 356.

32. E. M. Gyorgy, D. B. McWhan, J. R. Dillon, Jr., L. R. Walker, and J. V. Waszczak, Magnetic Behavior and Structure of Compositionally Modulated Cu-Ni Thin Films, Phys. Rev. B $\underline{25}$, 6739 (1982).

33. J. E. Cunningham and C. P. Flynn, Growth of Bicrystal Superlattices: Ru-Ir," J. Phys. F. L221-L226 (1985).

34. A. T. Macrander, R. D. Dupuis, J. C. Bean, and J. M. Brown, X-ray Characterization of Heteroepitaxial Structures with Large Mismatches, Semiconductor-Based Heterostructures Conf. Proc. Met. Soc. AIME, May 1986.

35. W. J. Bartels, Characterization of Thin Layers on Perfect Crystals with a Multipurpose High Resolution X-ray Diffractometer, J. Vac. Sci. Tech. B$\underline{1}$, 328 (1983).

RHEED OSCILLATIONS IN MBE AND THEIR APPLICATIONS TO PRECISELY CONTROLLED CRYSTAL GROWTH

Tsunenori Sakamoto

Electrotechnical Laboratory

1-1-4 Umezono Sakura-mura, Niihari-gun, Ibaraki, 305 Japan

INTRODUCTION

It is generally recognized that molecular beam epitaxy (MBE) technology has a number of advantages over conventional growth methods. For example, the real-time in-situ analysis of surface structures during growth is very attractive. For this purpose, reflection high energy electron diffraction (RHEED) has been commonly used, since the geometry of this technique is ideally suited for MBE, and the diffraction patterns contain considerable information about the surface.

One of the most exciting observations of recent years in MBE was the intensity oscillation of the specular beam spot in the RHEED pattern during MBE growth[1]. That is, just after the start of MBE growth, the intensity of the specularly reflected beam (00 spot) changes periodically, and one period of the oscillations corresponds to the growth of one monolayer thickness. The observation of this phenomenon opens the door for the dynamic monitoring of the layer-by-layer growth during MBE. This technique has been routinely used by many MBE research groups[2-6] to study initial growth mechanisms and to precisely calibrate growth rate and alloy composition. The computer controlled phase-locked epitaxy (PLE) method using RHEED intensity oscillations has been applied to the growth control of superlattice structures such as GaAs/AlAs[6-8], Ge_xSi_{1-x}/Si[9] etc., at an atomic scale.

In this paper, we will describe observations of the RHEED intensity oscillations of various materials, and their applications to precisely control MBE growth.

INTERPRETATIONS OF RHEED INTENSITY OSCILLATIONS

Figure 1 is a schematic illustration of the relationship between the incident electron beam and the diffraction patterns, taking an example of the zeroth Laue-zone for a GaAs (001) surface. [110] and [100] are two typical azimuths which are used to observe the diffraction patterns, and their oscillatory behaviors are, in general, different between them as will be shown later. During MBE growth, the intensity oscillations can be observed not only on the specular beam spot but also on the other diffraction spots and the Kikuchi lines. However, the specular beam is usually monitored because it has a strong intensity variation. Though it is generally recog-

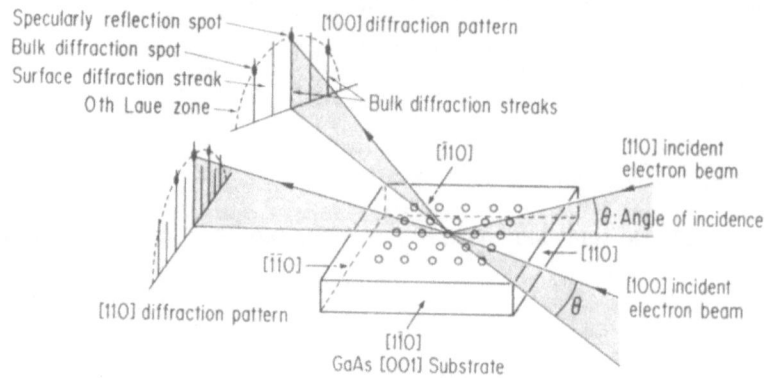

Fig.1. Schematic illustration of the relationship between incident electron beams and diffraction patterns on a GaAs (001) surface.

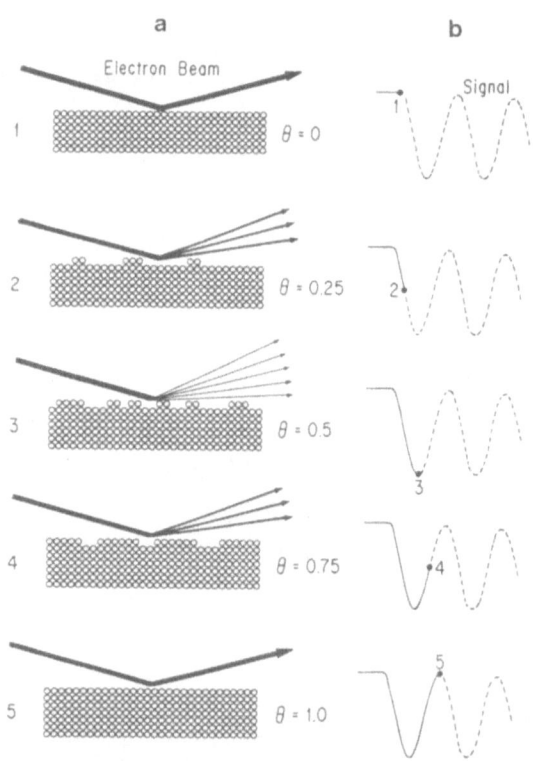

Fig.2. Single-scattering optical analogue model for RHEED intensity oscillations: (a) relationship between surface coverage and reflected electron beam; (b) specular beam intensity variation obtained on RHEED screen.

nized that the RHEED intensity oscillations originate from the layer-by-layer growth mechanisms in MBE[2-4], the growth model to explain the phenomenon has not yet been well established.

A single-scattering optical analogue model to interpret the RHEED intensity oscillations is illustrated in Fig.2. In this model, changes in intensity of the specular beam are directly related to the changes in the surface roughness at an atomic scale[3]. Since the de-Broglie wavelength of the RHEED electron (0.06 Å at V_a = 40 keV) is one order of magnitude less than the step height of the surface (2.83 Å for GaAs), diffuse reflectivity will result. Figure 2(a) shows the relationship between the surface coverages and the reflected electron beams, and Fig.2(b) shows the specular beam intensity variations obtained on a RHEED screen. The oscillations are observed when growth is initiated on a smooth surface, which is usually obtained after annealing of a buffer layer. At first, a maximum reflectivity is obtained when the surface is very smooth. After initiation of the growth, the surface becomes relatively rough giving rise to diffuse scattering of the reflected electron beam. When the surface coverage becomes half, the minimum reflectivity is obtained. After this, the surface smoothness recovers and the reflectivity increases. Finally, when the first monolayer growth is complete, the maximum intensity is again obtained. This model is based on a "rough-smooth transition" during the growth, where one period of the oscillations precisely corresponds to the growth of one monolayer. In the case of GaAs (001), it corresponds to the growth of $a_0/2$ (a_0: lattice constant of GaAs), i.e., a Ga layer plus an As layer. The optical analogue model based on the single-scattering theory is useful in most cases to explain the oscillatory behavior.

Recently, however, we reported bilayer-mode oscillations during Si MBE growth on a (001) substrate[10]. It is difficult to interpret the result using the single-scattering model shown in Fig.2. This peculiar oscillatory behavior is interpreted as being the result of alternating surface reconstructions of the "single-domain" surface during the growth[11-13]. This effect will be discussed later.

RHEED INTENSITY OSCILLATIONS IN GaAs

Experimental

A schematic diagram of the phase-locked epitaxy (PLE) system used for GaAs growth is shown in Fig.3. The experiments were performed using a RIBER 2300R/D MBE system. The base pressure and the pressure during growth in the growth chamber are 5×10^{-10} and 1×10^{-9} Torr, respectively. The acceleration

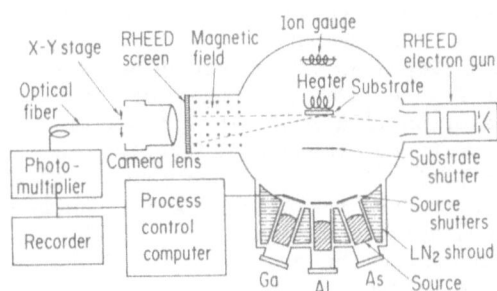

Fig.3. Schematic diagram showing the experimental arrangement for phase-locked epitaxy of GaAs

voltage of the RHEED system was 40 keV. The full width at half maximum (FWHM) of an electron beam spot on the RHEED screen was about 200 μm, and the glancing angle of the electron beam was less than 1 degree (off-Bragg condition). The intensity of the specular beam was measured via a plastic optical fiber by a photomultiplier. The output of the photomultiplier was monitored by a computer and at a pre-determined phase of the oscillations, the molecular beam cell shutters were controlled. A scanning magnetic field was applied on the diffracted electron beam to measure the rocking profile of the diffracted spots. After heat treatment of the substrate in the growth chamber, unintentionally doped buffer layers with a thickness of about 3000 Å were grown so that the initial surface roughness would not affect the oscillation results. Typical growth rates in this experiment were 0.3-0.5 Å/sec.

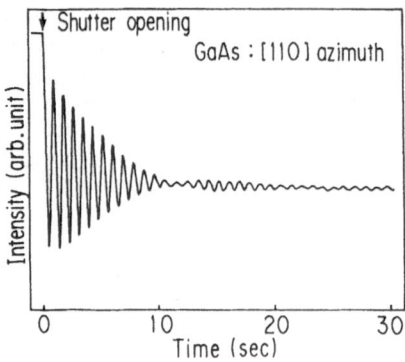

Fig.4. RHEED intensity oscillations of the specular bem during growth of GaAs; [110] azimuth

Results and Discussions

After growth of the GaAs buffer layer, the Ga source shutter was closed while maintaining the As flux. This process produces a well-ordered surface through the statistical migration of the surface Ga atoms to "kink" or "edge" sites on the as-grown surface. This surface smoothening effect will be discussed later.

Typical RHEED intensity oscillations of the specular beam observed during the growth of GaAs on a (001) 2x4 reconstructed surface in [110] azimuth are shown in Fig.4. The growth was initiated by impinging Ga atoms on the substrate surface. The equivalent vapor pressure ratio P(Ga)/P(As) was about 1/10. In the figure, the amplitude of the oscillations damped rapidly after a few tens of oscillations. This behavior is considered to be due to the presence of a three-dimensional growth. That is, although the growth takes place principally via a two-dimensional layer-by-layer process, a new layer starts growing before the preceding one has been completed.

Oscillations in [100] azimuth were also observed[6]. Typical oscillations obtained in this azimuth during the growth of $Al_x Ga_{1-x} As$ (x=0.41), where the Al mole fraction x was determined by the measurement of the oscillation frequency, which is discussed later, are shown in Fig.5. Even though a relatively rapid damping was observed just after the start of the growth, the oscillation persisted more than 700 periods in this azimuth. The reason for the differences in behavior of these oscillations between the [110] and the [100] azimuths has not yet been clearly understood. However, it may be related to an interference between the elastic specular scattering and the diffuse scattering, or the penetration depth differences of the electron beams between the two azimuths in the substrate surface.

Typical RHEED intensity oscillations obtained in the [100] azimuth during heteroepitaxial growth of GaAs/Al$_x$Ga$_{1-x}$As/AlAs are shown in Fig.6. In this experiment, 10 monolayers of GaAs, Al$_x$Ga$_{1-x}$As and AlAs layers were grown continuously. Since the oscillation periods are exactly proportional to the net growth rate, we can estimate an Al mole fraction x in the Al$_x$Ga$_{1-x}$As layer by measuring the frequency change of the oscillations between GaAs and Al$_x$Ga$_{1-x}$As layers as follows:

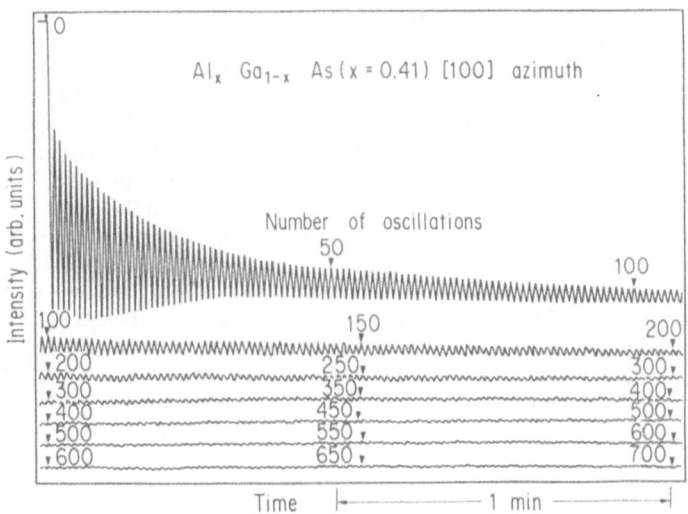

Fig.5. RHEED intensity oscillations of the specular beam during growth of Al$_x$Ga$_{1-x}$As (x = 0.41); [100] azimuth

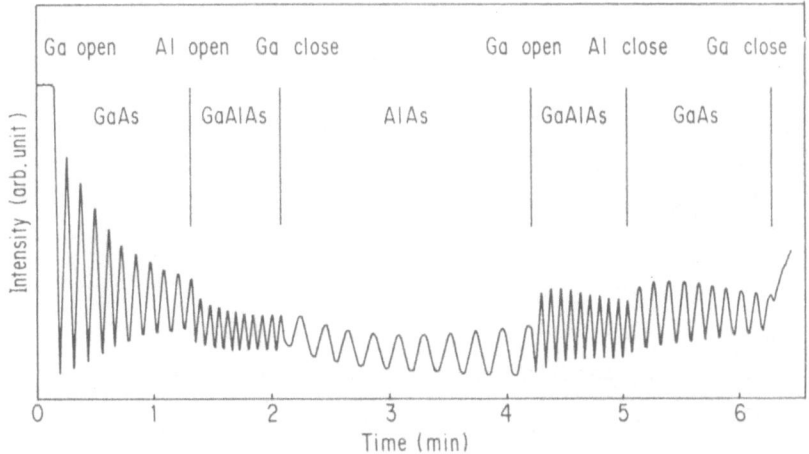

Fig.6. RHEED intensity oscillations during heteroepitaxial growth of GaAs/Al$_x$Ga$_{1-x}$As/AlAs; [100] azimuth

$$x = \frac{f(AlGaAs) - f(GaAs)}{f(AlGaAs)}$$

where f(GaAs) and f(AlGaAs) are the frequencies of the RHEED intensity oscillations obtained during the growth of GaAs and $Al_xGa_{1-x}As$ layers, respectively. The Al mole fractions x thus obtained are in good agreement with expected values from an impinging rate measurement method within an error of ±3%. The capability of the in-situ measurement of alloy composition is another significant advantage of the RHEED intensity oscillation technique.

One of the important observations obtained by the study of the RHEED intensity oscillation may be the surface smoothening effect during a growth interruption. RHEED intensity oscillations obtained during the heteroepitaxial growth of $Al_xGa_{1-x}As$/GaAs are shown in Fig.7. A 20-second growth interruption was made at the heterointerface, i.e., the Al and the Ga source shutters were closed while leaving the As source shutter open. During the growth interruption period, the specular beam intensity increased rapidly. This process is attributed to the recovery of the surface smoothness, which in turn is caused by the surface migration or the sublimation of adatoms. This effect has been studied in detail by Tanaka et al.[14] by a photoluminescence measurement using GaAs/$Al_xGa_{1-x}As$ quantum well structures.

To control crystal growth within an atomic-order precision it is very important to fabricate the quantum effect devices such as a quantum well device or a superlattice device. From this point of view, the conventional MBE growth method is not enough.

We have recently developed a new technique named "phase-locked epitaxy (PLE)", in which the RHEED intensity oscillations are analyzed by a computer and the source cell shutters are driven at a pre-determined phase of the oscillations[6,7]. The oscillations thus obtained during the growth of $(GaAs)_3(AlAs)_3$ tri-layer superlattice structure are shown in Fig.8. In this experiment, the Al source shutter was opened first and three monolayers of

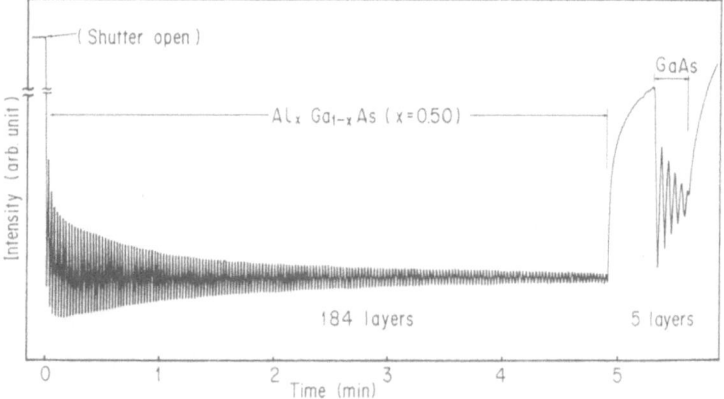

Fig.7. RHEED intensity oscillations during heteroepitaxial growth of $(Al_xGa_{1-x}As)_{184}$/$(GaAs)_5$; [100] azimuth. Note that by 20 seconds growth interruption at the heterointerface, oscillations can be recovered

AlAs were grown. Then the Al source shutter was closed and after a short growth interruption period (6 sec) the Ga source shutter was opened and three monolayers of GaAs were grown. After this, the short growth interruption period followed again before resuming the AlAs layer growth, and so on. Thus, by including a suitable growth interruption time at the heterointerfaces, the oscillations persist for a long time.

The PLE method has advantages over the conventional non-PLE method (i.e., a constant-time shutter controlling method) in growing superlattice structures. Simplified models of the cross-sectional view of superlattice structures grown by these two methods are shown in Fig.9.

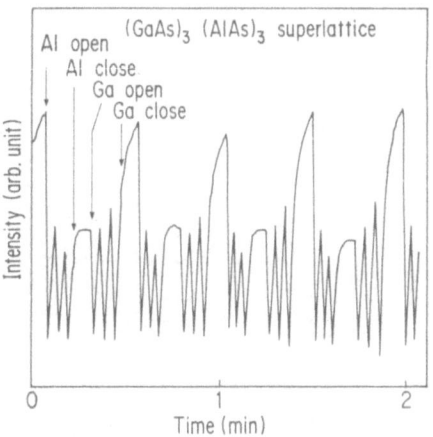

Fig.8. RHEED intensity oscillations during computer-controlled phase-locked epitaxy of $(GaAs)_3(AlAs)_3$ tri-layer superlattice structure

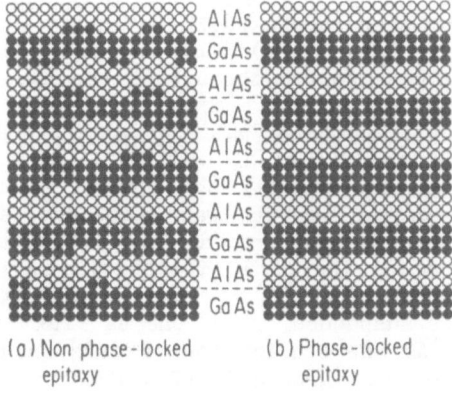

Fig.9. Simplified models of the cross-sectional view of $(GaAs)_3(AlAs)_3$ tri-layer superlattice structures grown by two methods: (a) conventional non-PLE method (constant-time shutter controlling method); (b) PLE method

In the case of the non-PLE method shown in Fig.9(a), the source shutters are driven at a random phase of the surface molecular coverage, and as a result, the heterointerfaces between GaAs and AlAs have roughness. On the other hand, in the case of the PLE method, the interfaces should be smooth as shown in Fig.9(b), since the molecular beam shutters are driven at the maximum amplitude of the oscillations, which occurs at the completion of the layer growth.

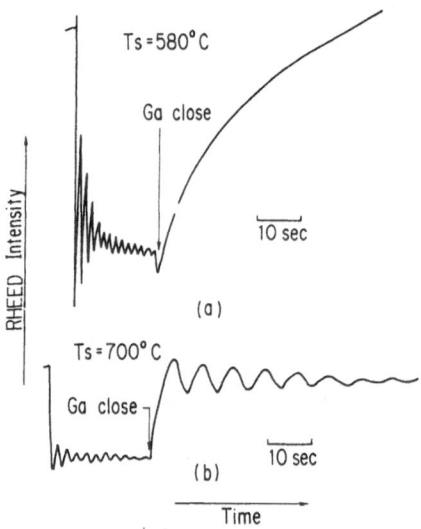

Fig.10. Growth and post-growth RHEED intensity oscillations for GaAs on a GaAs(001) surface: (a) at low temperature (T_S = 580°C); (b) at high temperature (T_S = 700°C). Arsenic flux is applied throughout

Recently, RHEED intensity oscillations of the post-growth process of the GaAs have been observed in the high temperature range (T_S > 690°C) by our group[15,16] and the University of Minnesota group[17], independently. Figure 10 shows growth and post-growth RHEED intensity oscillations: (a) at low substrate temperature (T_S = 580°C); (b) at high substrate temperature (T_S = 700°C). The As flux was applied throughout the experiments. The RHEED intensity oscillations for the growth started when a Ga flux was applied. During the growth, As-stabilized RHEED patterns gave the surface reconstruction of 2x4 at T_S = 580°C and 3x1 at T_S = 700°C. As mentioned previously, one period of the oscillations exactly corresponds to the growth of one monolayer of GaAs. After crystal growth was suspended, the surface reconstruction was essentially unchanged both at T_S = 580°C and T_S = 700°C. At the low substrate temperature the RHEED intensity recovers monotonically. However, at the higher substrate temperature, a new mode of the oscillations starts after the growth is suspended, i.e., the Ga flux is turned off while maintaining the As flux. As the sublimation of GaAs is reported at higher substrate temperatures, the post-growth RHEED intensity oscillations can be attributed to the sublimation of GaAs. In order to confirm this idea, we employed the new "deposition-sublimation" method utilizing the AlAs layer as a sublimation stopper. In this method, as shown in Figs.11(a) and (b),

GaAs of 5 and 10 monolayers are deposited by the PLE method on an AlAs sublimation-resistant buffer layer. When the PLE is finished, vacuum sublimation of GaAs is observed until the AlAs buffer is exposed and the post-growth RHEED intensity oscillation is terminated. The number of the oscillation periods for the sublimation process always equals the number of layers deposited on the AlAs within experimental error. This result indicates that the post-growth oscillation period corresponds to one monolayer sublimation of GaAs. It is also concluded that the vacuum sublimation should occur likewise in layer-by-layer fashion as long as the post-growth oscillations continue as in the case of the growth process. The mechanism of the post-growth oscillations is considered to be the reverse process of the MBE growth. The difference between them lies in the controlling parameters of the phenomena. In the case of deposition, the oscillations are determined by the intensity of the Ga flux while the sublimation is controlled by the substrate temperature and the impinging As flux. The post-growth RHEED intensity oscillation method is a new microscopic probe for the vacuum sublimation process of crystalline materials and can be used for etching with precise depth resolution.

RHEED INTENSITY OSCILLATIONS IN Si MBE

In this section the observations of the RHEED intensity oscillations during Si MBE growth and their applications to $Ge_x Si_{1-x}$/Si strained-layer superlattices are discussed.

Experimental

The experiments were performed in an ion-pumped Si MBE system. The base pressure and the pressure during growth in the growth chamber were 1×10^{-9} and 1×10^{-8} Torr, respectively. The acceleration voltage of the RHEED system was 40 keV. The incident angle of the electron beam on the substrate surface was less than 1° (off-Bragg condition). As a Si molecular beam

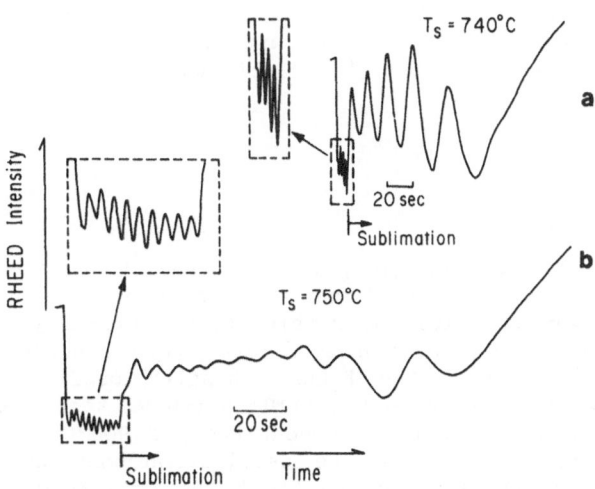

Fig.11. Growth and post-growth RHEED intensity oscillations for GaAs on an AlAs sublimation stopper layer: (a) GaAs (5 monolayers) on AlAs at $T_s = 740°C$; (b) GaAs (10 monolayers) on AlAs at $T_s = 750°C$

source, high purity crystalline Si was evaporated by a 2 kW electron-gun. Well-oriented Si (001) wafers were used. The substrates were degreased and then boiled in a HCl:H_2O_2:H_2O (1:1:4) solution to form a protective thin oxide film, which was removed by heating the substrate up to 800°C. Undoped buffer layers 1000-2000 Å thick were grown at a substrate temperature of 700°C prior to the RHEED intensity oscillation monitoring. The growth rate, typically 0.5 Å/sec, was monitored by a quartz thickness monitor.

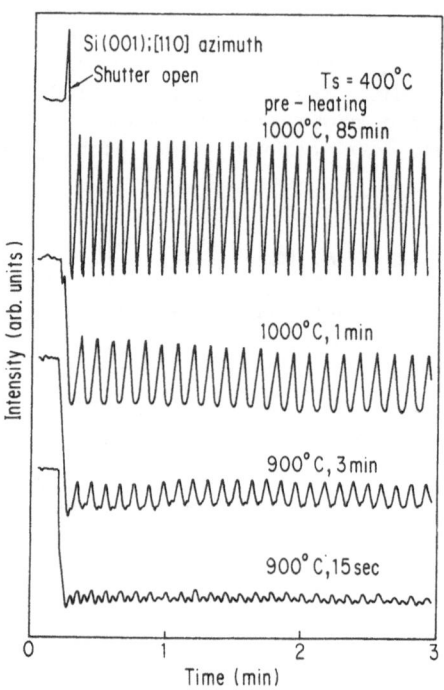

Fig.12. Intensity oscillation of the specular beam during homoepitaxial growth of Si on Si(001) surface with various preheating conditions after buffer layer growth; [110] azimuth, T_s = 400°C

Results and discussions

In case of Si growth, unlike GaAs, it was found that substrates must be annealed prior to growth at a temperature much higher than the growth temperature in order to obtain a stable RHEED intensity oscillation[10]. After the buffer layer growth, the substrates were annealed under various conditions. Figure 12 shows the RHEED intensity oscillations after the annealing, which were observed in the [110] azimuth at a substrate tempera-ture of 400°C. With a weak annealing condition of 900°C for 15 seconds, a small asymmetric monatomic-layer-mode oscillation was observed. With an increase of annealing temperature and time, the amplitude of the oscilla-tions increased rapidly and their period doubled. One period of the oscil-lations corresponds to a biatomic-layer growth ($a_0/2$=2.72 Å) on the Si (001) surface. This drastic change in the RHEED intensity oscillations after high temperature annealing means some constructive change in the annealed surface. We checked the RHEED patterns after the annealing to confirm the idea.

Figure 13 shows the RHEED patterns observed from the [010] azimuth at room temperature after subjecting the substrate to two different annealing conditions. After the annealing at 900°C for one minute, the two kinds of half-order diffraction spots originating from the 2x1 and the 1x2 reconstructed surfaces were both observed on the half-order Laue zone as shown in Fig.13(a). However, as shown in Fig.13(b), the annealing at 1000°C for 20 minutes made one of the series of the half-order spots (indicated by short arrows) disappear. We therefore concluded that by annealing at 1000°C for 20 minutes, the surface reconstruction of the Si(001) surface changed from the (2x1=1x2) double-domain structure to the 2x1 single-domain structure, i.e., the surface with only the 2x1 reconstruction, or at least with a preponderance of the 2x1 reconstruction. This means that small domains separated by monatomic-layer height steps in the as-grown Si (001) surface were developed to larger domains separated by biatomic-layer height steps by high temperature annealing. To the authors' knowledge, this is the first observation of the single-domain (001) 2x1 structure on a well-oriented (001) surface.

The peculiar biatomic-layer-mode oscillations on the Si (001) surface can be interpreted considering that the growth starts on the single-domain structure. A schematic model for the layer-by-layer growth on the Si (001) 2x1 single-domain structure is shown in Fig.14.

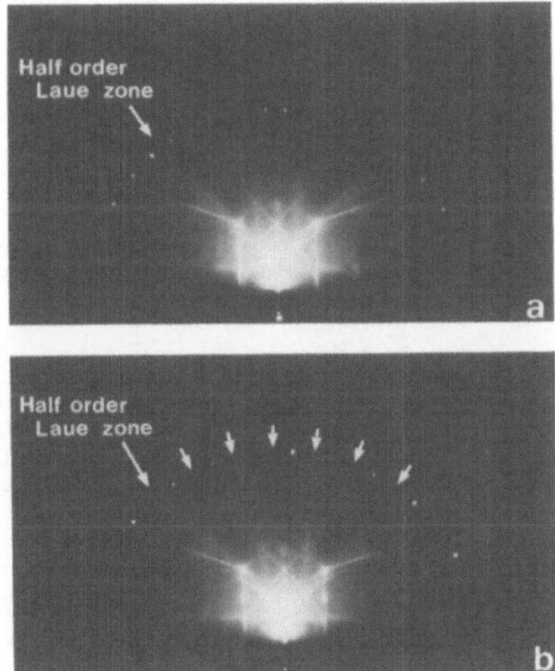

Fig.13. RHEED patterns taken from Si(001); [010] azimuth: (a) after the annealing at 900°C for 1 minute; (b) after the annealing at 1000°C for 20 minutes. The small arrows indicate vanished diffraction spots

Although the two-dimensional growth model of Si shown in this figure is analogous to that of GaAs[3], the difference is the fact that the elongated islands of monatomic-layer height with the longest dimension parallel to the [110] and the [1$\bar{1}$0] directions appear alternately for Si, while islands of biatomic-layer height with the longest dimension parallel to only one direction appear for GaAs. In the following discussion the propriety of this model is examined.

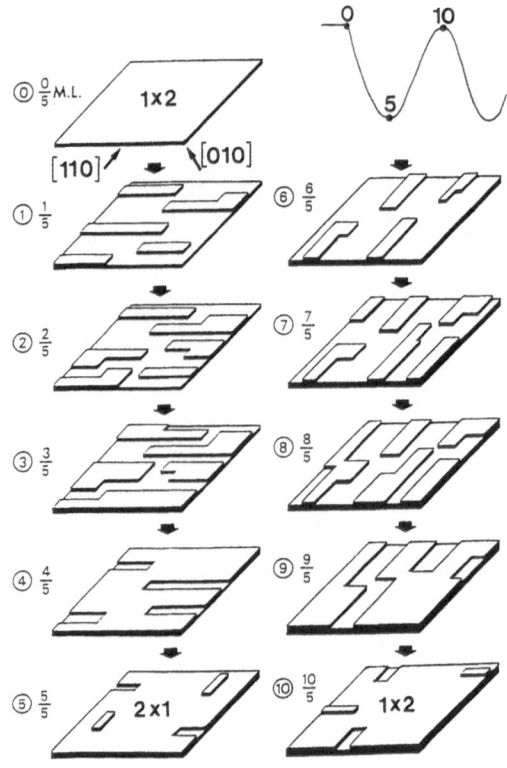

Fig.14. Schematic picture of Si(001) surfaces during layer-by-layer growth on the 2x1 single-domain structure. 1x2 and 2x1 reconstructed surfaces appear alternately

To investigate the alternating surface reconstructions in more detail, the RHEED intensity experiments were done during the growth using three sets of optical fiber monitoring systems simultaneously. Figure 15 shows the oscillations taken from three different diffraction spots in the RHEED patterns of [010] azimuth at a substrate temperature of 500°C. The upper inset figures show a reciprocal lattice of (001) 2x1+1x2 reconstructed surface and a corresponding diffraction pattern, schematically: (a) large solid circles; (b) small solid circles; (c) small open circles represent the bulk diffraction spots, 2x1 and 1x2 reconstruction related spots, respectively. Contrary to [110] azimuth, the oscillations of the specular beam in this azimuth is a monatomic-layer mode, i.e., one period of the oscillations shown in trace (a) corresponds to the monatomic-layer height ($a_0/4$). However, in both traces (b) and (c), one period of the oscillations corresponds to a biatomic-layer height. The phase difference between (b) and (c) is almost 180°C, and each oscillation maximum of (b) and (c) coincides with that of (a). The results indicate that though the 2x1 and 1x2

surface reconstructions appear alternately during growth, the specular
beam in the [010] azimuth is not affected by them. This fact is explained
as follows. The specular beam intensity obtained from the [110] azimuth in
the 1x2 reconstructed pattern is stronger than that in the 2x1 reconstructed
pattern at the glancing of 0.9°, which is used in the experiments, i.e., it
can be said that the 1x2 is a bright and the 2x1 is a dark reconstructed
surface. If the above-mentioned intensity difference between the 1x2 and
the 2x1 reconstructed surface is larger than that between the rough and the
smooth surface, which also appear alternately during the growth as shown in
Fig.14, the maximum intensity should be obtained at the completion of the
1x2 reconstructed surface and the minimum intensity at the completion of
the 2x1 reconstructed surface. Thus we can observe the biatomic-layer-mode
oscillations in the [110] azimuth. In the [010] azimuth, however, the two
surface reconstructions should not cause any intrinsic differences upon the
specular beam intensities. Then the monatomic-layer-mode oscillation, which
is caused by the usual "rough-smooth transition" of the surface, can be
observed.

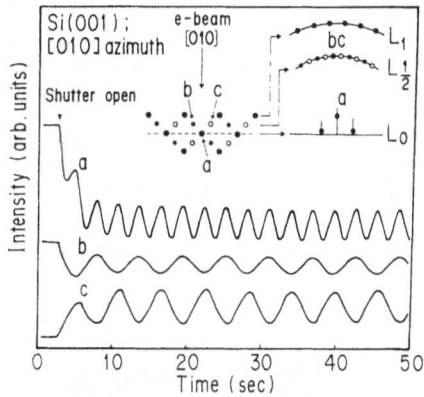

Fig.15. RHEED intensity oscillations of three different diffraction
        spots taken in the [010] azimuth pattern simultaneously:
        (a) specular beam; (b) 2x1 reconstruction spot; (c) 1x2 re-
        construction spot. Inset figures show a reciprocal lattice
        of (100) 2x1+1x2 and a corresponding diffraction pattern
        schematically: (a) large solid circles, (b) small solid
        circles, and (c) small open circles represent the bulk dif-
        fraction, 2x1 and 1x2 reconstruction related spots, respec-
        tively

    The bilayer-mode oscillations were also explained theoretically using
the calculation based on the multiple scattering theory. Details have been
described elsewhere [18,19].
    Figure 16 shows examples of the RHEED intensity oscillations during
the homoepitaxial growths on the single-domain Si (001) 2x1 reconstructed
surface at various substrate temperatures in the [110] azimuth. More than
ten periods of the oscillations are observed at room temperature which suggests that the layer-by-layer growth fashion in Si homoepitaxy persists

in more than 20 atomic layers even at room temperature. Oscillations of more than 2200 periods (which correspond to more than 6000 Å thickness) were observed at T_S = 500°C. Thus, the very stable oscillations can be observed on the single-domain structures.

RHEED intensity oscillations during the growth of $Ge_x Si_{1-x}$/Si strained-layer superlattice structures on Si (001) substrate

$Ge_x Si_{1-x}$ is a promising material for the fabrication of heterojunction devices of Si-based structure. Although the lattice mismatch between Si and Ge is relatively large ($\sim 4\%$), recent progress in Si MBE made it possible to achieve the pseudomorphic growth of $Ge_x Si_{1-x}$/Si on a Si substrate up to 1 μm in thickness[20]. However, to the authors' knowledge, no report on the RHEED intensity oscillations during the growth of the $Ge_x Si_{1-x}$/Si strained-layer heteroepitaxy has appeared prior to our recent work[9].

$Ge_x Si_{1-x}$ films with various Ge mole fractions were grown by controlling the evaporation rates of a Si and a Ge source. Typical RHEED intensity oscillations of a specular beam observed from the [110] azimuth at T_S = 450°C are shown in Fig.17. The growth of $Ge_x Si_{1-x}$ was preceded by Si growth for several oscillations in order to determine the alloy composition. While the oscillation amplitude and the average intensity were almost unchanged during the Si growth, they began to decrease after the $Ge_x Si_{1-x}$ growth started. The decrease took place more rapidly with an increase in the Ge mole fraction in the grown film. In the growth of Ge on Si at T_S = 450°C, i.e., x = 1.0, only two periods of the oscillations were observed[9]. The

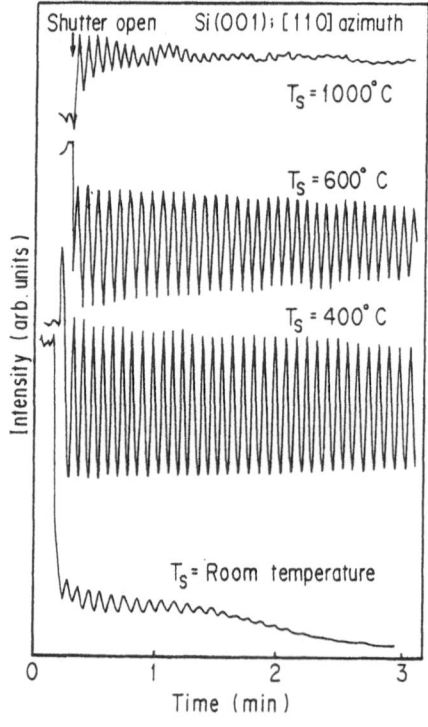

Fig.16. Examples of the RHEED intensity oscillations during growth of Si on the single-domain Si (001) 2x1 reconstructed surface at the various substrate temperatures; in the [110] azimuth

damping behavior of the oscillations could be explained by clustering during the Ge_xSi_{1-x} growth, as follows. When the Ge mole fraction was small, the RHEED pattern during Ge_xSi_{1-x} growth was almost the same as that during the Si growth.

However, with an increase of the Ge mole fraction, the RHEED pattern rapidly changed from the initial streaky type to the final spotty type. It was concluded from these observations of RHEED patterns that the decrease of the oscillation amplitude during Ge_xSi_{1-x} growth was a result of a change in the growth mode from a two-dimensional layer-by-layer type to a three-dimensional type which is caused by the clustering of Ge_xSi_{1-x}. The reason is not clear why three-dimensional growth was promoted by an increase in the Ge mole fraction. Considering the fact, however, that more than ten periods of the oscillations were observed during the homoepitaxial growth of Ge at $T_s = 350°C$, it cannot be said that Ge itself has a strong tendency to be a cluster formation.

These results suggest that the motive force for clustering has some relation to a lattice mismatch between a Si substrate and a Ge_xSi_{1-x} epilayer.

We fabricated Ge_xSi_{1-x} strained-layer superlattices on Si(001) substrate while monitoring the RHEED intensity oscillations. Figure 18 shows RHEED intensity oscillations during the growth of $Ge_{0.25}Si_{0.75}$ /Si strained-layer superlattices at $T = 450°C$; [110] azimuth. The Ge source shutter was opened and closed alternately at every fifth peak of the oscillations; thus, the strained-layer superlattice comprised 10 atomic layers of Si and 10 atomic layers of Ge_xSi_{1-x}, since one period of oscillations corresponds to one bilayer growth in this azimuth. Although it is needless to say that the precise controllability of the layer thickness is one of the advantages of the PLE method, a more important point is that we can also control the

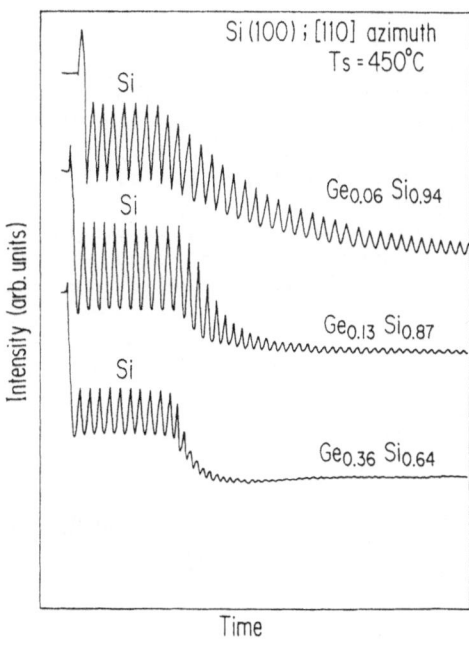

Fig.17. RHEED intensity oscillations of specular beam during Ge_xSi_{1-x} growth on Si(001) at 450°C taken from [110] azimuth for various Ge mole fractions. Si growth preceded the Ge_xSi_{1-x} growth to determine alloy compositions

interface roughness. The oscillation amplitude during the Ge_xSi_{1-x} growth rapidly decreased but recovered to some extent during the Si growth. This surface-smoothening by Si-overgrown layers assisted the longer observation of the RHEED intensity oscillations during the growth of a strained-layer superlattice much better than that during a single strained-layer growth. We can obtain 26 periods of a $(Ge_{0.25}Si_{0.75})_{10}/Si_{10}$ strained-layer superlattice, i.e., 520 atomic layers (70 nm), while monitoring the RHEED intensity oscillations. The total thickness of the $Ge_{0.25}Si_{0.75}$ layers (35 nm) was much larger than the critical thickness (8 nm at x = 0.25), at which the RHEED intensity oscillations almost faded away[9] The Ge_xSi_{1-x}/Si strained-layer superlattice structures obtained using the PLE technique have been characterized by TEM and EXAFS. Details have been reported elsewhere.

CONCLUSIONS

More than 700 periods of the intensity oscillations in a specular beam in the RHEED pattern were obtained in the [100] azimuth of the (001) GaAs substrate during the growth of $Al_xGa_{1-x}As$. Using these persisting oscillations, very accurate growth rate measurements up to about 2000 Å in thickness were performed. By measuring the oscillation frequency shifts, the precise Al mole fraction x of $Al_xGa_{1-x}As$ was obtained during the growth. A precisely defined $(GaAs)_3(AlAs)_3$ tri-layer superlattice was grown by computer-controlled phase-locked epitaxy using the RHEED intensity oscillations. It was shown that the PLE method has an advantage over the conventional MBE growth method for the precise control of very thin films and superlattice structures, because the PLE method is invulnerable to fluctuations of molecular beam flux.

Post-growth RHEED intensity oscillations were observed for GaAs at high substrate temperatures. One period of the oscillation precisely corresponds to one monolayer "sublimation" of GaAs from the surface. It is also concluded that the sublimation occurs in layer-by-layer fashion. The sublimation rate accurately determined from these post-growth oscillations is a function of the substrate temperature and the impinging As flux. The AlAs layer was found to act as a sublimation stopper. These post-growth oscillations can be used for etching with precise depth resolution as well as sublimation mechanism analyses.

The RHEED intensity oscillations were also observed during the growth of Si. We found that the RHEED intensity oscillations of the Si (001) substrate are strongly dependent on the annealing conditions prior to the

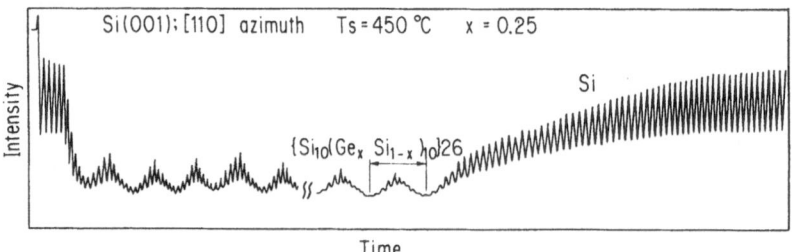

Fig.18. RHEED intensity oscillations during growth of
$(Ge_{0.25}Si_{0.75})_{10}/Si_{10}$ strained-layer superlattice structure on Si(001 at 450°C. using
PLE, 26 periods of straihned-layer superlattice
was grown without any growth interruption

growth. The high-temperature annealing changes the substrate surface not only into an atomically smooth one but also into the "single-domain" Si (001) 2x1 structure, which is essential for the observation of a stable oscillation. The alternating surface reconstructions of 2x1 and 1x2 were observed using the RHEED during Si growth on the single-domain structure. Using this result, the peculiar bilayer-mode oscillations in the [110] azimuth reported previously can be interpreted.

In the case of the strained-layer heteroepitaxy of Ge_xSi_{1-x} on the Si (001) substrate, the rapid decrease of the oscillations was observed just after the start of the Ge_xSi_{1-x} growth due to three-dimensional growth, however, the roughened Ge_xSi_{1-x} surface was found to be smoothened during the Si overlayer growth. Ge_xSi_{1-x}/Si strained-layer superlattices were grown using the PLE method without any growth interruption.

Although the RHEED patterns contain a lot of information about the surface, we can, at present, analyze only a few of them. Further development is expected with the progress of the dynamic signal processing technique and the theoretical analyses.

ACKNOWLEDGEMENT

The authors would like to thank Drs. N. Hashizume, K. Ohta, N.J. Kawai, T. Nakagawa, T. Kojima, K. Sakamoto, K. Miki, Prof. T. Kawamura, Prof. K. Tomizawa and Dr. Y. Bando for their helpful discussions. Thanks are due to H. Funabashi, S. Nagao, G. Hashiguchi, K. Kuniyoshi, N. Takahashi for their cooperative work and also to Dr. T. Tsurushima and Prof. S. Gonda for their continuous encouragment.

REFERENCES

1. J.J.Harris, B.A.Joyce and P.J.Dobson, Oscillations in the surface structure of Sn-doped GaAs during growth by MBE, Surf. Sci. 103: L90 (1981).
2. C.E.C.Wood, RED intensity oscillations during MBE of GaAs, Surf. Sci. 108: L441 (1981).
3. J.H.Neave, B.A.Joyce, P.J.Dobson and N.Norton, Dynamics of film growth of GaAs by MBE from Rheed observation, Appl. Phys. A31: 1 (1983).
4. J.M.Van Hove, C.S.Lent, P.R.Pukite and P.I.Cohen, Damped oscillation in reflection high-energy electron diffraction during GaAs MBE, J. Vac. Sci. & Technol. B1: 741 (1983).
5. B.F.Lewis, F.J.Grunthaner, A.Madhukar, T.C. Lee and R. Fernandez, Reflection high energy electron diffraction intensity behavior during homoepitaxial molecular beam epitaxy growth of GaAs and implications for growth kinetics and mechanisms, J. Vac. Sci. & Technol. B3: 1317 (1985).
6. T.Sakamoto, H.Funabashi, K.Ohta, T.Nakagawa, N.J.Kawai, T.Kojima and Y.Bando, Well defined superlattice structures made by phase-locked epitaxy using RHEED intensity oscillation, Superlattices and Microstructures 1: 347 (1985).
7. T.Sakamoto, H.Funabashi, K.Ohta, T.Nakagawa, N.J.Kawai and T.Kojima, Phase-locked epitaxy using RHEED intensity oscillation, Jpn. J. Appl. Phys. 23: L657 (1984).
8. N.Sano, H.Kata, M.Nakayama, S.Chika, H.Terauchi, Mono- and bi-layer superlattices of GaAs and AlAs, Jpn. J. Appl. Phys. 23: L640 (1984).
9. K.Sakamoto, T.Sakamoto, S.Nagao, G.Hashiguchi, K.Kuniyoshi and Y.Bando, Reflection high-energy electron diffraction intensity oscillations during Ge_xSi_{1-x} MBE growth on Si(001) substrates, Jpn. J. Appl. Phys. 26: 666 (1987).

10. T.Sakamoto, N.J.Kawai, T.Nakagawa, K.Ohta and T.Kojima, Intensity oscillations of reflection high-energy electron diffraction during silicon molecular beam epitaxial growth, Appl. Phys. Lett. 47: 617 (1985).

11. T.Sakamoto and G.Hashiguchi, Si(001)-2x1 single-domain structure obtained by high temperature annealing, Jpn. J. Appl. Phys. 25: L78 (1986).

12. T.Sakamoto, T.Kawamura and G.Hashiguchi, Observation of alternating reconstructions of silicon(001) 2x1 and 1x2 using reflection high-energy electron diffraction during molecular beam epitaxy, Appl. Phys. Lett. 48: 1612 (1986).

13. T.Sakamoto, T.Kawamura, S.Nagao, G.Hashiguchi, K.Sakamoto and K. Kuniyoshi, RHEED intensity oscillations of alternating surface reconstructions during Si MBE growth on single-domain Si(001) 2x1 surface, J. Cryst. Growth 81: 59 (1987).

14. M.Tanaka, H.Sakaki and J.Yoshino, Atomic-scale structures of top and bottom heterointerfaces in GaAs-Al$_x$Ga$_{1-x}$As (x=0.2-1) quantum wells prepared by molecular beam epitaxy with growth interruption, Jpn. J. Appl. Phys. 25: 155 (1986).

15. T.Kojima, N.J.Kawai, T.Nakagawa, K.Ohta, T.Sakamoto and M.Kawashima, Layer-by-layer sublimation observed by reflection high-energy electron diffraction intensity oscillation in a molecular beam epitaxial system, Appl. Phys. Lett. 47: 286 (1985).

16. N.J.Kawai, T.Kojima, F.Sato, T.Sakamoto T.Nakagawa and K.Ohta, Layer-by-layer sublimation and AlAs sublimation stopper formation AlAs-GaAs system observed by RHEED intensity oscillation, in Proc. 12th Int. Symp. on GaAs and Related Compounds, in Karuizawa (1985).

17. J.M.Van Hove and P.I.Cohen, Mass-action controll of AlGaAs and GaAs growth in moleculer beam epitaxy, Appl. Phys. Lett. 47: 726 (1985).

18. T.Kawamura, T.Sakamoto and K.Ohta, Origin of azimuthal effect of RHEED intensity oscillations observed during MBE, Surf. Sci. 171: L409 (1986).

19. T.Kawamura, T.Natori, T.Sakamoto and P.A.Maksym, Calculated of RHEED from stepped Si(001) for interpretation of RHEED oscillation during MBE, Surf. Sci. 181: L171 (1987).

20. J.C.Bean, L.C.Feldman, A.T.Fiory, S.Nakahara and I.K.Robinson, Ge$_x$Si$_{1-x}$/Si strained-layer superlattice grown by molecular beam epitaxy, J. Vac. Sci. & Technol. A2: 436 (1984).

PART II

PHYSICS OF MULTILAYER STRUCTURE

THEORETICAL BACKGROUND OF INTERFACES AND MULTILAYERS

J.P. Vigneron

Institut de Recherche sur les Interfaces Solides
Facultés Universitaires Notre-Dame de la Paix
61, rue de Bruxelles, B-5000 Namur (Belgium)

The considerable progress in material technology has made possible the fabrication of stratified crystals with widely adjustable geometrical parameters [1]. It is generally stressed that the availability of such materials opens the way to condensed-matter physics with reduced dimensionality where quantum size effects become dominant. One thinks of course primarily of the behavior of electrons in these sharp multilayer structures, for which these quantum size effects typically appear between 10 to 1000 Å in most studied materials. It should be emphasized, however, that all kinds of waves traveling through multilayer structures will be influenced by the stratification of the medium, possibly for other geometrical sizes than those considered for electrons. Photons propagating in heterostructures are not the least interesting of these wave phenomena, from the point of view of the applications.

This lecture will introduce very basic concepts used in the physics of multilayer systems. The presence of sharp or less sharp interfaces induces basically two types of new states in the vibration and electronic spectra of solids: the interface states and the confined states. Interface states are modes which peak at the location of the interface and keep localized around it, usually behaving as a traveling wave in directions parallel to the interface. Surface vibrations, like Rayleigh waves or Fuchs-Kliewer modes [2], are examples of such localized interface states, but more energetic excitations, like interface plasmons, also exist. Confined states, on the other hand, can be viewed as volume modes squeezed within a thin layer of material, because adjacent materials oppose the propagation of these excitations. Here also, one is led to a two-dimensional propagative behavior along any direction parallel to the confining layer. Thus, in both cases, one observes an energy localization in the direction normal to the interfaces and a bidimensional wave behavior along the layer planes.

When studying surfaces and interfaces in multilayer structures, one usually wishes to understand the effect of stratification on photon, phonon and electron states. Today, a fully microscopic description of these states is not yet available, except in a few specific systems, and most of the theoretical interpretations of data involves more or less macroscopic approaches. This is why, in this introductory lecture, we shall more precisely focus on these macroscopic descriptions. These involve explaining the origin of long-wavelength collective excitations like interface phonons and plasmons, the formation of electronic quantum-well states in reasonably wide semiconducting layers, in the framework of the envelope function approach, and using the same argument, tunneling effects involved in the electronic transport across barrier layers.

1. ELECTROMAGNETIC INTERFACE MODES

Electromagnetic interface modes play a substantial role in problems as different as those involved in optical properties of multilayered materials or in physisorption of highly polarizable molecules at surfaces.

1.1 Non-Retarded Case

In the non-retarded approximation, and discarding the details of the crystal structure on an atomic scale, the origin of these modes is easily understood in terms of the dielectric response of a stratified material to an external oscillatory electromagnetic field [3]. This response can be described by means of a frequency-dependent dielectric function $\varepsilon(\omega)$. The link between microscopic and macroscopic descriptions of the dielectric response is easily explained in the case of polar crystals with a simple crystal structure like zinc blende materials with only two distinct atoms per unit cell. The structure of the dielectric response function of a polar material reflects the presence of long-wavelength transverse and longitudinal vibrations modes (see Figure 1). According to the frequency, a polar material is able or unable to propagate electromagnetic energy, in a similar way as a periodic solid can or cannot propagate electron Bloch waves according to the value of the electron energy.

a. Long-Wavelength Transverse Optical Phonons

Let us consider the infrared dielectric response of a cubic polar material. We retain the simple case of a unit cell containing only two distinct atoms (s=2). In such a crystal, one observes the formation 3s-3 = 3 optical phonon branches. These branches correspond to the three possible polarizations of the atomic displacements with respect to the propagation direction, indicated by the orientation of the wavevector $\mathbf{k}$. The transverse optic (TO) mode is twofold degenerate because one assumes a perfect anisotropy. One can easily construct the Hamiltonian dynamics of these modes. Let $\mathbf{u_+}$ and $\mathbf{u_-}$ represent the respective displacements of the positive and negative ions, in terms of which the relative displacement associated with an optical mode can be defined:

$$\mathbf{w} = \frac{\mathbf{u_+} - \mathbf{u_-}}{\sqrt{Mn}} \qquad . \qquad (1)$$

Here M is the reduced mass of the ions in the unit cell and n represents the density of ion pairs in the solid. For long-wavelength vibration modes, this reduced coordinate becomes a continuous function of position, and the Hamiltonian constructed using this coordinate can be expressed in terms of a hamiltonian density. The kinetic energy density is easily written in terms of this relative displacement:

$$T = \frac{1}{2} \left(\frac{d\mathbf{w}}{dt} \right)^2 \quad , \qquad (2)$$

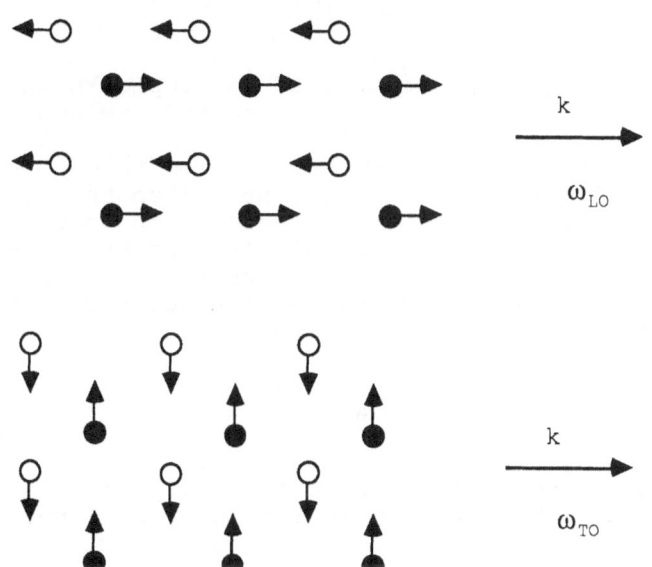

Figure 1. Long-wavelength longitudinal (ω_{LO}) and transverse (ω_{TO}) optical phonons

while the potential energy can be expressed as

$$U = \frac{1}{2} a \, w^2 - b \, \mathbf{w} \cdot \mathbf{E} - \frac{1}{2} c \, E^2 \quad . \qquad (3)$$

The elastic restoring energy (first term) is completed by two terms arising from the macroscopic electric field $\mathbf{E}$ which eventually accompanies the optical mode. The second term in (3) accounts for the dipole interaction with the macroscopic field $\mathbf{E}$, and the third term represents the field energy density. The meaning of the constants a, b, and c will be made clear in a moment. The equation of motion obtained from this potential energy are

$$\frac{d^2 \mathbf{w}}{dt^2} + a \, \mathbf{w} - b \, \mathbf{E} = 0 \quad , \qquad (4)$$

and the corresponding Hamiltonian density is

$$H = \frac{1}{2} \left(\frac{d\mathbf{w}}{dt} \right)^2 + \frac{1}{2} a \, w^2 - b \, \mathbf{w} . \mathbf{E} - \frac{1}{2} c \, E^2 \quad . \qquad (5)$$

The polarization, or dipole moment density is given by

$$\mathbf{P} = - \frac{dH}{d\mathbf{E}} = b \, \mathbf{w} + c \, \mathbf{E} \quad . \qquad (6)$$

Equations (4) and (6) can be applied in simple known cases, in order to assess the value of the constants a, b and c.

The first situation of interest is the case of a transverse mode, with the displacement $\mathbf{w}$ normal to the propagation direction. In this case, the ionic deformation corresponds to a shear motion unable to produce a polarization of the continuous medium. This of course means that no electric field develops in this situation ($\mathbf{E}=0$) and the equation of motion (4) reduces to

$$\frac{d^2 \mathbf{w}}{dt^2} + a \, \mathbf{w} = 0 \quad , \qquad (7)$$

which shows that the coefficient a is simply related to the frequency of the transverse optical phonon, the "reststrahlen" frequency

$$a = \omega_{TO}^2 \quad . \qquad (8)$$

The next situation of interest involves the application of a static external electric field $\mathbf{E_s}$. In this case, the equation of motion (4) gives a static deformation

$$\mathbf{w_s} = \frac{b}{a} \, \mathbf{E_s} \quad , \qquad (9)$$

which produces a polarization

$$\mathbf{P_s} = b \, \mathbf{w_s} + c \, \mathbf{E_s} = \left(\frac{b^2}{a} + c \right) \mathbf{E_s} \quad . \qquad (10)$$

This must be compared to the constitutive equation of the dielectric material

$$\mathbf{P_s} = \varepsilon_0 \, (\varepsilon_s - 1) \, \mathbf{E_s} \qquad (11)$$

in order to obtain a second relation linking the values of b and c:

$$\frac{b^2}{a} + c = \varepsilon_0 (\varepsilon_s - 1) \quad . \tag{12}$$

ε_s is the static (zero-frequency) dielectric constant, and ε_0 is the vacuum permittivity (SI units are used throughout this lecture).

Finally, the last situation that needs to be studied corresponds to the limiting case of an external high-frequency electric field $\mathbf{E}_\infty$. This field is assumed to vary so rapidly that no motion of the ions can take place:

$$\mathbf{w}_\infty = 0 \quad . \tag{13}$$

In this case, the expression of the crystal polarization reduces to

$$\mathbf{P}_\infty = c \, \mathbf{E}_\infty \quad , \tag{14}$$

which should be compared to the high-frequency response equation

$$\mathbf{P}_\infty = \varepsilon_0 (\varepsilon_\infty - 1) \, \mathbf{E}_\infty \tag{15}$$

to provide an explicit value for the coefficient c

$$c = \varepsilon_0 (\varepsilon_\infty - 1) \quad . \tag{16}$$

This result, combined with (12) and (8), gives the value of the last coefficient to be determined:

$$b = \sqrt{\varepsilon_0 (\varepsilon_s - \varepsilon_\infty)} \; \omega_{TO} \quad . \tag{17}$$

The consideration of some special cases allowed to express the unknown coefficients a, b and c in terms of more easily measurable quantities, the transverse optical frequency ω_{TO}, the static dielectric constant ε_s, and the high-frequency dielectric constant ε_∞. By "high frequency" we mean a frequency large enough so that no excitation of the crystal vibration modes can take place. Of course, at this high frequency, the electronic polarization can already be effective, and ε_∞ is different from the vacuum value ($\varepsilon_\infty \neq 1$).

b. Long-Wavelength Longitudinal Phonons

Longitudinal modes are characterized by a polarization field parallel to the direction of propagation. Without any free external charge, the displacement field vanishes and the macroscopic electric field is proportional to the polarization:

$$E = - \frac{P}{\varepsilon_0} \qquad \cdot \qquad (18)$$

Using this in (6), one can link the macroscopic electric field to the relative displacement $\mathbf{w}_L$

$$E = - \frac{b}{c + \varepsilon_0} \mathbf{w}_L \qquad , \qquad (19)$$

and eliminate this field from the equation of motion (4):

$$\frac{d^2 \mathbf{w}_L}{dt^2} + \left(a + \frac{b^2}{c + \varepsilon_0} \right) \mathbf{w}_L = 0 \qquad . \qquad (20)$$

The frequency of the longitudinal optical phonon is then readily obtained as

$$\omega_{LO}^2 = a + \frac{b^2}{c + \varepsilon_0} \qquad , \qquad (21)$$

which reduces to the well-known Lyddane-Sachs-Teller relation when the known values of a (8), b (17) and c (16) are properly substituted.

$$\omega_{LO}^2 = \frac{\varepsilon_s}{\varepsilon_\infty} \omega_{TO}^2 \qquad . \qquad (22)$$

Finally, from these relations, it is possible to give an explicit form of the dielectric function at all frequencies in terms of the fundamental optical constants ε_s, ε_∞, and ω_{TO}.

c. Frequency-Dependent Dielectric Function in the Infrared Energy-Range

By definition, the dielectric function $\varepsilon(\omega)$ relates the macroscopic oscillatory electric field and the sample polarization:

$$\mathbf{P}(\omega) = \varepsilon_0 \left[\varepsilon(\omega) - 1 \right] \mathbf{E}(\omega) \qquad . \qquad (23)$$

Having reached an oscillatory regime for the relative displacement,

$$\mathbf{w} = \mathbf{w}(\omega) e^{i\omega t} \qquad , \qquad (24)$$

one can solve the equation of motion (4) under the explicit form

$$\mathbf{w}(\omega) = \frac{b}{a - \omega^2} \mathbf{E}(\omega) \qquad (25)$$

and express the polarization directly in terms of the electric field:

$$P(\omega) = \left(\frac{b^2}{a - \omega^2} + c \right) E(\omega) \quad . \qquad (26)$$

Comparing this expression with the definition (23), one obtains the dielectric function $\varepsilon(\omega)$ in the form

$$\varepsilon(\omega) = \varepsilon_\infty + \frac{(\varepsilon_s - \varepsilon_\infty)\, \omega_{TO}^2}{\omega_{TO}^2 - \omega^2} \qquad (27)$$

or

$$\varepsilon(\omega) = \varepsilon_\infty \frac{\omega^2 - \omega_{LO}^2}{\omega^2 - \omega_{TO}^2} \quad . \qquad (28)$$

It is seen that the dielectric function vanishes at the longitudinal optical phonon frequency and has a pole at the transverse optical phonon frequency. In the frequency range between ω_{TO} and ω_{LO}, the dielectric function gets negative. The index of refraction

$$n(\omega) = \sqrt{\varepsilon(\omega)} \qquad (29)$$

is then imaginary and the wave cannot propagate through the crystal without decay. This form of the dielectric function then provides a kind of band structure for the propagating waves in the infrared frequency range. Propagation is always allowed, except in the reststrahlen gap, between ω_{TO} and ω_{LO}. In this region, the reflectivity would reach 100%, if the optical phonon lifetime were infinite. In practice imperfect reflectivity is still observed in the reststrahlen band mostly because anharmonicities in the ion-ion potential allow for a decay of the optical phonons into other types of vibration modes. This effect can be accounted for in a phenomenological way by introducing in the equations of motion a dissipating term, proportional to the first derivative of the relative displacement and constantly opposed to the ion motion:

$$\frac{d^2 w}{dt^2} + \gamma \omega_{TO} \frac{dw}{dt} + a\,w - b\,E = 0 \quad . \qquad (30)$$

This actually does not change the solution (25), except that the frequency ω^2 is now replaced by the complex number $\omega^2 + i\gamma\omega\omega_{TO}$. The dielectric function then takes the form (see in Figure 2 the dielectric function of GaAs, using parameters obtained from infrared reflectance measurements)

$$\varepsilon(\omega) = \varepsilon_\infty + \frac{(\varepsilon_s - \varepsilon_\infty)\, \omega_{TO}^2}{\omega_{TO}^2 - \omega^2 - i\,\gamma\,\omega\,\omega_{TO}} \quad . \qquad (31)$$

For more than two ions per unit cell, several transverse optical phonon frequencies and several reststrahlen bands are found. In that case, more complex expressions of the dielectric function must be used, accounting for more than one oscillator:

$$\varepsilon(\omega) = \varepsilon_\infty + \sum_{j=1}^{n} \frac{\rho_j \, \omega_{TO,j}^2}{\omega_{TO,j}^2 - \omega^2 - i\gamma_j \omega \omega_{TO,j}} \quad . \tag{32}$$

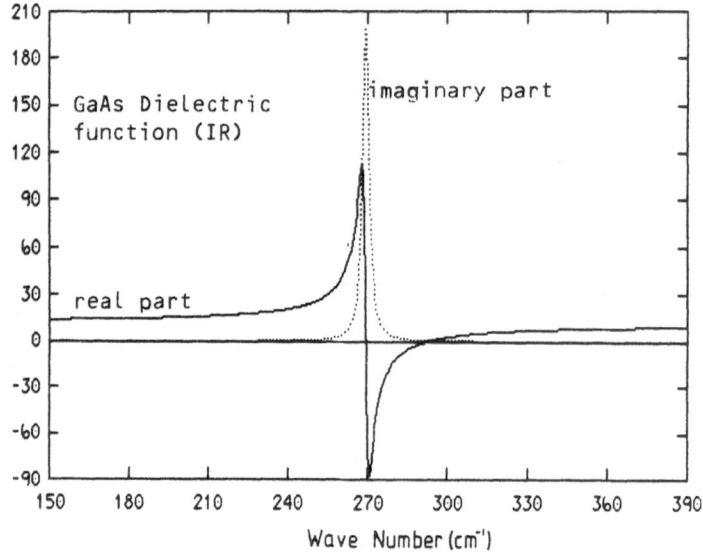

Figure 2. GaAs infrared dielectric function, including damping due to phonon finite lifetime

d. Infrared Interface Electromagnetic Modes

We now come to the central part of this section. When considering several regions described by different dielectric functions, can new states of the electromagnetic field appear? In the most general situation, several regions of space coexist, with different dielectric responses. This can be viewed as defining a variable local dielectric function $\varepsilon(r,\omega)$. Of course, the spatial variation does not describe the response on an atomic scale: all wavelengths involved in the problem must be much larger than the interatomic spacing in any region. If an external oscillating charge distribution $\rho(r,\omega)$ is established in this region, it will determine a potential which can be calculated by solving a Poisson equation

$$\nabla . \varepsilon(r,\omega) \nabla V(r,\omega) = -\frac{\rho(r,\omega)}{\varepsilon_0} \quad . \tag{33}$$

The electromagnetic eigenmodes of the system are oscillating potentials which develop without the need for any external charge density. They are given by non-trivially null potentials which verify

$$\nabla . \varepsilon(r, \omega) \nabla V(r, \omega) = 0 \quad . \quad (34)$$

This is only possible at precise frequencies, which come out of the above non-linear eigenvalue problem (34).

As a specific example, let us consider the simple case of a planar interface separating two media of different dielectric responses [4]. Medium 1 is described by the dielectric function $\varepsilon_1(\omega)$ and medium 2 is described by the dielectric function $\varepsilon_2(\omega)$ (see Figure 3).

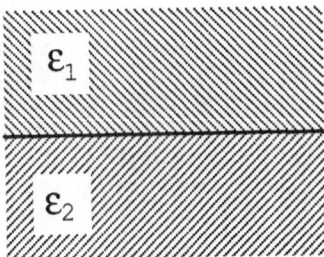

Figure 3. Two media joining at a planar interface may determine interface electromagnetic modes appearing in the reststrahlen regions of both materials

The eigenmodes are described by a potential which verifies

$$\nabla^2 V(r, \omega) = 0 \quad (36)$$

in both regions. Due to the translational invariance of the system, the potential

$$V(r, \omega) = e^{i\mathbf{k} . \rho} v(z, \omega) \quad (37)$$

separates into a two-dimensional wave travelling parallel to the layers, and a localized envelope decaying away from the interface. The two-dimensional **k** vector fixes the propagation direction and the wavelength of the mode along the interface. It also controls its decay length in the orthogonal (z) direction. The function $v(z, w)$ satisfies the one-dimensional differential equation

$$\frac{d^2 v}{dz^2} - k^2 v = 0 \quad , \quad (38)$$

which accepts solutions with a finite limit for large $|z|$ in the form

$$v(z,\omega) = A\,e^{-kz} \qquad \text{for } z>0$$

$$= B\,e^{kz} \qquad \text{for } z<0 \qquad . \qquad (39)$$

The matching conditions at the interface (continuous normal displacement **D** and tangential electric field **E**) simply reduce, in this geometry, to the continuity of v and $\varepsilon(dv/dz)$. These give the constraints

$$A = B \qquad\qquad\qquad (40)$$

$$\varepsilon_1(\omega) + \varepsilon_2(\omega) = 0 \qquad . \qquad (41)$$

Equation (40) shows that the potential is symmetrical with respect to the interface plane, while (41) fixes the frequency of the interface mode. It is interesting to note that such a mode will only appear at specific frequencies such that at least one of the dielectric constants ε_1 or ε_2 is positive, the other being negative. At an interface mode frequency, one of the materials must be able to propagate the field, while the other one must be reflecting. When medium 1 and medium 2 (assuming two atoms per unit cell) both present reststrahlen bands and if those bands do not overlap, there are always two distinct solutions, one in each reststrahlen region. There is no k-dependence of those interface mode frequencies. This is related to the absence of any unit of length in this geometry.

The case of free surface has been specially studied. The dielectric function of vacuum is simply $\varepsilon=1$ and the surface magnetic mode is obtained from $\varepsilon(\omega_{so})=-1$. This mode always appear inside the reststrahlen region of surfaces and interfaces, usually closer to the LO frequency when the matrix is less polar.

The next important question that must be asked is: How can we observe those interface electromagnetic modes and have they been observed?

The simple electromagnetic eigenmodes described above are in fact accessible to electron energy loss measurements [5-10], for which, in contrast to the case of reflectivity, retardation effects are negligible. When an electron with a few electron volts of kinetic energy is reflected by the external surface of the layer it will lose energy, coupling its own Coulomb field to the decaying field associated with the surface or shallow interface modes. Figure 4 shows the reflection energy-loss spectrum of a semi-infinite GaAs-Al$_{0.3}$Ga$_{0.7}$As superlattice with about 200 Å layer thickness [11-12]. This spectrum can be interpreted as follows. The main peak, labeled A, on the right (loss) and left (gain) of the elastic peak is associated with an evanescent surface mode attached to the external GaAs surface, while the weaker structure, labeled B, is due to a propagating interface mode having primarily an Al-As stretching character. This spectrum can be quantitatively accounted for by the dielectric approach described above. The full line in Figure 4 is a simulation based on this approach.

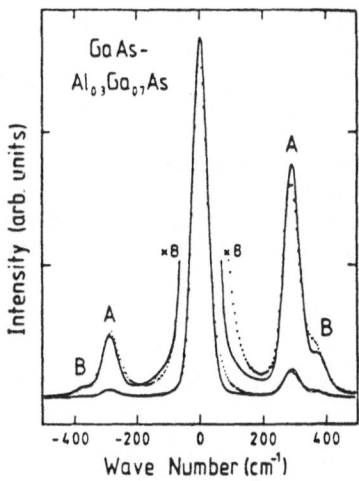

Figure 4. HREELS spectrum from a GaAs/AlGaAs
superlattice. Dots are experimental points. The
full line is a simulation spectrum based on a
dielectric approach

e. Surface and Interface Plasmons

The frequency-dependent dielectric function for the plasmon system (collective excitation of the electron gas), i.e.
in the electron-volt energy range, is usually obtained from
semi-empirical band structures. State-of-the-art self-consistent band structure calculations start being able today to
provide such data, but little work has yet been done in this
direction. The difficulty lies essentially in the fact that
the frequency-dependent dielectric function is not a ground-
state property and advanced quasi-particle descriptions of the
crystal electronic structure are necessary to predict the correct distances in energy between occupied and unoccupied
bands. For metals, plasma oscillations can be considered as
the analogous of the optical modes in ionic crystals. Here,
positive ion cores play a role similar to positive ions in
ionic crystals and the delocalized electron cloud that of negative ions [13-18]. Here again, a "macroscopic" view of this
system, forgetting about the details of the crystal structure,
will be adopted. It leads to consider a homogeneous medium
made of two uniformly dense gases: the negatively charged
electron plasma in a uniform positively charged background.
The electronic analogous of the transverse modes corresponds
to a shearing motion of the electron gas, which cannot produce
a charge density change (nor any electric field) and develops
without experiencing any resistance. The corresponding force
constant is then zero, which implies a frequency

$$\omega_T = 0 \quad . \tag{42}$$

The longitudinal modes give charge density fluctuations
accompanied by an electric field which induces a restoring
force leading a finite frequency. This classical plasma fre-

quency can easily be determined by the following simple argument. A parallelepipedic volume of rather large section S filled with a uniform electron gas with density n is displaced rigidly along a small distance x away from their original position. A planar hole of positive charge (containing $n|e|Sx$ units of charge) appears on one side of the box, and the corresponding excess of negative charge appears on the other side (see Figure 5.) The electric field which develops inside the volume is that found in a charged plane capacitor:

$$E = \frac{\sigma}{\varepsilon_0} = \frac{n|e|}{\varepsilon_0} x \quad , \tag{43}$$

which shows that each electron in the box experiences a restoring force

$$F = - \frac{ne^2}{\varepsilon_0} x \quad . \tag{44}$$

This force provides an oscillation frequency which is nothing else than the classical plasma frequency

$$\omega_L = \omega_p = \sqrt{\frac{ne^2}{m\varepsilon_0}} \quad . \tag{45}$$

If one further assumes that the ionic cores are not deformable, ε_∞ is set to the vacuum value 1, and the dielectric function takes the form

$$\varepsilon(\omega) = 1 - \frac{\omega_p^2}{\omega^2} \quad . \tag{46}$$

The static dielectric constant of the metal is infinitely large and negative, as it should be for a perfect conductor.

The matching conditions described above for infrared d electric response will still apply here and give rise to interface plasmons, with a frequency obtained from $\varepsilon_1 + \varepsilon_2 = 0$. For surface plasmons, for instance, $\varepsilon_2 = 1$ and the frequency is given by

$$\omega_s = \frac{\omega_p}{\sqrt{2}} \quad . \tag{47}$$

Interface plasmons such as surface plasmons can also be detected by electron energy loss spectroscopy. They play a central role in physisorption because their zero point energy is sensitive to the geometry of the surfaces on which they develop. The Van der Waals energy between two macroscopic bodies is essentially due to this effect.

The finite lifetime of plasmons can be accounted for, as was done for the infrared excitations by replacing the square of the frequency ω^2 by a complex quantity, here $\omega^2 - i\omega/\tau$.

1.2 Retarded Case: Interface Polaritons

Multilayer optics requires a full treatment of the electromagnetic excitations, including the retardation effects and require solving more completely the Maxwell's equations [19-22]. For stratified materials, which keep translational invariance with respect to any displacement perpendicular to the growth direction, two independent solutions of Maxwell's equations are the well-known s and p electromagnetic waves. In the former case (s), the radiation is polarized in a direction parallel to the surface and perpendicular to the direction of incidence (x). In the latter case (p), the electric field is located in the plane of incidence (y,z). s modes do not

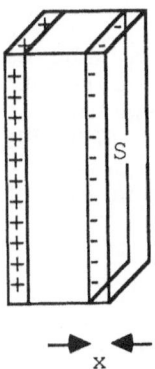

Figure 5. The charge density fluctuations oscillate with the classical plasma frequency. Rigidly displacing a finite volume of electrons with respect to the rest of the cloud and with respect to the background positive charge generates a restoring force. This force explains the finite frequency of longitudinal plasma waves

contribute a charge accumulation corresponding to any interface polarization, since the corresponding motions of matter remain strictly parallel to the surface. No interface mode can develop for this polarization. p modes, by contrast, induce a superficial charge density which may react on the ions in the volume of the material and can give rise to interface excitations. Again because of the translational symmetry parallel to the interface, the fields can all be described as a product of a wave travelling parallel to the interface and an envelope function varying in the normal (z) direction:

$$\mathbf{H}(\mathbf{r},t) = e^{iky-i\omega t} \, U_p(z) \, \mathbf{e}_x \quad . \tag{50}$$

If the wave travels in a medium with a stratified structure, described by a local dielectric function which varies only in the z direction, $\varepsilon(z,\omega)$, Maxwell's equations reduce to a modified wave equation in one dimension:

$$\varepsilon(z,\omega) \; \frac{d}{dz} \left[\frac{1}{\varepsilon(z,\omega)} \frac{dU_p}{dz} \right] + \left[\left(\frac{\omega}{c}\right)^2 \varepsilon(z,\omega) - k^2 \right] U_p(z) = 0 \; ,$$

(51)

the other fields being given by

$$D(r,t) = \varepsilon_0 \varepsilon(z,\omega) \; E(r,t)$$

$$= e^{iky-i\omega t} \left(i\frac{dU_p}{dz} e_y + kU_p(z) e_z \right)/\omega \; .$$ (52)

From these general expressions, the case of a single interface can easily be obtained. It requires the solution of the wave equation in both regions of space, with dielectric constants ε_1 and ε_2, and matching of U_p and $(1/\varepsilon) dU_p/dz$ at the boundary.

In this geometry, the solutions are still exponential

$$\begin{aligned} U_p(z) &= A_1 \; e^{-\alpha_1 z} & z>0 \\ &= A_2 \; e^{\alpha_2 z} & z<0 \end{aligned}$$ (53)

with

$$\alpha_i = \sqrt{k^2 - \varepsilon_i \left(\frac{\omega}{c}\right)^2} \; , \quad i=1,2 \; .$$ (54)

The decays are now different in both media: The mode asymmetry is a result of the retardation effect. If α_i is imaginary, we have a radiative type of mode in the corresponding medium i. These modes cannot sustain themselves at the interface, as the radiation will drive away its energy. We shall rather seek for solutions in the regions where both α_i are real. The matching conditions lead to the following constraints:

$$A_1 = A_2$$ (55)

and

$$\alpha_2 \varepsilon_1 + \alpha_1 \varepsilon_2 = 0 \; .$$ (56)

These conditions become identical to the non-retarded interface mode defining equation as k becomes much larger than ω/c. The basic difference introduced by retardation is the dispersion (the frequency is different for each wavelength). For a surface polariton (one of the joined media is vacuum) one obtains the dispersion shown in Fig. 6.

126

An important feature of the surface and interface modes is
that they determine an electric field outside the volume of
the crystal. An electron in the vacuum, outside the crystal
will experience this field and will not have an infinite mean
free part. This is the main idea pursued in using high resolution electron energy loss spectroscopy in the detection of
these excitations [23].

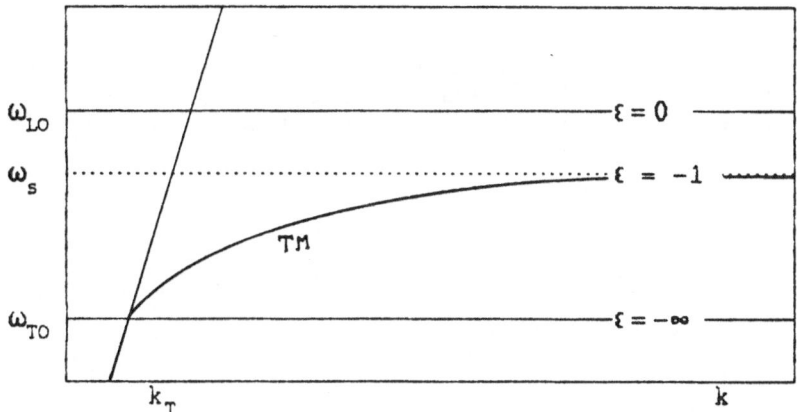

Figure 6. Surface polariton. Dispersion appears at
wavelengths close to that of light at comparable
frequencies. At short wavelengths (large k) the
mode frequency approaches the value found for the
electrostatic interface mode found in neglecting
all retardation effects

2. COUPLING OF INTERFACE OR CONFINED MODES

Interface modes are peaked at interfaces and decay into the
joining materials. When another interface is formed close to the
first one the interface modes couple to give new "slab" modes with
different frequencies. This mechanism is, in essence, similar to
that brought about by the tight-binding description of the molecular states starting from the atomic states: tunneling takes place
to couple similar interface modes. This, as we will see in a moment, is also true for electromagnetic waves. Tunnel effect is in
fact a wave phenomenon, not necessarily connected to a quantum-
mechanical system. When a superlattice is considered, the large
number of interfaces can give rise to the formation of bands, in a
way very similar to the formation of bands in solids from the periodic repetition of atoms in the crystal structure. This band formation concerns both interface modes and confined states. The development of minibands in a composition superlattice can be viewed as
the building of Bloch combinations of coupled quantum-well confined
states.

To study the formation of multiple states by coupling of interface modes, we shall again consider the simple example of the

classical electromagnetic interface modes. The simplest compound
state is that provided by the electromagnetic excitation of a thin
plane slab, of thickness L. Let us here consider first the non-
retarded modes of a self-supported slab: the slab, with dielectric
constant $\varepsilon(\omega)$, is surrounded by vacuum ($\varepsilon=1$). The z=0 plane (see
Fig.7) is a plane of symmetry, so that the p-polarized modes can be
readily classified as symmetric or antisymmetric. The quasistatic
potential is of the form:

$$V(z) = A\ e^{kz} + B\ e^{-kz} \tag{57}$$

and can be explicitly written inside and outside the slab. The
matching conditions may only be expressed at the boundary z = L/2,
since they will automatically be satisfied at z=-L/2. One gets the
following relations:

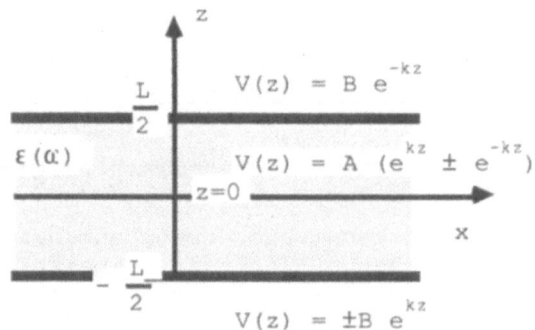

+ : symmetric mode
− : antisymmetric mode

Figure 7. The slab geometry and the
solutions of the Laplace equation
defining the coupled interface modes.
Symmetric and antisymmetric modes develop

$$A\ (e^{k\frac{L}{2}} \pm e^{-k\frac{L}{2}}) = B\ e^{-k\frac{L}{2}} \tag{58}$$

and

$$\varepsilon(\omega)\ k\ A\ (e^{k\frac{L}{2}} \overset{-}{+} e^{-k\frac{L}{2}}) = -\ B\ k\ e^{-k\frac{L}{2}} \quad, \tag{59}$$

which give a non-trivially null potential if

$$\varepsilon(\omega) = -\ \text{th}^{\pm 1}\ k\frac{L}{2} \quad. \tag{60}$$

The corresponding dispersion relation is shown in Fig.8 for the
symmetric (ω_+) and the antisymmetric (ω_-) modes. At short wavelengths (k large), both frequencies degenerate into a single one
equal to the surface mode. In this situation, the potentials as-

sociated with the interface excitation do not overlap, which means
here that the ionic motions at either surfaces do not influence
each other. For wavelengths of the order of the slab thickness, the
interface modes couple together and one obtains two well-separated
frequencies.

When retardation effects are taken into account, the dispersion relation is obtained from the modified equation

$$\varepsilon(\omega) = -\frac{\alpha_2}{\alpha_1} \, th^{\pm 1}\left(\alpha_1 \frac{L}{2}\right) \quad , \tag{61}$$

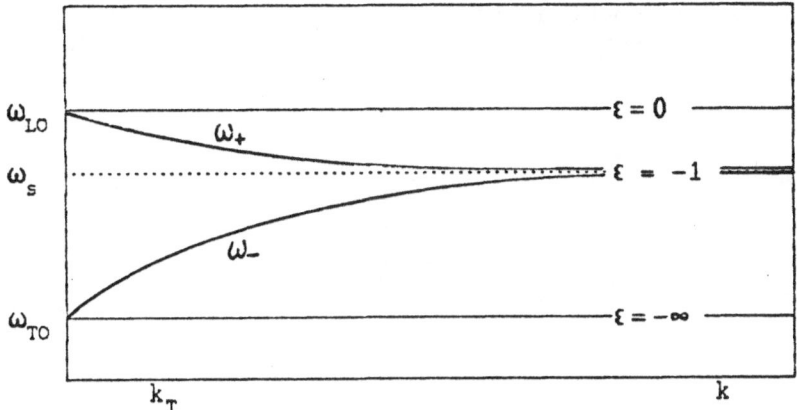

Figure 8. Electromagnetic modes in a
slab. The two modes can be interpreted as
coupled interface modes

where

$$\alpha_1^2 = k^2 - \left(\frac{\omega}{c}\right)^2 \tag{62}$$

and

$$\alpha_2^2 = k^2 - \varepsilon(\omega)\left(\frac{\omega}{c}\right)^2 \quad . \tag{63}$$

Here again, one sees that retardation effects rapidly become negligible for $k \gg k_T$, the wavelength of light with a frequency of the
order of ω_T.

When several interfaces are present, the number of levels increases and in the case of an infinite or a semi-infinite superlattice, continuous bands are formed, with propagating waves of a
new kind. Each wave is formed by a Bloch combination of interface
modes. Figure 9 shows the result of a calculation of coupled interface polaritons in a GaAs-AlAs semi-infinite superlattice, that
is a superlattice terminated by a free surface with vacuum. One ob-

serves the formation of three kinds of modes that can be labelled according to the short-wavelength limit of their frequencies. Modes of the first group converge, for large k, to the interface state located in the AlAs reststrahlen frequency. The isolated mode near 290 cm^{-1} approaches the GaAs surface mode frequency, while the modes of the next group converging near 280 cm^{-1} reproduce the independent interface modes located in the GaAs restrahlen band. Some of these modes are part of a continuum, which shows that they can propagate energy, while some others are isolated, a sign that these modes are localized near the surface of the superlattice. These isolated surface modes are also built from a combination of surface and interface polaritons, but the corresponding potential amplitude is evanescent in both directions away from the free surface of the superlattice.

The formation of interface polaritons in the electromagnetic excitation spectrum of a superlattice, while electron minibands are rather interpreted as combinations of electron quantum-well states, are clear illustrations of the concept of interface states, and

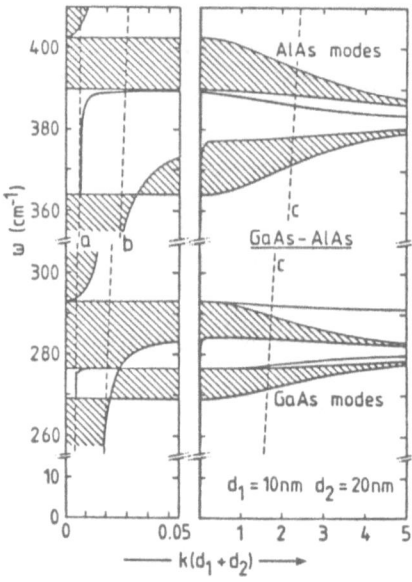

Figure 9. Polariton structure (p polarization) for a GaAs-AlAs superlattice, represented on two different scales. The layer thicknesses are 10 and 20 nm for GaAs (first layer) and AlAs, respectively. The dashed lines visualize the lines of equation (a) w=kc, (b) w=kc/n$_{prism}$ and (c) w=kv$_{//}$ relevant for IR, ATR and EELS spectroscopies, respectively

confined states, respectively. It should be emphasized that interface or confined states can be found in both systems. Confined modes can be found in volume polaritons such as those located in a layer of material which behaves as a wave guide, when total reflec-

130

tion occurs against the neighboring layers. Also, in superlattices like HgTe-CdTe, where the envelope function formalism with the simple boundary conditions described here most probably fail, the electronic states may rather take the form of interface states.

3. ELECTRON STATES IN QUANTUM WELLS AND MULTILAYERS

Diffraction of electrons by a crystal structure takes place at energies which, translated into wavelength, adapts roughly to the lattice parameter

$$\lambda \ (\text{Å}) \ \approx \ \frac{12}{\sqrt{E \ (\text{eV})}} \ \approx \ a \ . \tag{64}$$

Because of this diffraction effect, electrons will propagate or be highly reflected when entering the material through a given interface, according to their energy. The regions of high reflection coincide with the forbidden energy gap, where no bulk electron states are found, and the regions of free propagation correspond to unoccupied allowed bands. In a quantum well structure, energy regions show up where propagation is allowed within the central thin layer and is not allowed outside. This condition may lead to states confined to the central layer and which eventually propagate parallel to the interfaces. These states bear similarity with the states developing in one-dimensional potential wells, with energies determined by the constraining effect of the Heisenberg uncertainty principle, basic to the quantum nature of electrons. It should be emphasized that the similarity between the quantum-well layer in a solid and the potential well of elementary quantum mechanics lies basically in the geometrical ordering of propagating/not propagating layers. It is not at all easy, from a theoretical point of view, to determine the equivalent potential step height to be used to simulate the quantum-well electronic structure. This theoretical issue is the one involved in determining the band offset which establishes itself in a heterostructure. One should be conscious that ab initio modelisation of a quantum well of reasonable thickness is still a formidable task when dealt with from the microscopic point of view. However, much progress is being made on the experimental characterization of quantum wells using perpendicular transport and, in view of the importance of these studies, it seems advisable to introduce here the theoretical framework in which most tunneling experiments are looked at. We shall focus on the effective mass (or envelope function) approach since this is the starting point of most quantum-well and tunnelling experiments in multilayer structures.

The envelope function approximation [24-27] results from an application of the effective mass theory to the electronic structure of multilayers. It can be considered as a macroscopic approach, in the sense that the details of the atomic structure are hidden in parameters like effective masses, barrier heights, etc… In essence, the envelope function approximation bears resemblance with the dielectric approach described above for photon states. Each layer is characterized by macroscopic parameters taken from known bulk properties, and appropriate matching conditions are defined. In what follows, we shall recall elementary results of the k.p perturbation theory used to describe perfect crystal states, rederive effective mass equations in a form appropriate for studying the wave-function matching at interfaces, and derive the matching conditions basic to the envelope function treatment.

3.1 k.p Perturbation Expansion

The k.p perturbation expansion aims at describing perfect-crystal Bloch states at the vicinity of specific Brillouin zone points. The Bloch wavefunctions of the periodic crystal satisfy the Schrödinger equation

$$\left[-\frac{\hbar^2}{2m} \nabla^2 + v(r) \right] \psi_{nk}(r) = \varepsilon_{nk} \psi_{nk}(r) \quad , \qquad (65)$$

where $v(r)$ is the common effective periodic potential seen by all electrons in the crystal and $\psi_{nk}(r)$ is the nth Bloch state at the point $\mathbf{k}$ in the Brillouin zone. m is the free-electron mass. The Bloch theorem sets the analytic form of these wavefunctions as

$$\psi_{nk}(r) = u_{nk}(r) \frac{e^{i\mathbf{k}.\mathbf{r}}}{\sqrt{8\pi^3}} \quad , \qquad (66)$$

where $u_{nk}(r)$ has the same periodicity as $v(r)$ and is normalized over the volume v_c of the crystal unit cell.

We consider the most simple situation of a single isotropic and parabolic conduction band in a direct band gap semiconductor (the case of GaAs, and many other III-V compounds). More complex situations, like non-parabolic or degenerate bands, can be handled using similar arguments, but we shall not describe these in detail. We assume knowing the full set of k=0 energies and wavefunctions and ask about the specific lower conduction band for k≈0. Using the Bloch form of the wavefunction, we can transform the original Schrödinger equation into an equation for the periodic function u_{nk},

$$\left[-\frac{p^2}{2m} + v(r) + \frac{\hbar}{m} \mathbf{k}.\mathbf{p} \right] |u_{nk}\rangle = (\varepsilon_{nk} - \frac{\hbar^2 k^2}{2m}) |u_{nk}\rangle, \qquad (67)$$

with periodic boundary conditions. In the $\mathbf{r}$ representation, $\mathbf{p}$ is the differential operator $-i\hbar\nabla$. The states $|u_{n0}\rangle$ and energies ε_{n0} are assumed to be explicitly known. Using standard perturbation theory for eigenvalue problems, we can obtain the first-order correction to the periodic part of the Bloch function

$$|u_{nk}\rangle = |u_{n0}\rangle + \sum_{j \neq n} \frac{\frac{\hbar}{m} \mathbf{k}.\langle u_{j0}|\mathbf{p}|u_{n0}\rangle}{\varepsilon_{n0} - \varepsilon_{j0}} |u_{j0}\rangle \qquad (68)$$

and the second-order correction to the energy

$$\varepsilon_{nk} - \frac{\hbar^2 k^2}{2m} = \varepsilon_{n0} + \frac{\hbar}{m} \, \mathbf{k} . <u_{n0}|\mathbf{p}|u_{n0}> +$$

$$\sum_{j \neq n} \frac{\left(\frac{\hbar}{m}\right)^2 |<u_{n0}|\mathbf{k}.\mathbf{p}|u_{j0}>|^2}{\varepsilon_{n0} - \varepsilon_{j0}} \quad . \tag{69}$$

Defining the vector

$$\mathbf{P}_{nj} = <u_{n0}|\mathbf{p}|u_{j0}>$$

$$= i \, \frac{m}{\hbar} \, (\varepsilon_{j0} - \varepsilon_{n0}) \, <u_{n0}|\mathbf{r}|u_{j0}> \tag{70}$$

(the last expression results from the identity $\mathbf{p} = -i(m/\hbar)[H,\mathbf{r}]$),
one sees immediately that $p_{nn} = 0$ and that there is no first-order
correction to the energy. Also, $p_{mn} = p^*_{nm}$. With these notations,
one gets a dispersion relation valid at the vicinity of k=0 in the
form

$$\varepsilon_{nk} = \varepsilon_{n0} + \frac{\hbar^2 k^2}{2m^*} \tag{71}$$

with an effective mass m* defined as

$$\frac{1}{m^*} = \frac{1}{m} + \frac{2}{m^2} \sum_{j \neq n} \frac{|\mathbf{k}_0 . \mathbf{P}_{jn}|^2}{\varepsilon_{n0} - \varepsilon_{j0}} \quad . \tag{72}$$

$\mathbf{k}_0$ is a unit vector in the direction of $\mathbf{k}$. In an isotropic medium,
the effective mass is independent of the direction of motion and $\mathbf{k}_0$
can be given any direction.

3.2 Effective Mass Theory (Envelope Function Approximation)

We now depart from the case of a perfect crystal to consider the
electronic states induced by a "macroscopic" perturbation. By
"macroscopic" we mean a change in potential that takes place over a
large number of the crystal unit cells. This corresponds to rather
thick layer heterostructures, and the envelope function approach
will almost certainly fail for layers reduced to a few atomic
planes. These ultra-thin structures should rather be treated as a
new crystal lattice, using microscopic approaches. The new
Schrödinger equation for the perturbed crystal reads

$$\left[\frac{p^2}{2m} + v(r) + U(r)\right] \psi(r) = \varepsilon \psi(r) \quad , \qquad (73)$$

where $U(r)$ is the difference between the potential energy of an electron in the perturbed system and in the reference perfect crystal. The exact wavefunction will be expanded in series of the perfect crystal Bloch states

$$\psi(r) = \sum_{n'} \int_{BZ} dk'^3 \, F_{n'k'} \, \psi_{n'k'}(r) \quad , \qquad (74)$$

which provides a representation for the perturbed Schrödinger equation

$$\varepsilon_{nk} F_{nk} + \sum_{n'} \int_{BZ} dk'^3 \, U_{nk\,n'k'} \, F_{n'k'} = \varepsilon F_{nk} \quad , \qquad (75)$$

where

$$U_{nk\,n'k'} = \int \psi^*_{n'k'}(r) \, U(r) \, \psi_{nk}(r) \, dr^3 \quad . \qquad (76)$$

There is no approximation at this point. Using the explicit form of the Bloch functions one can still write this matrix element under the form

$$U_{n'k'nk} = \sum_G C^G_{n'k'\,nk} \, \hat{U}(k'-k+G) \quad , \qquad (77)$$

if one makes use of the Fourier expansion of the periodic function

$$u^*_{n'k'}(r) \, u_{nk}(r) = \sum_G C^G_{n'k'\,nk} \, e^{iGr} \qquad (78)$$

and if one defines the Fourier transform of the perturbing potential

$$\hat{U}(q) = \frac{1}{8\pi^3} \int U(r) \, e^{-iqr} \, dr \quad . \qquad (79)$$

If the perturbing potential is "macroscopic", that is slowly varying on the length scale of the crystal unit cell, its Fourier transform $\hat{U}(k)$ is sharply peaked around k=0 and vanishes inside most of the volume of the Brillouin zone. Due to this, we can approximate the above matrix element of the perturbing potential by retaining only the **G**=0 contribution

$$U_{n'k'\,nk} \approx C^0_{n'k'\,nk} \, \hat{U}(k'-k) \quad , \qquad (80)$$

while considering only those terms for which **k**≈**k**'. Moreover, since we are only interested in states close to the direct minimum of the perfect-crystal conduction band, we shall only consider those states for which **k** and **k**' keep close to the Brillouin zone center.

134

As a consequence, we arrive at

$$C^0_{n'\mathbf{k}'\,n\mathbf{k}} \approx \frac{1}{v_c} \int_{v_c} u^*_{n\mathbf{k}}(r)\, u_{n'\mathbf{k}}(r)\, dr = \delta_{nn'} \quad . \qquad (81)$$

The hypothesis of a macroscopic perturbing potential then leads to a complete decoupling of the bands and a partial decoupling of the points of the Brillouin zone. The Schrödinger equation describing the perturbed states to be associated with the point **k** is then

$$\varepsilon_{n\mathbf{k}}\, F_{n\mathbf{k}} + \int d\mathbf{k}\, \hat{U}(\mathbf{k}-\mathbf{k}')\, F_{n\mathbf{k}'} = \varepsilon F_{n\mathbf{k}} \quad . \qquad (82)$$

Focussing over the states appearing near the band extremum, at k=0, and making use of the k.p expansion of the energy around this point, we obtain

$$\frac{\hbar^2 k^2}{2m^*}\, F_{n\mathbf{k}} + \int d\mathbf{k}'\, \hat{U}(\mathbf{k}-\mathbf{k}')\, F_{n\mathbf{k}'} = (\varepsilon - \varepsilon_{n0})\, F_{n\mathbf{k}} \quad , \qquad (83)$$

i.e. a Schrödinger equation, written in momentum space, for a free particle of mass m*, with an energy counted from the k=0 level. With a standard Fourier transform of $F_{n\mathbf{k}'}$ one can translate this equation into its real space equivalent,

$$\left[-\frac{\hbar^2}{2m^*}\, \nabla^2 + U(r) \right] F_n(r) = (\varepsilon - \varepsilon_{n0})\, F_n(r) \quad , \qquad (84)$$

where

$$F_n(r) = \int d\mathbf{k}\, F_{n\mathbf{k}}\, e^{i\mathbf{k}r} \quad . \qquad (85)$$

This equation describes the motion of a particle which only feels the influence of the perturbing potential. It is important to notice that in this expression the material is solely described by two parameters, namely the effective mass m* and the energy of the band edge ε_{n0}. The wavefunction $F_n(r)$ representing the states of this fictitious particle is related in a simple way to the perturbed wavefunction of the crystal electrons

$$\psi(\mathbf{r}) = \sqrt{\frac{1}{8\pi^3}} \int d\mathbf{k}\, F_{n\mathbf{k}}\, u_{n\mathbf{k}}(\mathbf{r})\, e^{i\mathbf{k}r} \quad . \qquad (86)$$

This representation of the wavefunction can be made more explicit using the k.p first-order expression for the periodic part of the Bloch function. One writes

$$\psi(\mathbf{r}) = F_n(\mathbf{r})\, \psi_{n0}(\mathbf{r}) + \frac{\hbar}{m} \sum_{j \neq n} \frac{\left[-i\nabla F_n(\mathbf{r}) \right] \cdot \mathbf{p}_{jn}}{\varepsilon_{n0} - \varepsilon_{j0}}\, \psi_{j0}(\mathbf{r}) \quad .$$

$$\qquad (87)$$

In most derivations of the effective mass approximation only the first term is retained and $F_n(r)$ appears as an envelope function multiplying the k=0 Bloch state to give the real electron wavefunction. In this sense, the particles of mass m^* are fictitious "envelope-driven" particles not identical to the electrons whose state is defined by the full wavefunction ψ. It should be emphasized that the second term involving the gradient of the envelope function must be included if the analytical order of all k-wise expansions we used in this section is to be maintained (second order in the energies, first order in the wavefunction). There would be a problem interpreting the current of particles if this term was omitted. The current probability of the envelope-driven particles would be different from the current probability of the real crystal electrons. If we average the current on a unit cell located at the point r, properly including all terms to first order, one gets the following average current $[\Phi_n(r) = F_n(r)/(8\pi^3)^{1/2}]$

$$\bar{J}_z = \hbar \left(\frac{1}{m} + \frac{2}{m^2} \sum_{j \neq n} \frac{|p_{nj}^z|^2}{\varepsilon_{n0} - \varepsilon_{j0}} \right) \text{Im } \Phi_n^*(r) \frac{\partial}{\partial z} \Phi_n(r) \quad (88)$$

or

$$\bar{J}_z = \left(\frac{\hbar}{m^*} \right) \text{Im } \Phi_n^*(r) \frac{\partial}{\partial z} \Phi_n(r) \quad . \quad (89)$$

This result shows that the current can be calculated by using the envelope function as if it were a proper wavefunction. The current is in fact carried by the effective-mass charge carriers.

Since the current carried by real electrons is conserved in space, the current carried by the effective-mass particles will also be conserved. This suggests that the following boundary conditions be applied at a plane interface at z=0 between two different semiconducting compounds

$$\Phi_n(0^+) = \Phi_n(0^-) \quad (90)$$

$$\frac{1}{m^*(0^+)} \frac{\partial \Phi_n(0^+)}{\partial z} = \frac{1}{m^*(0^-)} \frac{\partial \Phi_n(0^-)}{\partial z} \quad . \quad (91)$$

The effective mass approach assumes that we apply a perturbation potential to a perfect crystal. In the case of a quantum well or a multilayer structure this means that the periodic parts of the Bloch functions remain identical throughout the crystal. Thus, one should keep in mind that the envelope function approach is actually designed for heterostructures in which the joining materials are not "too different". This will not be the case, for instance, when the conduction band in one layer lines up with a valence band in the next layer as it seems to happen in InAs-GaSb heterojunctions. Also, as said before, the layer thickness has to be large enough and the perturbing potential weak enough so that the states we describe have Bloch function distribution coefficients F_{nk} sharply peaked in the Brillouin zone. In principle, this means that the quantum well should not be very deep, with respect to the top of the well barrier. One may however be astonished that even rather deep levels like those found in some GaAs-GaAlAs quantum wells can be correctly described by the effective mass approach. Such a deep level decays rapidly in the GaAlAs layer, and one may think that the $F_{n\mathbf{k}}$ coefficients would extend too much in k space to provide enough accuracy on the description of the perturbed wavefunction. Errors on the energy position are, however, moderated by the fact that the charge carriers spend only a very short time in the barrier region, which is not accurately described. It should also be mentioned that the above derivation is restricted to the case of a single band at the zone center. In the case of several degenerate bands, the effective-mass procedure can still be used, on the basis of a multiple envelope representation, and using the Kane model to represent off-center Bloch functions. See, for instance, Gerald Bastard [25] and references therein.

The consideration of heterostructures with a variable effective mass for the purpose of studying the one-dimensional confinement or tunneling requires to construct a new effective-mass Hamiltonian which guarantees at the same time hermiticity and current conservation. The following form of the effective mass equation

$$
-\frac{\hbar^2}{2} \frac{d}{dz} \left(\frac{1}{m(z)} \frac{d\Phi(z)}{dz} \right) + \left[\frac{\hbar^2}{2m_{//}} k_{//}^2 + V(z) \right] \Phi(z) = \varepsilon \, \Phi(z)
$$

(92)

fulfills these prescriptions. Hermiticity of the Hamiltonian $[<\Phi_1|H|\Phi_2> = <\Phi_2|H|\Phi_1>]$ can easily be insured when Born-Von Karman conditions are fulfilled, while for a stationary state, the current of particles

$$
J_z = \frac{\hbar}{2i} \left(\Phi^* \frac{1}{m(z)} \frac{d\Phi}{dz} - \Phi \frac{1}{m(z)} \frac{d\Phi^*}{dz} \right)
$$

(93)

is independent of z, for any parallel wavevector $k_{//}$. Also, for a superlattice [V(z) periodic] the current provided by a state n$\mathbf{k}$ is given by the simple relation

$$
J_z = \frac{1}{\hbar} \frac{\partial \varepsilon_n(k_z, k_{//})}{\partial k_z}
$$

(94)

similar to that found for a constant effective mass.

References

1. For a wide review on the physics and applications of multilayer structures, see IEEE Journal of Quantum Electronics, Vol QE, No. 9, September 1986.
2. R. Fuchs and K. L. Kliewer, Phys. Rev. 140, 2076 (1965).

3. M. Born and K. Huang, Dynamical Theory of Crystal Lattices (Clarendon, Oxford, England, 1966).
4. K. L. Kliewer and R. Fuchs, Adv. Chem. Phys. XXVII, 358 (1974)
5. H. Ibach, Phys. Rev. Lett. $\underline{24}$, 1416 (1970).
6. A.A. Lucas and M. Sunjic, Phys. Rev. Lett. 26, 229 (1971).
7. H. Ibach and D. L. Mills, Electron Energy Loss Spectroscopy and Surface Vibrations, pp. 63-126, Academic Press, New York (1982).
8. A.A. Lucas, J.P. Vigneron, Ph. Lambin, P.A. Thiry, M. Liehr, J.J. Pireaux, and R. Caudano, Intern. J. Quantum Chem.: Quantum Chem. Symposium $\underline{19}$, 687 (1986).
9. A. A. Lucas and J.P. Vigneron, Solid State Commun. $\underline{49}$, 327 (1984).
10. M. Liehr, P. A. Thiry, J.J. Pireaux and R. Caudano, J. Vac. Sci. Technol. $\underline{A2}$, 1079 (1984).
11. Ph. Lambin, J.P. Vigneron, A.A. Lucas, P. A. Thiry, M. Liehr, J.J. Pireaux, R. Caudano, and T. J. Kuech, Phys. Rev. Lett. $\underline{56}$, 1842 (1986).
12. Ph. Lambin, J.P. Vigneron, and A. A. Lucas, Phys. Rev. B 32,8203 (1985).
13. E.N. Economou and K. L. Ngai, Adv. Chem. Phys. XXVII, 265 (1974).
14. H. Raether, Excitation of plasmons and Interband Transitions by Electrons (Springer-Verlag, Berlin, 1980).
15. E. P. Pokatilov and S.L. Beril, Phys. Stat. Sol. (b) 118,567 (1983).
16. G.F. Giuliani and J.J. Quinn, Phys. Rev. Lett. 51, 919 (1983).
17. G. F. Giuliani and J.J. Quinn, and R. F. Wallis, J. Phys. (Paris) Colloq. 45, C5-285 (1984).
18. J.J. Quinn, Solid State Commun. 52, 607 (1984).
19. K. Huang, Proc. Roy. Soc. London A208, 352 (1951).
20. E.N. Economou, Phys. Rev. 182, 539 (1969).
21. G. Borstel and H. J. Falge, in Electromagnetic Surface Modes (Edited by A. D. Boardman), p. 219, John Wiley & Son Ltd., New-York (1982).
22. Ph. Lambin, J.P. Vigneron, A.A. Lucas, and A. Dereux, Physica Scripta, 35, 343 (1987).
23. see, for instance, A. A. Lucas, Physica Scripta, T13, 150 (1986).
24. S. White and L. J. Sham, Phys. Rev. Lett., 47, 879 (1981).
25. G. Bastard, Phys. rev. B, 25, 7584 (1982); in Molecular Beam epitaxy and Heterostructures, (Edited by L.L. Chang and K. Ploog) NATO ANSI Series E, vol 87, Martinus Nijhoff, Dordrecht, p. 381 (1985).
26. M. Altarelli, in Proceedings of Les Houches Winterschool Semiconductor Superlattices and Heterojunctions (Edited by G. Allan, G. Bastard, N. Boccara, M. Lannoo and M. Voos) Springer Verlag, 1987.
27. M. F. H. Schuurmans and G.W.'t Hooft, Phys. Rev. B, 31, 8041 (1985).

THE PHYSICS OF METALLIC SUPERLATTICES: AN EXPERIMENTAL POINT OF VIEW

Ivan K. Schuller

Materials Science Division

Argonne National Laboratory, Argonne, Illinois   60439 USA

I. INTRODUCTION

Metallic multilayers offer an ideal test ground for many physical phenomena occurring at different length scales.[1-2] For the purpose of these lectures it is convenient to categorize the physical phenomena under study as short (below ~ 50 Å) and long (above ~ 50 Å) length scales.[3]

The structural properties that have to be controlled are of course related to the length scale which governs the physical property being investigated. Because of this, for some studies structural control almost at the atomic level is necessary whereas in other cases disorder, interdiffusion and roughness at much larger scales can be tolerated. Since multilayers exhibit a preferred direction (the growth direction) a large amount of work has been invested in characterizing their structure along this direction.[4] If the atomic planes are coherently stacked it is customary to label the material as a "superlattice" whereas if this additional ordering is absent the term "multilayer" is used.[5] The physical properties of multilayers and superlattices can be strongly modified because of the peculiarities of this unique geometry. It is convenient to categorize the physical phenomena in increasing order of complexity as single film, proximity, coupling and superlattice effects. Single film effects are caused by the fact that the dimension of the material is constrained in one direction. An example of this type of phenomena is the increase of the film resistivity with decreasing thickness.[6] Proximity effects occur due to existence of a boundary between two materials, for instance magnetic-normal[7-8] or superconducting-normal[9-10] interfaces. Proximity effects can modify the physical properties of the materials in contact due to electron transfer, interfacial strain, a variety of interactions (dipolar, exchange, etc.) and due to long range effects such as superconductivity. Coupling effects occur in general between two like materials across an unlike material, for instance magnetic or superconducting coupling across normal materials. The coupling lengths can vary from interatomic distances in cases such as direct exchange or Rudman-Kittel-Kasuya-Yoside(RKKY) to long length scales such as superconducting or dipolar coupling. In principle in the former three cases, multilayers are not necessary since the effects can be studied in "isolated," single, double, or triple films, respectively. In practice however, multilayers offer a significant advantage over isolated film studies. The main advantages are that large volumes of the material can be prepared which makes many studies easier and surface contamination can be avoided since only a small fraction of the material is contaminated, unlike in isolated film studies where in situ probes are necessary. One additional advantage which is not commonly recognized is that many structural probes especially ("superlattice") diffraction[4] studies rely on the multilayer property in order to provide information about structure and chemical composition at interfaces.

One type of effect which only exists in multilayers and cannot even in principle be applied to a few layers, is the so-called "superlattice" effect. In this case, the effects crucially depend on the repetitive nature of the multilayer just as the development of electronic bands in a solid depends on the repetitive (i.e., crystalline) nature of materials. Three examples of these type of "superlattice effects" are: the observation of phonon folding in semiconductor superlattices,[11] the development of magnon bands in magnetic superlattices and the enhancement of critical fields due to matching of the vortex lattice to the superperiodicity in superconducting superlattices.[3]

In the present set of lectures we will describe the normal state, magnetic and superconducting properties of metallic superlattices with special emphasis on the relationship between structure and physical properties.

II. NORMAL STATE PROPERTIES: Electrical Resistivity, Magnetotransport, and Elastic Constants

a) Transport

The electrical resistivity and the magnetotransport are one of the very first measurements that are commonly applied to multilayered materials. Due to the structural anisotropy it is expected that the resistivity will be considerably anisotropic. Although a variety of interesting theoretical predictions[12-14] have been made regarding the transport in the perpendicular (to the layers) direction, no clear-cut experimental studies[15] have emerged yet. The main reason for this lack of studies is the fact that serious technical difficulties are present. Due to the small thickness of the films very sensitive measurements have to be performed and it is hard to isolate effects coming from the multilayer from effects produced by the contact resistance.

The parallel resistivity on the other hand, has been extensively studied in a variety of multilayers and superlattices.[16] In principle, superlattices should exhibit anomalous transport properties due to electronic folding and the opening of minigaps (at the zone boundary) in the electronic band structure.

Experimentally however, the situation looks quite different. Figure 1 shows the electrical resistivity as a function of inverse layer thickness $(1/t)$ for Nb/Cu superlattices.[6,17] The resistivity scales with the inverse thickness until reaching a value close to 150 $\mu\Omega$cm and then it saturates. The saturation value is close to the Ioffe-Reggel limit[18] which occurs when the electronic mean free path becomes comparable to the layer thickness. This, together with the fact that the resistivity for thicker layers scales with $1/t$, is a clear indication of the existence of strong interfacial boundary scattering. The origin of this scattering is not clear at the present time; defects and lattice mismatch certainly contributes to this, in addition the difference between the electronic properties of the materials might also add to this scattering through contact potentials, for instance. There is one report in the literature in which extremely long mean free paths (spanning several layer thicknesses) have been reported for Nb/Ta superlattices.[19] The mean free paths in these studies were <u>calculated</u> based on free electron theory and were later corrected and reduced by a further refinement of the calculation.[20] At the present time it is not clear to us whether such long mean free paths are present in Nb/Ta superlattices or whether these calculations point towards failures of free electron theory when applied to transition metal superlattices. This

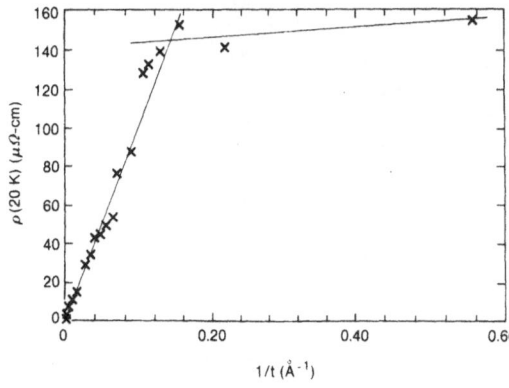

Fig. 1. Low temperature (20 K) resistivity as a function of inverse layer thickness for Nb/Cu super-lattices (after refs. 6, 17).

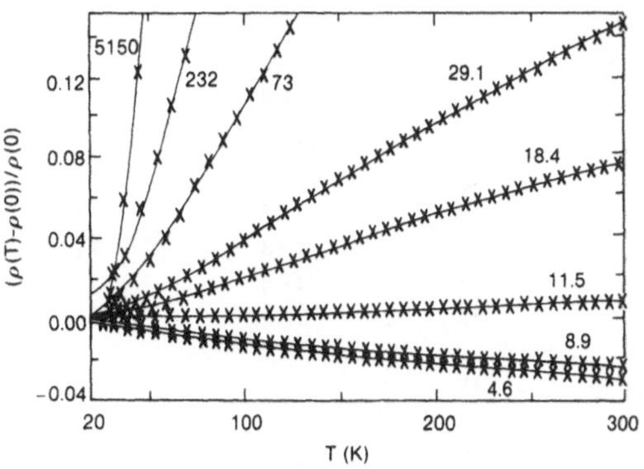

Fig. 2. Temperature dependent part of the resistivity versus tempera-ture for a series of Nb/Cu super-lattices. Note the metal-nonmetal transition between 11.5 Å and 8.9 Å (after refs. 6, 17).

is specially true since even disordered Nb–Ta alloys[21] have a very long mean free path if free electron theory is used for these calculations. It would be of considerable interest to study the behavior of the resistivity as a function of layer thickness in this system in order to understand the origin of the scattering mechanism.[16] If interfacial scattering is unimportant in the Nb/Ta system, the resistivity should not scale with 1/t as observed in Nb/Cu superlattices, for instance.

The temperature dependence of the resistivity has been measured in a variety of systems, Nb/Cu[6,17] and Nb/Al[22] superlattices being most extensively studied. The temperature dependence of the resistivity shows a clear metal-nonmetal transition in the temperature coefficient of resistiv-ity (TCR). Figure 2 shows this behavior for a series of sputter deposited Nb/Cu samples.[6,17] As the layer thickness decreases there is a systematic change in the temperature coefficient of resistivity and at about t ~ 10 Å the material exhibits a negative "non-metallic" TCR. This behavior is quite similar to that observed in other short mean free path materials, such as single thin films,[23] amorphous,[24] granular[25] metals and low dimen-sional semiconductor devices.[26] The origin of this type of behavior is

Table I. Experimental measurements of elastic constants in metallic super-lattices. Code: Ev = Evaporation, Sp = Sputtering, R = Rolling; SS = Self Supporting; Y = Young, Y_B = Biaxial, F = Flexural, T = Torsional, BS = Breaking Stress; E = Enhancement, N = No Change, S = Softening; BT = Bulge Tester, CR = Cantilever Reed, BS = Brillouin Scattering, SAW = Surface Acoustic Waves, LVDT = Linear Voltage Displacement Transducer, RT = Rigid Tensile Machine.

System	Prep Method	Removal from Substrate	Constant	Result	Measurement Method	Ref.
Au/Ni	Ev	Yes	Y_B	E ($\sim$ 225%)	BT	33
Cu/Pd	Ev	Yes	Y_B	E ($\sim$ 470%)	BT	33
	Ev	Yes	Y_B	N	BT	34
Cu/Ni	Ev	Yes	F	E ($\sim$ 225%)	CR	35
	Ev	Yes	Y	E ($\sim$ 170%)	LVDT	35
	Ev	Yes	T	E ($\sim$ 170%)	CR	35
	Ev	Yes	Y	E ($\sim$ 200%)	CR	36
	Ev	Yes	T	E ($\sim$ 180%)	CR	36
	Ev	Yes	Y_B	E ($\sim$ 200%)	BT	37
	Ev	Yes	Y	N	CR	38
	Ev	Yes	BS	E ($\sim$ 175%)	LVDT	35
Ag/Pd	Ev	Yes	Y_B	E ($\sim$ 230%)	BT	39
Cu/Au	Ev	Yes	Y_B	N	BT	39
Cu/Al	R	SS	BS	E ($\sim$ 300%)	RT	40
Nb/Cu	Sp	No	C_{44}	S ($\sim$ 35%)	BS	41
Mo/Ni	Sp	No	C_{44}	S ($\sim$ 40%)	BS	42
V/Ni	Sp	No	C_{44}	S ($\sim$ 40%)	BS	43
	Sp	No	C_{44}	S ($\sim$ 40%)	SAW	43
Au/Cr	Ev	No	c_{44}	E ($\sim$13%)	BS	44
Fe/Pd	Sp	No	C_{11}	S ($\sim$20%)	BS	45

thought to be related to the existence of localization[27] and correlation[28] effects which have been the subject of much experimental and theoretical research. In a recent theoretical analysis of the temperature dependent resistivity Gurvitch[22] was able to extract mean free paths for the indivi-dual constituents of the superlattice. The general conclusions of this work is that the mean free path is smaller than the layer thickness except for the case of Cu (in Nb/Cu) where the mean free path is layer thickness limited. This is quite interesting in view of the fact the in plane crystallographic coherence of Nb/Al has been claimed to be higher than for Nb/Cu.[29] This implies that the structural properties obtained from dif-fraction data do not determine uniquely electronic properties such as the mean free path.

It is perhaps useful to consider thermodynamic properties such as the binary phase diagram. In general, two materials that form solid solutions in their thermodynamic phase diagram are perhaps expected to have similar electronic structures. Because of this, the interface between these kinds of materials will scatter less (Nb/Ta for instance) than if the constituents exhibit a eutectic phase diagram (Nb/Cu for instance). Therefore, when studying the effect of interfaces on the electrical resistivity, it is desirable to compare the resistivity to that of the alloys, not only to the individual constituents.

A variety of other isolated transport measurements have been performed in a number of systems, however, a clear-cut picture has not emerged. These studies include thermopower,[30-31] magnetoresistance and Hall effect[30] and are included here for the sake of completeness.

To summarize, the parallel resistivity of metallic multilayers has been the object of considerable study. Its behavior in systems studied extensively as a function of thickness and temperature indicate that substantial scattering originates from the interfaces. The temperature coefficient of resistivity changes sign at the structural order-disorder transition. At present it is not clear whether the interfacial scattering originates from defects or the electronic differences at the interfaces. The experimental results seem to imply the absence of extended electronic states. This fact has serious implications for theories (especially band structure calculations) which rely on the existence of extended electronic states perpendicular to the layers. There is a great lack of further thermal and magnetotransport studues and in general transport perpendicular to the layers.

b) <u>Elastic Constants</u>

One of the most interesting and extensively studied phenomena in metallic multilayers has been related to the elastic behavior.[32] Major interest has been motivated by the original report of enormous enhancements of the biaxial modulus of Au/Ni superlattices.[33] In general the observation of anomalous elastic behavior occurs in two distinct categories. Superlattices whose constituents form solid solutions in their thermodynamic phase diagram ("solid solution" superlattices) exhibit enhancements whereas others exhibit softenings. In Table I we have listed the experimental measurements and the most important conclusion claimed regarding the elastic constants.

1) <u>"Solid Solution" Superlattices</u>

Superlattices that form solid solutions in their thermodynamic phase diagram[46] generally are lattice matched. The archetypal system of this type, Cu/Ni, shows large enhancements on a number of elastic constants[35-37].

Figure 3 shows the stress strain relationship obtained from Bulge test measurements for three Au/Ni superlattices of various modulation wavelengths λ.[33] The thinnest λ sample exhibits a marked nonlinearity and an enhanced slope at the origin. The biaxial modulus extracted from these slopes is shown in Figure 4. For these Au/Ni superlattices an enormous enhancement of about factors of 2.5 are shown in Fig. 4. Even larger enhancements (~ factors of 8) have been claimed for Cu/Ni superlattices.[37] Enhancements have also been reported in the Young, Flexural and Torsional moduli and in the breaking stress of a number of solid solution superlattices.[33,35-37,39] These enhancements are known in the literature as the

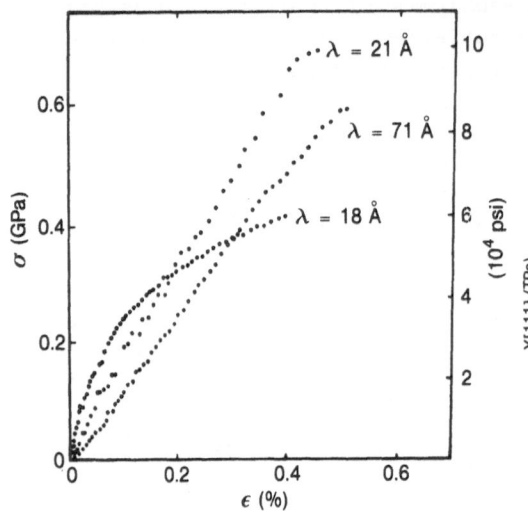

Fig. 3. Stress–strain relation
for a series of Au/Ni superlattices
(after ref. 33).

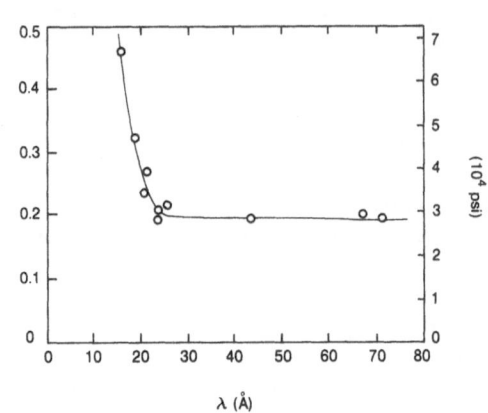

Fig. 4. Biaxial modulus Y(111)
versus modulation wavelength (λ)
for Au/Ni superlattices extracted
from the slopes at the origin of
Fig. 3 (after ref. 33).

"supermodulus effect".[32] An early report[38] claimed a negative result in
Cu/Ni superlattices although the structure was not reported. Not long ago
a criticism was raised to these measurements,[34] mainly because of the
observed nonlinearities in the stress–strain curves. This work claims that
the nonlinear stress–strain relation is due to a plastic deformation of the
sample when removed from the substrate. Because of this deformation, the
elastic constant is seemingly enhanced although this would be only an
experimental artifact. The reason for these discrepancies has not been
solved at the present time.

The theoretical origin of the supermodulus effect has been the subject
of a number of investigations.[47-54] It was first suggested by Henein[47]
that the supermodulus effect occurs at a thickness for which the Fermi
surface first touches the Brillouin zone. In spirit, this explanation is
similar to the Hume-Rothery effect in which a structural change is induced
at an electron concentration for which the Fermi sphere touches the
Brillouin zone.[55] This idea has been used to explain the existence of the
supermodulus effect in Cu/Ni and Ag/Pd and its absence in Cu/Au.[47-48] In a
similar formulation, the elastic constant has been related to singularities
in the second derivatives of the dielectric function.[49] It was claimed
that due to the new superlattice periodicity the Fermi surface develops
flat portions ("Fermi surface nesting") which in turn produces singularities in the dielectric function. Of course, in order for this effect to be
operative, the existence of extended electronic states perpendicular to the
layers is necessary. Further transport measurements of solid solution
superlattices are necessary in order to prove the existence of extended
electronic states as explained above in Section IIa. All the early
theories were of a qualitative nature and no attempts at quantitative
comparisons with experiments were made.

A molecular dynamics calculation[50] has also been performed which
claims a quantitative agreement with the enhanced modulus in Au/Ni.

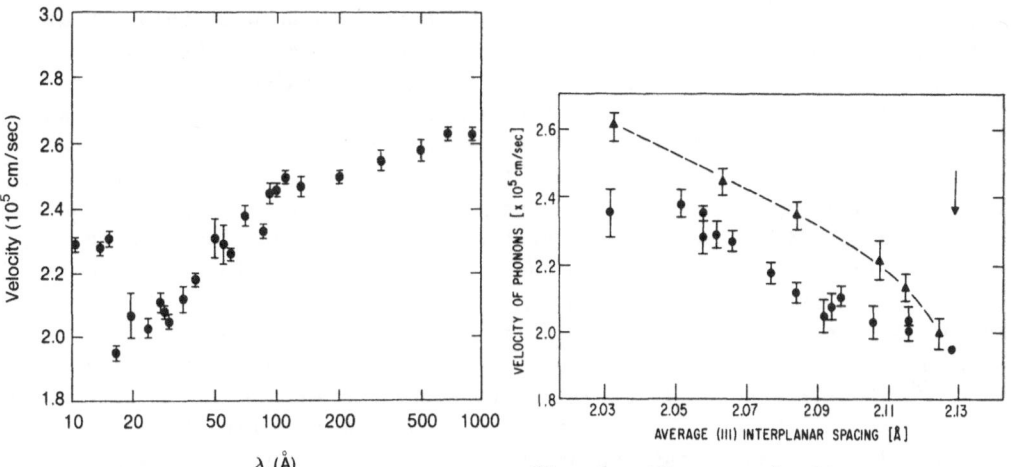

Fig. 5. Acoustic phonon velocity
for equal thickness Mo/Ni superlattices (after ref. 42).

Fig. 6. Phonon velocity versus
Ni interatomic spacing (after ref.
51).

However, this calculation was performed for a perfectly segregated
superlattice and the supermodulus effect was computed for a strain along
the perpendicular to the film whereas experimentally the strain is applied
in the plane of the film. A recent calculation claims that coherency
strains alone cannot be responsible for the elastic constant enhancements.[52] The main idea behind this calculation is that the enhancement
obtained in one part of the superlattice is lost in the other, if coherency
strains are present. It is therefore suggested that other effects beyond
coherency strains are responsible for the supermodulus effect.

2) Non "Solid Solution" Superlattices

The study of the elastic constants for nonlattice matched combinations
that do not form solid solution in their binary phase diagram[46] has been
reported recently.[41-43,56] The shear elastic constant (C_{44}) was measured
using Brillouin scattering techniques in which a monochromatic laser beam
is incident on the sample (without removal from the substrate) and the
light frequency, shifted by the thermally excited lattice vibrations
("phonons"), is detected using a very sensitive Fabry-Perot interferometer.
The phonon velocity is simply related by conservation equations to scattering angles, and frequencies of the incident and shifted light. The
phonon velocity in turn is directly proportional to the square root of the
elastic constant. Figure 5 shows such a measurement for the Mo/Ni
system.[42] A softening is observed at around a modulation wavelength of
20 Å. Similar behavior has also been observed in Nb/Cu[41] and V/Ni[43]
superlattices.

To our surprise, there has only been one measurement of elastic
constants using Surface Acoustic Waves (SAW).[43] The difficulty with these
measurements is that the sound wavelength (~20 μm) is usually longer than
the overall thickness ($\lesssim$ 2 μm)of the superlattice film. Because of this a
correction has to be applied to take into account the contribution from the
substrate. SAW and Brillouin scattering measurements have been claimed to
be in excellent quantitative agreement, in Ni/V superlattices.

The only report in which a quantitative comparison between theoretical calculations and elastic anomalies observed in metallic superlattices (Ni/Mo) has been performed using molecular dynamics techniques.[51] In this calculation the softening of the elastic constant has been attributed to an observed expansion in the lattice spacing perpendicular to the layers. Figure 6 shows a comparison of the experimentally measured shear elastic constant as a function of the experimentally observed lattice expansion and the molecular dynamics calculation described earlier, with no adjustable parameters. Since the experiment measures a surface wave velocity which is related to the bulk wave velocity through a constant ~0.8-1.0, the agreement between theory and experiment is considered excellent.

At this stage we would like to call attention to several theoretical papers[57-60] which specifically address many questions related to surface localized phonons, the appearance and dispersion of waves which have no counterpart in the homogeneous medium and the phonon structures in multi-layered structures.

In summary, elastic constant anomalies have been observed in many superlattice systems. The experimental measurements listed in Table I can be summarized as follows; combinations that form solid solutions and are lattice matched exhibit large enhancements whereas systems that do not form solid solutions and are not lattice matched exhibit softenings. Although several qualitative, theoretical explanations have been advanced there is only one quantitative calculation which compares favorably to experiments. The elastic enhancements observed in solid solution lattice matched super-lattices are claimed to be caused by an electronic instability when the Fermi surface first touches the Brillouin zone boundary, whereas in non-lattice matched superlattices the elastic constant softenings are thought to be caused by structural changes particularly an expansion observed for decreasing layer thickness. Clearly extensive additional work, experimental and especially theoretical, is needed to further clarify the situation.

III. MAGNETIC PROPERTIES

Magnetic superlattices have been an extremely fruitful area of research. The main reasons are that a variety of interesting effects can be artificially engineered into these types of materials. In addition, a number of important applications rely on the possibility of preparing magnetic multilayers with well-defined structural and magnetic properties. A number of causes have been invoked to explain unusual effects in magnetic superlattices and multilayers. These include: changes in the surface and electronic structure, the development of interfacial states, proximity effects and electron transfer, changes in the thickness when compared to a characteristic magnetic length (RKKY, dipolar, exchange, etc.) and coupling across nonmagnetic layers.

Table II summarizes the experimental magnetic measurements performed to date. The large number of systems explored to date shows the growing importance these type of materials are acquiring in the study of a variety of basic problems in magnetism. It is not possible in the present set of lectures to properly review the large body of work that has been done especially since the field is still in development. We will only illustrate some of the most important physical phenomena by choosing arbitrarily an example from each category. Most properties observed to date can be classified as: thin film, two dimensional, interfacial, proximity, coupling and superlattice effects.

Table II. Magnetic Measurements on Metallic Superlattices. Code:
KE = Magneto-optical Kerr Effect, ACS = A-C Susceptibility,
FMR = Ferromagnetic Resonance, M = Magnetization, BS = Brillouin
Scattering, ND = Neutron Diffraction, MS = Mössbauer Spectroscopy,
MXS = Magnetic X-ray Scattering, LM = Lorentz Microscopy, Ev = Evaporation,
Sp = Sputtering, El = Electrolitic Method

System	Magnetic Measurement	Preparation Method	Reference
Ni/Cu	ACS	Ev	61,62
	M	Ev	61,63,64,65,66,67
	FMR	Ev	68,69
	M	SpDC	70,71
	ND	Ev	72
Ni/Mo	FMR	SpDC	73
	BS	SpDC	74-76
	M	SpDC	77
	ND	SpDC	78
Ni/Cr	M	Ev	79
Ni/C	M,FMR	Ev	80
Ni/V	R,M	SpDC	99
NiFe/TiN	M	SpDC	100
Co/Cu	KE	SpRF	81
	M,FMR	Ev	82
Co/Au	KE	SpRF	81
Co/Nb	M,FMR	Ev	83-85
	BS	Ev	86
	NMR	Ev	87
Co/Sb	NMR	Ev	88,89
Co/P	ACS	El	90
Co/Pd	M	SpRF	91,92
	M	Ev	93,94
Co/Cr	M	Ev	95
Co/Mn	FMR	Ev	63
	M,FMR	Ev	96
Co/Gd	M	SpDc	97
CoNb/CoTi	M	Sp	98
CoSiBi/CoTi	M	Sp	98
Fe/Cu	KE	SpRF	81,101,102
	LM	Sp	103
	M	SpRF	104,
Fe/Au	KE	SpRF	81
Fe/Sb	MS	Ev	105,106,107
	FMR,ND,MS	Ev	108
Fe/Sn	MS	Ev	109
Fe/Mg	MS	Ev	110,111
	M,MS,ND	Ev	112
Fe/V	MS	Ev	111,113
	NMR	Ev	88,114,115
	M	Ev	116
Fe/W	BS	SpDC	117,118
Fe/Ta	M	SpDC	119
Fe/Y	M	Ev	120
Fe/Pd	MS	Ev	121
	M	Ev	93,94
	BS	SpDC	117,118
Fe/Cr	M	Ev	120
Fe/FeO	M	Sp	121
Fe/Mn	NMR	Ev	88
Fe/Nd	M	SpDC	123
Fe/Gd	M	Ev	124

(continued)

System	Magnetic Measurement	Preparation Method	Reference
Fe/Tb	M	Sp	125
FeB/Ag	M	SpDC	126
FeCo/Si	M	Sp	127
FeCo/Tb	FMR	Ev	128
Mn/Sb	M	Ev	129
Dy/Y	M	Ev	130,131
	MXS	Ev	132
	ND	Ev	133-135
Gd/Y	M	Ev	136,137
	ND	Ev	138,139
	MXS	Ev	140
Tm/Lu	M	Sp	141

a) Thin Film Effects

Many properties observed in metallic superlattices can be explained as due to thin film effects, i.e., the reduction of the sample dimension in the growth direction. Extensive interest on the properties of magnetic super-lattices was motivated by the original report of an enhanced magnetization of Ni in Cu/Ni superlattices, extracted from FMR measurements.[68] The determination of the magnetization using FMR requires a knowledge of the anisotropy field. Because of this a number of investigators[64-67] (including the original ones) performed magnetization measurements in order to check directly the possibility of an enhanced magnetization. Figure 7 shows the magnetization (M) as a function of thickness in equal thickness Cu/Ni superlattices.[64,65] The magnetization is independent of the Cu layer thickness, however it depends linearly on the inverse layer thickness. Therefore, these properties are determined by the Ni area-to-volume ratio indicating that thin film effects are responsible in grand part for the magnetic effects. It is quite interesting that these effects extend to very small layer thickness, even though the structure is not expected to be well controlled for the thinnest layers. A manifestation of structural disorder is perhaps the origin for the tendency to saturation observed in the thinnest samples. Similar conclusions (i.e., decreasing M for decreasing Ni thickness) were reached simultaneously[66,67] by other investigators and therefore the early results claiming an enhanced magnetization are probably a consequence of the assumptions regarding the anisotropy used in the interpretation of the FMR data. Many superlattice systems studied to date exhibit very similar behavior in the magnetization, namely a decreasing magnetization with decreasing layer thickness. The detailed functional dependence, in systems that do not form solid solutions, can be fitted using "dead layer" models. Further detailed magnetic work, combined with extensive structural studies and careful comparison with theoretical models, is necessary to ascertain whether dead magnetic layers are indeed present at the interfaces. The dependence of the magnetization in solid-solution superlattices is complicated by the fact that considerable interdiffusion is expected for the thinnest layers and therefore the magnetic behavior should be interpreted as for a compositionally modulated alloy.

The anisotropy is another property which, in many cases, has been shown to exhibit mostly thin film effects. For most systems the anisotropy is in the plane, as expected from pure shape anisotropies. In some exceptional cases, Co/Cr[95] and Tb/Fe[125] being the most studied, a perpendicular anisotropy is found. This property of course can have important applications in perpendicular recording devices where a perpendicular anisotropy is desirable.

b) Two-Dimensional Effects

Low dimensional magnetization has been the subject of considerable interest especially because clear cut theoretical predictions and numerical calculations are available for the behavior of well-defined model systems.[42] Metallic superlattices can be used as a testing ground for these theories especially because a direct measurement of quantities, such as the magnetization, can be performed and they do not have to be inferred from indirect measurements.

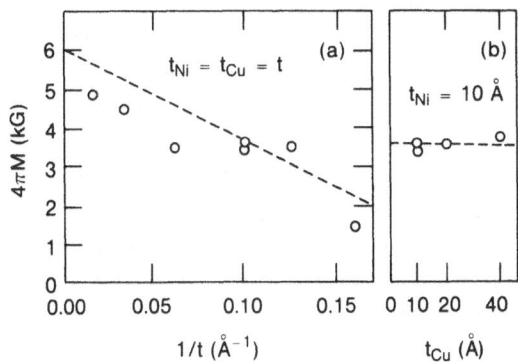

Fig. 7. Magnetization versus a) inverse layer thickness 1/t and b) normal metal thickness t_{Cu} in equal thickness Cu/Ni superlattices (after ref. 64).

The saturation magnetization of Fe in Fe/V superlattices as a function of temperature for a varying number of Fe atomic planes (n) in a layer is shown in Fig. 8.[116] As the layer thickness is decreased below about 10 atomic planes the magnetization deviates from the expected and observed magnetization in bulk Fe. This deviation increases in a systematic way with decreasing n. Interestingly the n = 1.43 sample shows a linear temperature dependence, in a very wide temperature range. A similar behavior was claimed earlier for n = 3 with thicker V layers (number of V atomic planes >9). A linear temperature dependence of the magnetization was also observed in Cu/Ni superlattices.[68,71] Theoretically a linear temperature dependence is expected from a two-dimensional (2)D spin wave contribution to the magnetism. In order to establish uniquely the 2D origin of these temperature dependences further structural work to clearly characterize the type of defects present in these very thin films is necessary. Moreover, it is necessary to rule out whether in principle defects such as steps, roughness, intermixing, island formation, etc. can give origin to such a remarkable linear temperature dependence.[7]

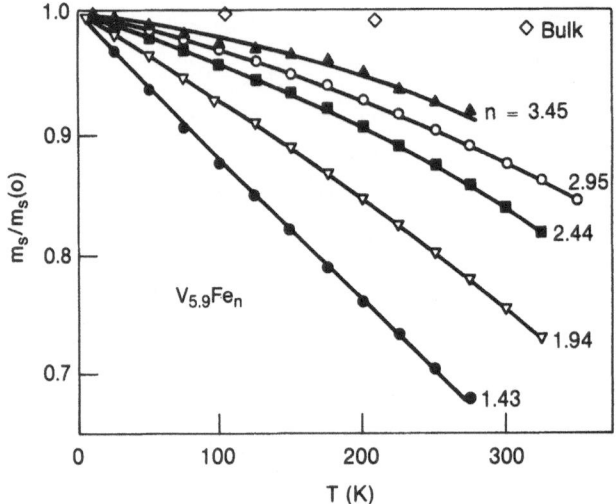

Fig. 8. Magnetization versus temperature for V/Fe superlattices. Note the linear dependence for n = 1.43 and the deviation from bulk for samples with n $\lesssim$ 3.5 (after ref. 116).

(c)   Interfacial and Proximity Effects

Interfacial[142-144] and proximity effects[145] can occur due to a modification of the properties of a material in contact with a different material. The origin of these type of effects can be due to the development of interfacial electronic states, electron transfer, modification of the interfacial structure etc. The existence of interfacial effects has been claimed in the Stoner enhancement of Pd in Au/Pd/Au trilayer films[146] and in Pd/Co,[91-94] Tb/Fe,[125] and Pd/Fe[93-94] multilayer films. These effects are especially pronounced in the dependence of the perpendicular anisotropy field H_K as a function of layer thickness in the multilayered films. Figure 9 shows the dependence of H_K on $1/t_{Co}$ for varying Pd thicknesses in Pd/Co superlattices.[91,92] "This linear dependence in single magnetic films is usually taken as evidence for the presence of surface anisotropy."[91,92] As in all thin films, the

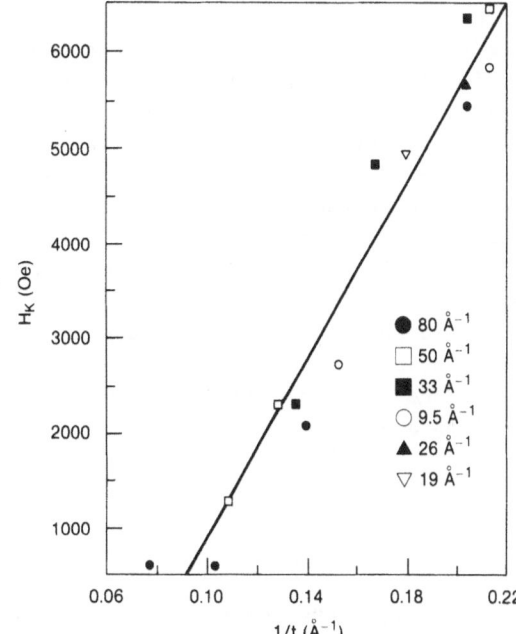

Fig. 9. Perpendicular anisotropy field H_K versus reciprocal Co layer thickness in Pd/Co superlattices (after ref. 91, 92).

measured magnetic anisotropy includes a contribution from a purely shape anisotropy. It is quite remarkable that even with the large shape anisotropies (which tend to keep the magnetization in the plane of the film) expected for the thinnest films, the total magnetic easy axis is in the perpendicular direction. In Pd/Co the transition from parallel to perpendicular magnetic anisotropy occurs at approximately 8 Å layer thickness. A detailed analysis of the structure in the thinnest (<13 Å) Pd/Co superlattices would be desirable in order to pin down uniquely the origin of the perpendicular anisotropy. This is especially true since 8 Å is the layer thickness for which a number of other superlattices (Mo/Ni for instance)[147] undergo a transition from layered to disordered growth.

The origin of the perpendicular anisotropy in the thinnest layers is not clear at present. It was claimed that this is possibly due to an "interface magnetization which favors a perpendicular magnetization". It has been suggested that the microscopic origin of this is a magneto-strictive effect due to the lattice mismatch at the interfaces. Similar perpendicular anisotropies observed in Tb/Fe amorphous multilayers have been assigned to the "anisotropic distribution of Tb-Fe pairs".[125] At this stage we should stress that the magnitude of this anisotropy is not a purely thin film effect since a comparison of Pd/Co and Pd/Fe grown under similar conditions shows the perpendicular anisotropy to be present in Pd/Co but not in Pd/Fe. To complicate matters even further recent work of ours[148] shows that the magnetic anisotropy measured by FMR and dc magneti-zation differ considerably.

All this clearly emphasizes that in order to clarify the situation in addition to magnetic measurements, extensive structural work is needed especially if a detailed understanding between magnetism and structure is to emerge in these very thin films.

d) Coupling Effects

Metallic multilayers are ideal systems for the study of magnetic coupling phenomena at various length scales.[3] The different types of

coupling mechanisms that are present at short length scales (< 50 Å)
include direct exchange, RKKY interaction, etc. At long length scales
(above 50 Å) the dipolar interaction is perhaps the most important coupling
mechanism. It is important to emphasize from the very beginning that the
structural perfection required for the observation of short length scale
phenomena is quite stringent. For instance for the observation of coupling
through direct exchange crystallographic perfection at the atomic level is
required, on the other hand dipolar coupling being very long ranged is more
"forgiving". Because of this, especially in the case of short coupling
phenomena, it is important to understand the effect slight interdiffusion,
pinholes, roughness, etc. would have on the magnetic properties.[7] In
addition, an extensive set of complementary structural probes shoud be
applied since in many cases one characterization tool cannot uniquely pin-
down all parameters of the structure.

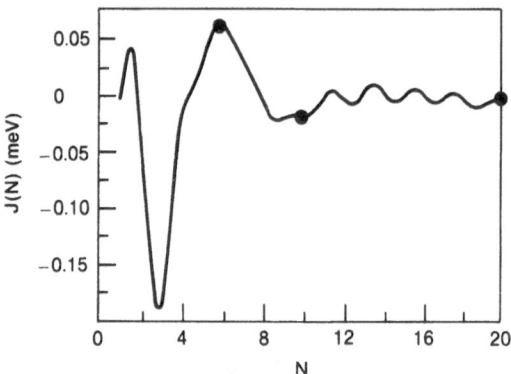

Fig. 10. Range function of the
RKKY interaction along the
modulation direction for Gd/Y
superlattices (N = number of Y
atomic layers). Dots indicate
experimentally measured data
(after ref. 149).

The first observation of magnetic coupling across nonmagnetic Cu was
claimed from measurements of the ac susceptibility in Cu/Ni superlattices.[61] The Curie temperature was found to increase with decreasing Cu
thickness in a functional form expected from RKKY coupling. However, no
detailed characterization was presented in this case. Recently similar
ideas have been applied to Gd/Y superlattices which were characterized
using x-ray, neutron and high energy electron diffraction[136-140]. The dependence of the calculated indirect exchange interaction between Gd ion as
a function of the number of Y layers in Gd/Y superlattices is shown in Fig.
10.[149] The data points extracted from experimental results are also shown
in this graph. It has also been claimed that for $6 < N < 10$ successive Gd
layers order parallel and antiparallel to each other, in agreement with the
calculation shown in Fig. 10. The same type of approach also has been
applied to a helical configuration of spins as observed in Dy/Y superlattices.[130-135] In this case, the helical spin configuration in each Dy
layer locks together across Y layer. The type of indirect exchange
described above is capable of providing a coupling mechanism in this case
also,[149] although a detailed quantitative comparison between experiment and
theory has not yet been done. It should be stressed that for these types of
short range coupling effects to be operational atomically sharp interfaces
and no intermixing is necessary.

e) Superlattice Effects

Superlattice effects arise from the periodicity of the superlattice
and are unique to multilayers, i.e., they cannot be obtained in only a few
layers are present. In semiconductor superlattices, these types of effects
were observed in the phonon spectrum due to the "folding" of the phonon
spectrum by the periodicity, the so-called "phonon folding".[11] Similar
folding effects should also be present in the electronic spectrum, if
extended electronic states are present. Although considerable effort has
been devoted by a number of groups for the observation of phonon or
electronic folding in the Raman spectrum or the electronic transport, these
searches have been fruitless to date in metallic systems. Possibly the
lack of these types of effects in the phonon and electronic spectrum is due
to the structural imperfections existing at the interfaces, which can
destroy the short length scale (< 50 Å) coherence necessary for their
existence.

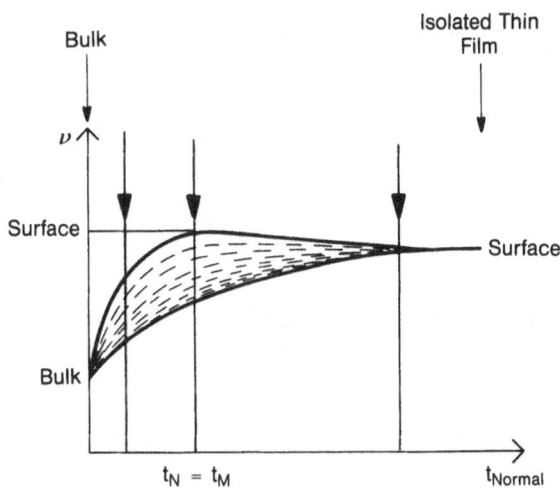

Fig. 11. Magnon frequency versus normal metal separator t_N in Mo/Ni
superlattices. The series of experimental measurements are indicated by
the arrows (after ref. 150).

At dipolar length scales (> 50 Å), however, superlattice effects have
been observed and quantitatively compared to theoretical calculations.
Theoretical calculations[150-151] imply that as a function of normal material
thickness (t_N) the discrete magnons existing in each magnetic layer of a
superlattice couple together to give rise to a band of magnons. Figure 11
shows qualitatively the expected dependence of the magnon frequency (ν) as

a function of t_N. If the layers are separate one excitation is present which spreads into a band with decreasing t_N. At the thickness for which the magnetic layer thickness $t_M = t_N$ an additional surface mode splits off. These theories[150-151] predict in detail the dependence of the magnon frequency as a function of t_N, t_M, saturation magnetization M_s, magnetic field H and scattering vector $Q_\parallel$.

Figure 12 shows the magnetic field dependence of the magnon frequency (measured using Brillouin scattering) together with theoretical fits using only the saturation magnetization as an adjustable parameter.[74] The saturation magnetization obtained in this fashion from Brillouin scattering and directly from magnetization measurements are in quantitative agreement.[75] In addition, the wave vector dependence is also found to be in agreement with theoretical predictions which implies that the existence of super-lattice magnon bands is on firm footing.[75]

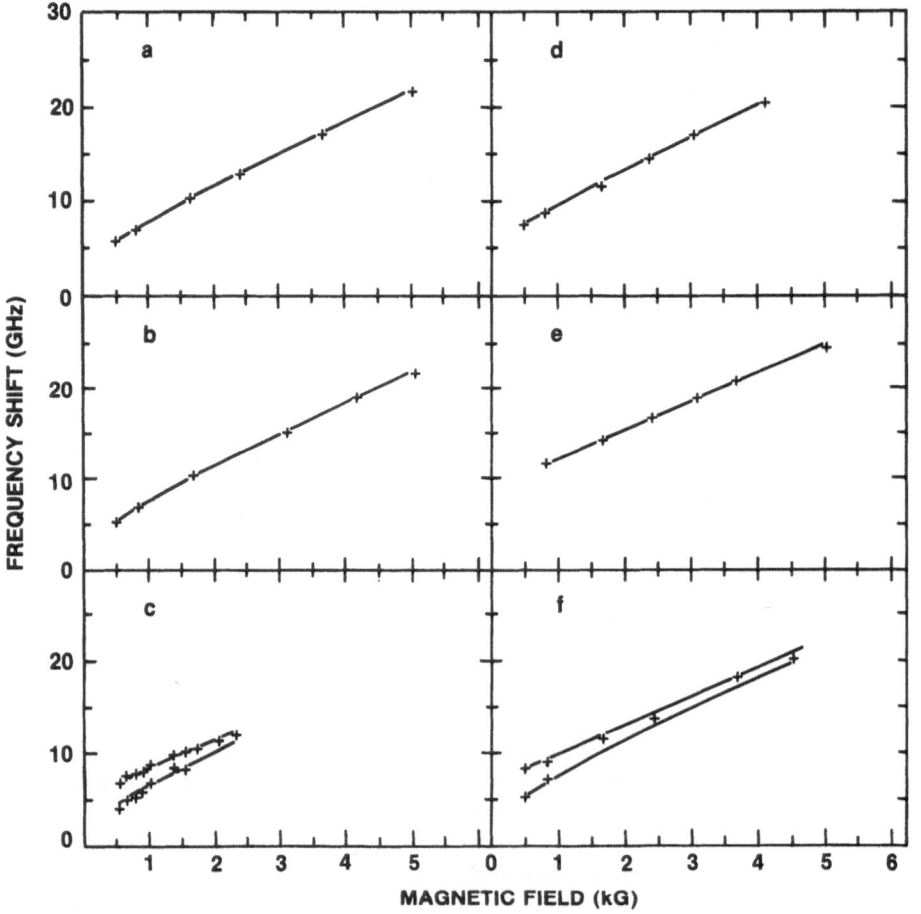

Fig. 12. Magnon frequency (crosses) for a series of Mo/Ni superlattices versus magnetic field together with theoretical fits (solid lines) (after ref. 74-76).

All these results imply that superlattice effects are observable in the magnetic properties of superlattices at long length scales. The reason for their absence at short length scales is not clear at the present time but it is probably related to the existence of structural defects at the interfaces.

In summary, the study of magnetic multilayers has been the subject of much exploratory work in a large number of systems. Although many properties are explainable as due to simple thin film effects, two dimensional, interfacial, proximity, coupling and superlattice effects have been claimed in a number of carefully constructed systems. A number of observations indicate that multilayers offer considerable technological advantages over single films. These phenomena include the observation of enhanced coercivities in Mo/Ni superlattices and the observation of perpendicular anisotropy in Pd/Co and Tb/Fe superlattices.

IV. SUPERCONDUCTING PROPERTIES

The study of superconductivity in layered materials has been a subject of intense interest especially in the naturally occurring intercalated graphite and layered dichalcogenides.[152] The recent discovery of high temperature superconductivity in metal oxides[153-154] will give a renewed impetus to the study of anisotropic superconductors, because in these materials as well, the basic building block consists of Cu-O layers.[155-156] Artificially prepared multilayers have been the subject of intense investigation for a number of years, especially in the search of novel superconducting mechanisms,[157-161] and in the study of dimensional effects. Superconductivity is a specially fruitful territory for exploration in connection with layered materials because the characteristic length which determines the physical properties (i.e., the superconducting coherence length) is much longer than any defects present in reasonably carefully prepared samples. Because of this, a wealth of physical phenomena can be studied without major concerns regarding slight roughness and disorder at the interfaces. It is however quite important to assure that the individual layers are not contaminated and that a minimal interdiffusion occurs at the interfaces. The superconducting behavior in metallic superlattices can be classified (as in the case of magnetic superlattices) as: thin film, two dimensional, interfacial, proximity, coupling and "superlattice" effects. Since the coherence length is very long, unlike in the case of magnetism, many studies have been performed in single, double and triple layered films, to clarify superconductivity issues. Here we will only concentrate on those issues which have been mostly addressed using multilayers. Clearly, problems such as the superconducting proximity effect[9,10] have been studied for a large number of years and are beyond the scope of this review.

Table III summarizes the experimental work on superconducting multilayers.

a) Thin Film, Interfacial and Proximity Effects

One of the early motivations for the interest in multilayers started because of the prediction of unusual pairing mechanisms in metal/semiconductor multilayers.[157,158,161] Some of the earliest experimental work in this field was performed in the search for this type of unusual pairing.[159-160] Anomalous enhancements of the transition temperatures were observed in metal/semiconductor multilayer as the number of layers (or

Table III. Experimental measurements in superconducting multilayers. Code: T_c = transition temperature, H_c = critical field, j_c = critical current, PD = penetration depth, Ev = evaporation, Sp = sputtering.

System	Superconducting Measurement	Prep Method	Ref.
Mo/Ni	T_c, H_c	Sp	162–163
Mo/Sb	T_c	Ev	164
Mo/Ge	H_c	Sp	165
Mo/Si	T_c, H_c	Sp	166–167
MoC/Si	T_c	Sp	168
Al/Ge	T_c	Ev	168
In/Ag	T_c	Ev	170
Sn/Ag	T_c	Ev	170
Pb/Bi	J_c	Ev	171–173
Pb/Fe	T_c, T	Ev	174
Pb/Cu	T_c, T	Ev	175
Pb/Ge	T_c, H_c	Ev	176
PbBi/Cr	J_c	Ev	177
V/Ni	T_c, H_c	Sp	178–179
V/Fe	T_c, H_c	Ev	180
V/Ag	T_c, H_c, PD	Ev	181–182
V/Mo	T_c, H_c	Sp	183–184
Nb/Al	T_c, T, H_c	Sp	185–188
Nb/Cu	T_c, T, H_c, PD	Sp	189–201
Nb/Ge	T_c, H_c	Sp	202–204
Nb/Ta	T_c, T	Ev, Sp	20,205–208
Nb/Ti	T_c, H_c	Sp	209–210
Nb/Zr	T_c, C_v	Sp	211–212
Nb/Si	T_c, H_c	Sp	166–167
Nb/CeCu$_6$	T_c, T	Sp	213
NbGe/NbGe	T_c	Sp	214
Nb$_3$Ge/Nb$_3$Ir	T_c	Sp	215
Nb$_3$Sn/	J_c	Ev	216
NbTi/Ge	T_c, H_c	Sp	217
NbN/AlN	T_c, H_c	Sp	218

number of interfaces) was increased. Originally it was thought that the enhancements are due to an excitonic mechanism, however, it seems that the reason for these enhancements is not uniquely established yet.

The transition temperature T_c of multilayered superconductors in many cases is found to decrease with layer thickness. Notable exceptions are the enhancements mentioned earlier[159–160] and the increase of T_c observed in Mo/Ni superlattices.[162–163] In general, the changes in T_c can be simply understood, like in A15 superconductors as due to a decrease in the mean free path (see Section IIa), and a consequent decrease in the density of states.[219–220] Figure 13 shows the change of the T_c of Nb in three different geometries, Nb/Cu[192] and Nb/Ge[204] multilayers and single Nb films.[221] The Nb T_c in the Nb/Ge is obtained by extrapolating the T_c of the multilayer as a function of Ge thickness (t_{Ge}) to $t_{Ge} = 0$ Å, whereas in

t (Å)

100.0 50.0 33.3 25.0 20.0 16.7

T_c of Nb (°K)

t^{-1} (10^{-3} Å^{-1})

Fig. 13. Nb transition temperature extracted from single Nb films (triangles, ref. 221), Nb/Ge multilayers (open squares, ref. 204) and Nb/Cu superlattices (closed circles, ref. 192).

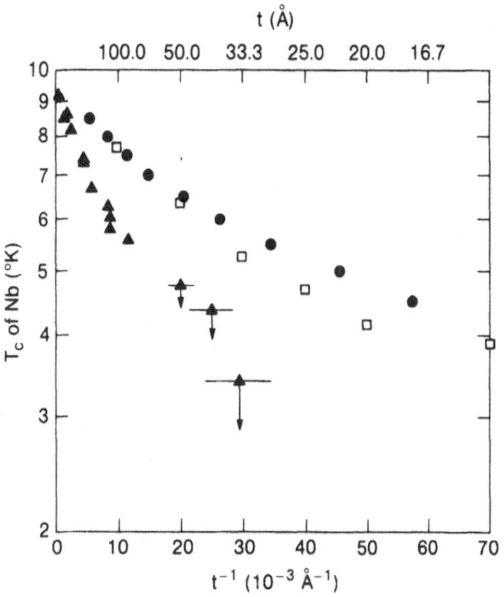

the Nb/Cu superlattices a more complicated analysis based on proximity effect theories is necessary.[192] The general decrease in multilayers is due to a smearing of the density of states caused by a decrease in the electronic mean free path. However, to explain the steep decrease observed in single Nb films an additional mechanism due to the proximity with a contaminated niobium oxide surface layer had to be invoked. A comparison of the T_c's obtained in multilayers and single layers illustrates beautifully the "technical" advantage that the multilayer geometry offers, in avoiding complications due to surface contamination. The T_c's obtained from both types of multilayers are in good agreement whereas in single films a major fraction of the sample is contaminated by surface oxidation, depressing the T_c further and complicating the analysis considerably.

The phonon spectrum obtained from tunneling data shows also a combination of thin film and proximity effect.[195] Figure 14 shows a series of tunneling curves obtained from Nb/Cu superlattices using a technique developed by Wolf, Zasadzinski, Osmund and Arnold.[222] In this technique the tunneling barrier is formed by evaporating and oxidizing a thin film of Al on top of the Nb/Cu superlattice. Although the Al film somewhat complicates the analysis of the tunneling data, the Al phonon at 37 meV serves as an internal calibration for the superconductivity in these multilayers. The superconducting energy gap measured in Nb/Cu superlattices decreases with layer thickness due to the reduced electronic mean free path and the proximity effect with the normal metal (Cu). This decrease in the energy gap in turn affects the amplitude of the Nb (transverse acoustic (TA) ~ 17 mV and longitudinal optic (LO) ~25 mV) and Al phonons. In good agreement with theoretical expectations the phonon peak <u>amplitude</u> decreases with the square of the energy gap. The <u>position</u> of the phonon peak on the other hand does not change in a perceptible manner, only a broadening is observed. A comparison of the phonon spectrum extracted from an inversion of the tunneling curves for Nb and a Nb/Cu superlattice shows that in effect to quite high accuracy there is only a broadening of the phonon spectrum with no shifts in peak positions.[223] Phonon folding as found in semiconducting GaAs/AlAs [11] has not been observed in any metallic superlattices. Similar behavior is observed in

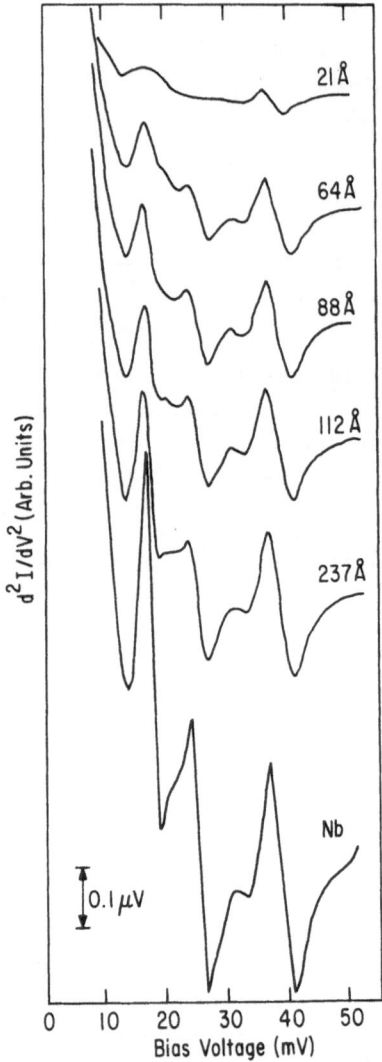

Fig. 14. Second derivative tunneling curve versus bias voltage for a series of equal thickness Nb/Cu superlattices showing Al (~37 meV, longitudinal optic Nb (~25 meV) and transverse acoustic Nb (~17 meV) phonons (after ref. 195).

Nb/Ta superlattices where for the thinnest layers the smearing of the phonon density of states is similar to that observed in Nb-Ta alloys and has been attributed to interdiffusion.[208] In Nb/Al superlattices again, no shift in the phonon spectrum is found although in this case irreproducible peaks are obtained at low energies.[185] At the present time, it is not clear whether these low frequency peaks are due to an experimental artifact or a consequence of the layering process.

b) Dimensional and Coupling Effects

The critical field of multilayered superconductors has been the most extensively studied property, especially in connection with dimensional and coupling effects[224].

Two-dimensional superconductivity is easily observable in thin superconducting films if their thickness t is smaller than the superconducting coherence length ξ. The critical field of a three-dimensional (3D) anisotropic superconductor is given by

$$H_{c2\parallel}(T) = \frac{\phi_o}{2\pi} \frac{1}{\xi_\parallel(T) \, \xi_\perp(T)} \tag{1}$$

and

$$H_{c\perp}(T) = \frac{\phi_o}{2\pi} \frac{1}{\xi_\parallel^2(T)} \tag{2}$$

where $\xi_\parallel(T)$ and $\xi_\perp(T)$ are the temperature dependent parallel and perpendicular coherence lengths and ϕ_o is the flux quantum. If a single superconducting thin film has a dimension $t \ll \xi_\perp(T)$, equation (1) becomes

$$H_{c\parallel}(T) = \frac{\phi_o}{2\pi} \frac{1}{t} \frac{1}{\xi_\parallel(T)} \tag{3}$$

in two dimensions (2D). Since the superconducting coherence length $\xi \propto 1/\sqrt{T_c - T}$, equations (1) and (3) imply that in 3D the temperature dependence of $H_{c2\parallel}$ is linear and in 2D it is square-root like. For a superconductor constructed from stacks of 2D superconductors separated by a semiconductor (Josephson coupled) or metal (proximity coupled), an interesting situation arises as a function of temperature. For low temperatures where the separation $t_N \gg \xi_\perp(T)$, the multilayer is expected to behave as a single 2D layer, i.e., square-root like in temperature, whereas at high temperature where $t_N \ll \xi_\perp(T)$, the layers strongly couple and the behavior is 3D like, i.e., linear in temperature.

This type of behavior was first observed in artificially prepared multilayers by Ruggiero, Barbee, and Beasley in Josephson coupled Nb/Ge multilayers.[202] They clearly showed that, indeed for thin Ge layers in the 3D regime, the behavior is linear in temperature whereas for thick Ge as the Nb layers the behavior becomes square root 2D like. This behavior was found to be in good agreement with theoretical expectations.[225,226] By properly changing the thickness of the films the regime of "dimensional crossover" can also be observed. A recent theoretical calculation[227,228] has shown that the temperature at which the dimensional crossover inflection occurs, depends on the ratio of the density of states of the superconductor and separator. Because of this it is considerably easier to

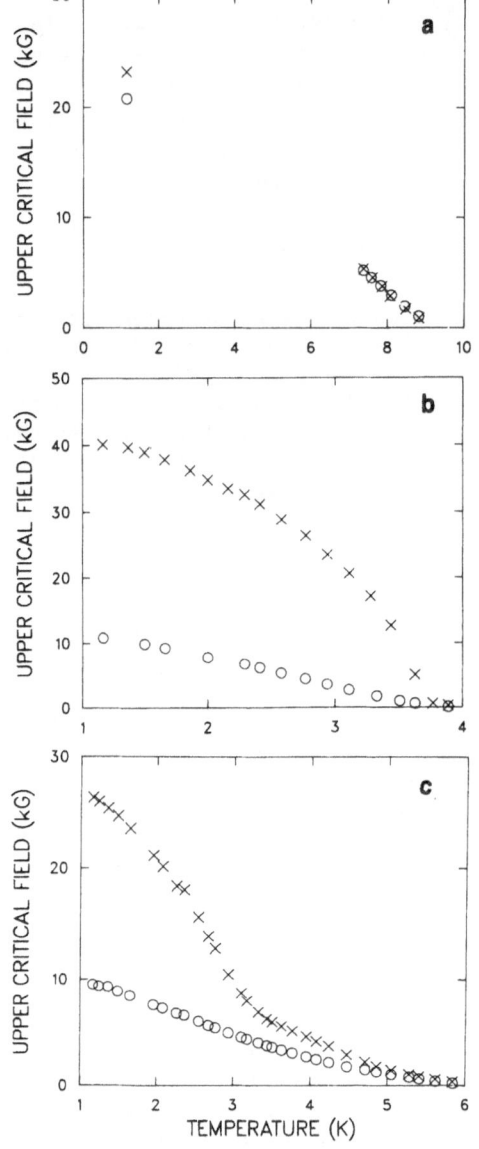

Fig. 15. Perpendicular (circles) and parallel (crosses) upper critical
fields as a function of temperature in a) three dimensions, b) two
dimensions, and c) crossover (after ref. 196).

observe dimensional crossover in proximity coupled superconductors, i.e.,
it occurs at larger thicknesses of the separator, so questions regarding
the perfection of the layers are not very critical.

Figure 15 shows the behavior of the critical fields in proximity
coupled Nb/Cu superlattices. If the Cu thickness is small the layers are
strongly coupled and the dependence of the <u>parallel</u> critical field is
linear, for thicker Cu layers the dependence of H_{c2} becomes square-root

like and for $t_{Cu} \sim 170$ Å a very clear dimensional crossover is found. In a variety of detailed studies it was shown that the dependence of $H_{c2\parallel}$ as a function of separator thickness, temperature, angle and ratio of density of states is well understood.[176,196] The perpendicular critical field is not sensitive to dimensional effects as expected from the qualitative description given above and therefore its temperature dependence is linear for all thicknesses. We would also like to point out at this time that it is of importance to carefully eliminate the effect of surface of supercon-ductivity[229] which also can give raise to anisotropic behavior in films and which however is not related to the dimensional behavior described above.

c) Superlattice Effects

Superlattice effects in layered superconductors have been predicted a number of years ago. Van Gelder[230] solved the Bogoliubov equations for the situation in which the superconducting pair potentials vary as a function of distance as in the Kronig-Penney model. These calculations have claimed the existence of states in the forbidden superconducting gap and mini gaps in the continuum. Two attempts[175,195] to observe these effects have been unsuccessful and it is not clear at the present time whether the condition under which the experiments were done do not agree with the theoretical model or if the theoretical model is incorrect. Very recently, tunneling data of high T_c superconductors[231] have claimed the observation of structure in the superconducting density of states which is reminiscent of the van Gelder prediction.

Superlattice effects have been observed in equal layer thickness Nb/Cu superlattices in the thickness dependence of the parallel critical field.[192,194,197] The perpendicular critical fields in these samples do not vary with varying layer thickness which together with equation (2) imply a coherence length of ~170 Å, independent of layer thickness. Figure 16 shows the parallel critical field as a function of layer thickness at

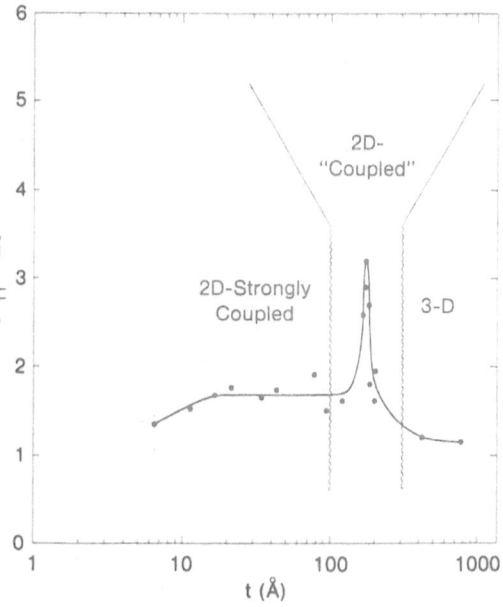

Fig. 16. Critical field anisotropy showing the enhancement caused by the matching of the vortex lattice to the superlattice period (after ref. 193).

low temperature (~1.1 K). If the layer thicknesses are small, the layers are coupled strongly and the critical field anisotropy is slightly greater than unity due to surface superconductivity.[229] At high thicknesses each individual layer is 3D in character, and the anisotropy is close to unity. If the vortex lattice spacing (given by the coherence length) is matched to the superperiodicity, the vortices can "comfortably sit" inside the normal metal, and therefore the critical field is enhanced for a layer thickness around 170 Å.

A number of other effects have been studied in superconducting superlattices including the interaction of superconductivity and competing mechanisms (localization[162-163] and magnetism[178-179]), the proximity effect with semiconductors[204] and the effect of crystallization on the proximity effect.[176]

V. THE ULTIMATE SUPERLATTICE: High T_c Oxide Superconductors

With the recent advent of high temperature superconductivity in metallic oxides,[153-154] it is perhaps appropriate to highlight here the fact that these materials also fall in the general category of layered superconductors[155,156] and therefore many of their anisotropic properties will possibly be similar to those of the artificially layered superlattices.

Figure 17 shows a plot of the crystal structure of $YBa_2Cu_3O_{7-x}$[156] obtained from Rietveld refinement of neutron diffraction scattering. The structure consists of dimpled Cu-O planes parallel to the <u>a-b</u> plane

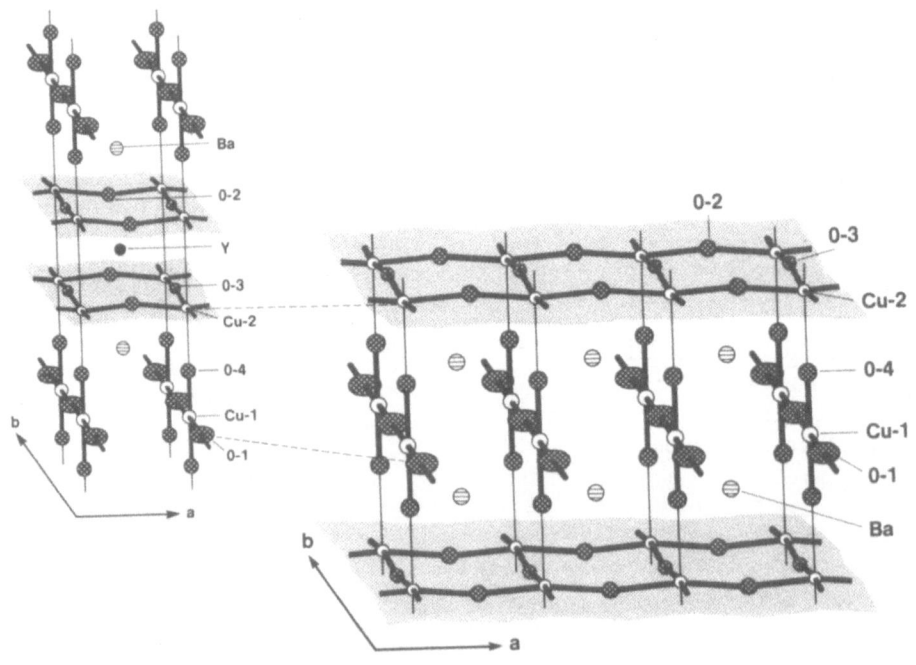

Fig. 17. Crystal structure of the 93°K superconductor $YBa_2Cu_3O_{7-x}$ showing Cu-O planes and chains (after ref. 156).

surrounding the Y atoms, weakly linked together with one-dimensional chains extending along the b axis.

It is a general consensus that most of the conduction occurs in the Cu-O planes although the detailed superconducting mechanism is not known at present. The anisotropic properties of this material have not been studied until the writing of this paper, however we expect that they should manifest the layered structure in a similar fashion as described above. Perhaps even the observation of superlattice effects in the density of states (à la van Gelder) might be possible.[230]

In concluding, the study of superlattices has produced and helped out the understanding of a number of physical phenomena at different length scales. It is quite important to note that many of the studies performed in structurally well-controlled artificial multilayers is of relevance to many new exciting developments in science and technology.

Acknowledgments: I would like to thank my many collaborators over a number of years. Their help, encouragement, and insight made it all possible and certainly lots of fun. Many thanks to my colleagues A. S. Arrott, M. B. Brodsky, R. C. Dynes, C. M. Falco, G. P. Felcher, C. P. Flynn, A. J. Freeman, E. M. Gyorgy, J. B. Ketterson, J. Kwo, C. F. Majkrzak, D. B. McWhan, G. P. Prinz, and S. K. Sinha for criticism and discussions which helped me dig deeper and work harder to try understanding. This work was supported by the U.S. Department of Energy, BES-Materials Sciences, under Contract #W-31-109-ENG-38 and the Office of Naval Research under Contract #N00014-83-F-0031.

REFERENCES

1. See various articles in "Synthetic Modulated Structures", L. L. Chang and B. C. Giessen, Academic Press, Inc., Orlando (1985).
2. See various articles in "Interfaces, Superlattices and Thin Films", J. D. Dow and I. K. Schuller, Material Research Society Publishers, Pittsburgh (1987).
3. I. K. Schuller and H. Homma, MRS Bulletin XII, 18 (1987).
4. See for instance, D. B. McWhan in Ref. 1, p. 43.
5. I. K. Schuller, Phys. Rev. Lett. 44, 1597 (1980).
6. T. R. Werner, et al., Phys. Rev. B26, 2224 (1982).
7. For an early review see A. Yelon, in Physics of Thin Films, M. H. Francombe and R. W. Hoffman eds., Academic Press, NY (1971), pg. 205.
8. J. C. S. Levy, Surf. Scie. Rep. 1, 39 (1981).
9. G. Deutscher and P. G. deGennes in Superconductivity, R. D. Parks ed., Marcel Dekker, Inc., NY, 1969, pg. 1005.
10. A. Gilabert, Ann. Phys. 2, 203 (1977).
11. C. Colvard, R. Merlin, M. V. Klein, and A. C. Gossard, Phys. Rev. Lett. 45, 298 (1980).
12. J. Sokoloff, Phys. Rev. B22, 5823 (1980).
13. J. Sokoloff, Sol. St. Commun. 40, 633 (1981).
14. M. Luban and B. N. Harmon, Sol. St. Commun. 51, 199 (1986).
15. J. Slaughter, W. P. Pratt Jr., H. Sato, and P. A. Schroeder, Bull. Am. Phys. Soc. 32, 693 (1987).
16. See for instance, C. M. Falco and I. K. Schuller in Ref. 1, pg. 339.
17. T. R. Werner, Ph.D. Thesis, Northwestern University, (1983) (unpublished).
18. A. Ioffe and A. Reggel, Prog. Semicond. 4, 237 (1960).

19. S. M. Durbin, J. E. Cunningham, and C. P. Flynn, J. Phys. F12, L75 (1982).
20. P. R. Broussard and T. H. Geballe, Phys. Rev. B35, 1664 (1987).
21. A. Calverly and A. C. Rose-Innes, Proc. R. Soc. London, 255A, 267 (1960).
22. M. Gurvitch, Phys. Rev. B34, 540 (1986).
23. R. C. Dynes, J. Garno, and J. Rowell, Phys. Rev. Lett. 10, 479 (1978).
24. R. W. Cochrane, R. Harris, J. O. Ström-Olson, and M. J. Zuckermann, Phys. Rev. Lett. 35, 676 (1975).
25. A. F. Hebard and J. M. Vandenberg, Phys. Rev. Lett. 44, 50 (1980).
26. D. J. Bishop, D. C. Tsui, and R. C. Dynes, Phys. Rev. Lett. 44, 1153 (1980).
27. P. W. Anderson, Phys. Rev. 109, 1492 (1958).
28. B. Altshuler and A.Aronov, Sov. Phys. JETP 50, 968 (1979).
29. D. B. McWhan, M. Gurvitch, J. M. Rowell and L. R. Walker, J. Appl. Phys. 54, 3886 (1983).
30. I. K. Schuller, et al., in "Modulated Structures", J. M. Cowley, J. B. Cohen, M. B. Salamon, and B. J. Wuensch, eds., AIP, NY (1979), pg. 417.
31. D. Baral and J. Hilliard, Appl. Phys. Lett. 41, 156 (1979).
32. J. E. Hilliard, AIP Conf. Proc. 53, 407 (1979).
33. W. M. C. Yang, T. Tsakalakos, and J. E. Hilliard, Jour. Appl. Phys. 48, 876 (1977).
34. Hideo Itozaki, Ph.D. Thesis, Northwestern University (1982) (unpublished).
35. D. Baral, J. B. Ketterson, and J. E. Hilliard, J. Appl. Phys. 57, 1076 (1985).
36. L. R. Testardi, et al., J. Appl. Phys. 52, 510 (1981).
37. T. Tsakalakos and J. E. Hilliard, J. Appl. Phys. 54, 734 (1983).
38. B. S. Berry and W. C. Pritchet, Thin Solid Films 33, 19 (1976).
39. G. E. Henein and J. E. Hilliard, J. Appl. Phys. 54, 728 (1983).
40. V. S. Kopan and A. V. Lysenko, Fiz. Metal. Metalloved. 29, 183 (1970).
41. A. Kueny, et al., Phys. Rev. Lett. 48, 166 (1982).
42. M. R. Khan, et al., Phys. Rev. B27, 7186 (1983).
43. R. Danner, et al., Phys. Rev. B33, 3696 (1986).
44. P. Bisanti, M. B. Brodsky, G. P. Felcher, M. Grimsditch, and L. R. Sill, Phys. Rev. B35, 7813 (1987).
45. P. Baumgart, et al., Phys. Rev. B34, 9004 (1986).
46. See for instance, K. P. Standhammer and L. E. Murr "Atlas of Binary Alloys - A Periodic Index", Marcell Dekker, Inc., NY, (1973).
47. G. E. Henein, Ph.D. Thesis, Northwestern University, (1979) (unpublished).
48. W. E. Pickett, J. Phys. F12, 2195 (1982).
49. T.-B. Wu, J. Appl. Phys. 53, 5265 (1982).
50. Y. Sasajima, M. Imafaku, R. Yamamoto, and M. Doyama, J. Phys. F14, L167 (1984).
51. I. K. Schuller and A. Rahman, Phys. Rev. Lett. 50, 1377 (1983).
52. A. Banerjea and L. R. Smith, Phys. Rev. B35, 5413 (1987).
53. V. G. Vaks and A. V. Trefilov, JETP Lett. 38, 453 (1983).
54. R. C. Camarata, Scripta Metallurgica 20, 479 (1986).
55. See for instance, C. Kittel "Introduction to Solid State Physics" 2nd edition, John Wiley and Sons, NY, 1957, pg. 325.

56. For a review see I. K. Schuller, "1985 Ultrasonics Symposium", B. R. McAvoy ed., IEEE Publishers, pg. 1093.

57. Shi-jie Xioug and Chien-hua Tsai, Acta Phys. Sinica 32, 1073 (1983).

58. B. Djafari-Rouhani, L. Dobrzynski, O. Harduin-Duparc, R. E. Camley, and A. A. Maradudin, Phys. Rev. B28, 1711 (1983).

59. B. Djafari-Rouhani, L. Dobrzynski, and O. Harduin-Duparc, J. Elec. Spec. Rel. Ph. 30, 119 (1983).

60. R. E. Camley, B. Djafari-Rouhani, L. Dobrzynski, and A. A. Maradudin, Phys. Rev. B27, 7318 (1983).

61. W.-S. Zhou, H. K. Wong, J. R. Owers-Bradley and W. P. Halperin, Physica 108B, 953 (1981).

62. A. A. Hirsch, N. Friedman and Z. Eliezer, Physica 30, 2314 (1964).

63. H. Sakakima, R. Krishnan and M. Tessier, J. Appl. Phys. 57, 3651 (1985).

64. E. M. Gyorgy, et al., Phys. Rev. Lett. 45, 57 (1980).

65. E. M. Gyorgy, et al., Phys. Rev. B25, 6739 (1982).

66. J. Q. Zheng, C. M. Falco, J. B. Ketterson and I. K. Schuller, Appl. Phys. Lett. 38, 424 (1981).

67. J. Q. Zheng, J. B. Ketterson, C. M. Falco and I. K. Schuller, J. Appl. Phys. 53, 3150 (1982).

68. B. J. Thaller, J. B. Ketterson and J. E. Hilliard, Phys. Rev. Lett. 41, 336 (1978).

69. R. Krishnan and M. Tessier, Sol. St. Commun. 60, 637 (1986).

70. J. Baszynski, L. Smardz and F. Stobiecki, Phys. St. Sol. 84, K129 (1984).

71. G. Xiao and C. L. Chien, J. Appl. Phys. 61, 4061 (1987).

72. G. P. Felcher, et al., J. Mag. Magn. Mat. 21, L198 (1980).

73. M. J. Pechan, M. B. Salamon and I. K. Schuller, J. Appl. Phys. 57, 3678 (1985).

74. M. Grimsditch, M. Khan, A. Kueny and I. K. Schuller, Phys. Rev. Lett. 51, 498 (1983)

75. A. Kueny, M. Khan, I. K. Schuller and M. Grimsditch, Phys. Rev. B29, 2879 (1984).

76. I. K. Schuller and M. Grimsditch, J. Appl. Phys. 55, 2491 (1984).

77. M. R. Khan, P. Roach and I. K. Schuller, Thin Solid Films 122, 183 (1985).

78. J. W. Cable, M. R. Khan, G. P. Felcher and I. K. Schuller, Phys. Rev. B34, 1643 (1986).

79. M. B. Stearns and C. H. Lee, J. Appl. Phys. 61, 4064 (1987).

80. R. Krishnan, K. B. Youn and C. Sella, J. Appl. Phys. 61, 4073 (1987).

81. T. Katayama, H. Awano and Y. Nishihara, Jour. Phys. Soc. Jpn. 55, 2539 (1986).

82. M. Dariel, et al., J. Appl. Phys. 61, 4067 (1987)

83. R. Krishnan and W. Jantz, Sol. St. Commun. 50, 533 (1984).

84. R. Krishnan and P. Gerard, Sol. St. Commun. 50, 529 (1984).

85. R. Krishnan, J. Mag. Magn. Mat. 50, 189 (1985).

86. R. Krishnan, W. Janz, W. Wettling and G. Rupp, IEEE Trans. Mag. MAG-20, 1264 (1984).

87. R. Krishnan, K. Le Dang and P. Veillet, Phys. Lett. 110A, 170 (1985).

88. K. Takanashi, H. Yasuoka and T. Shinjo, J. Mag. Magn. Mat. 54, 783 (1987).

89. K. Takanashi, et al., J. Phys. Soc. Jpn. 53, 2445 (1984).

90. J. M. Riveiro and G. Rivero, IEEE Trans. Mag. MAG-17, 3082 (1981).

91. P. F. Garcia, A. D. Meinhaldt and A. Suna, Appl. Phys. Lett. 47, 178 (1985).

92. P. F. Garcia, A. Suna, D. G. Onn and R. van Antwerp, Superlattices and Microstructures 1, 101 (1985).

93. H. J. G. Draisma, W. J. M. de Jonge, and F. J. A. den Broeder, Jour. Mag. Magn. Mat. 66, 351 (1987).

94. F. J. A. den Broeder, et al., J. Appl. Phys. 61, 4317 (1987).

95. M. B. Stearns, C. H. Lee and S. P. Vernon, J. Mag. Magn. Mat. 54, 791 (1986).

96. H. Sakakima, et al., J. Mag. Magn. Mat. 54, 785 (1987).

97. D. J. Webb, et al., Phys. Rev. B32, 4667 (1985).

98. N. S. Kazama and H. Fujimori, J. Mag. Magn. Mat. 54, 793 (1986).

99. H. Homma, C. S. L. Chun, G.-G. Zheng and I. K. Schuller, Phys. Rev. B33, 3562 (1986).

100. L. Maritato, and C. M. Falco, J. Appl. Phys. 61, 1588 (1987).

101. T. Katayama, H. Awano, J. Maedomari and Y. Nishihara (unpublished).

102. T. Katayama, Y. Nishihara and H. Awano, J. Appl. Phys. 61, 4329 (1987).

103. F. J. A. den Broeder, J. Appl. Phys. 60, 3381 (1986).

104. Y. Kozono, et al., J. Appl. Phys. 61, 4311 (1987).

105. T. Shinjo, H. Hosoito and T. Takada, J. Mag. Magn. Mat. (Proc. ICM '82, Kyoto).

106. J. M. Friedt, N. Hosoito, K. Kawaguchi and T. Shinjo, J. Mag. Magn. Mat. (ICMFS 10, Yokohama '82).

107. N. Hosoito, K. Kawaguchi, T. Shinjo and T. Takada, J. Phys. Soc. Jpn. 51, 2701 (1982).

108. T. Shinjo, et al., J. Phys. Soc. Jpn. 52, 3154 (1983).

109. T. Shinjo, N. Hosoito, S. Hine and T. Takada, Jap. Jour. Appl. Phys. 19, L531 (1980).

110. T. Shinjo, K. Kawaguchi, R. Yamamoto, N. Hosoito and T. Takada, Sol. St. Commun. 52, 257 (1984).

111. T. Shinjo, Hyperfine Interactions 27, 193 (1986).

112. T. Shinjo, et al., Jour. Mag. Magn. Mat. 54, 737 (1986).

113. T. Shinjo, et al., Proc. Inter. Conf. on the Dynamics of Interfaces, Lille, France, September 12-16, 1983.

114. K. Takanashi, et al., J. Phys. Soc. Jpn. 53, 4315 (1984).

115. K. Takanashi, et al., J. Phys. Soc. Jpn. 51, 3743 (1982).

116. H. K. Wong, et al., J. Appl. Phys. 55, 2494 (1984)

117. B. Hillebrands, et al., J. Appl. Phys. 61, 4309 (1987).

118. B. Hillebrands, et al., Phys. Rev. B34, 9004 (1986).

119. Z. S. Sham, Z. R. Zhao, J. G. Zhao and D. J. Sellmeyer, J. Appl. Phys. 61, 4320 (1987).

120. C. Sellers, et al., J. Mag. Magn. Mat. 54, 787 (1986).

121. N. Hosoito, T. Shinjo and T. Takada, Jour. Phys. Soc. Jpn. 50, 1903 (1981).

122. R. R. Ruf and R. J. Gambino, J. Appl. Phys. 55, 2628 (1984).

123. D. J. Sellmeyer, Z. R. Zhao, Z. S. Shan and S. Nafis, J. Appl. Phys. 61, 4323 (1987).

124. T. Morishita, Y. Toqami and K. Tsushima, J. Mag. Magn. Mat. 54, 789 (1986).

125. N. Sato, J. Appl. Phys. 59, 2514 (1986).

126. G. Xiao, C. L. Chien and M. Natan, J. Appl. Phys. 61, 4314 (1987).

127. N. S. Kazama and H. Fujimori, J. Mag. Magn. Mat. 35, 86 (1983).

128. S. C. Shin, M. L. Cofield and R. H. D. Nutall, J. Appl. Phys. $\underline{61}$, 4326 (1987).

129. T. Shinjo, N. Nakayama, I. Moritani and Y. Endoh, J. Phys. Soc. Jpn. $\underline{55}$, 2512 (1986).

130. S. Sinha, et al., J. Mag. Magn. Mat. $\underline{54}$, 773 (1986).

131. J. Borchers, et al., J. Appl. Phys. $\underline{61}$, 4049 (1987).

132. D. B. McWhan, et al., J. Mag. Magn. Mat. $\underline{54}$, 775 (1986).

133. M. B. Salamon, et al., Phys. Rev. Lett. $\underline{56}$, 259 (1986).

134. J. J. Rhyne, et al., J. Appl. Phys. $\underline{61}$, 4043 (1987).

135. M. Hong, et al., J. Appl. Phys. $\underline{61}$, 4055 (1987).

136. J. Kwo, et al., J. Mag. Magn. Mat. $\underline{54}$, 771 (1986).

137. J. Kwo, et al., Phys. Rev. Lett. $\underline{55}$, 1402 (1985).

138. C. F. Majkrzak, et al., J. Appl. Phys. $\underline{61}$, 4055 (1987).

139. C.F. Majkrzak, et al., Phys. Rev. Lett. $\underline{56}$, 2700 (1986).

140. C. Vettier, et al., Phys. Rev. Lett. $\underline{56}$, 757 (1986).

141. W. P. Lowe, et al., J. Appl. Phys. $\underline{58}$, 1615 (1985) and L. H. Greene, et al., Superlattices and Microstructures $\underline{1}$, 407 (1985).

142. See for instance, U. Gradmann, J. Mag. Magn. Mat. $\underline{6}$, 173 (1977).

143. C. L. Fu and A. J. Freeman, Jour. Mag. Magn. Mat. $\underline{54}$, 777 (1986) and references cited therein.

144. T. Oguchi and A. J. Freeman, Jour. Mag. Magn. Mat. $\underline{54}$, 797 (1986) and references cited therein.

145. R. M. White and D. J. Friedman, J. Mag. Magn. Mat. $\underline{49}$, 117 (1985).

146. M. B. Brodsky and A. J. Freeman, Phys. Rev. Lett. $\underline{45}$, 133 (1980).

147. M. R. Khan, et al., Phys. Rev. B$\underline{27}$, 7186 (1983).

148. M. Pechan and I. K. Schuller, Phys. Rev. Lett. (in press).

149. Y. Yafet, J. Appl. Phys. $\underline{61}$, 4058 (1987).

150. P. Grünberg and K. Mika, Phys. Rev. B$\underline{27}$, 2955 (1983).

151. R. E. Camley, T. S. Rahman and D. Mills, Phys. Rev. B$\underline{27}$, 261 (1983).

152. For a recent paper see M. Dresselhaus, Materials Research Society Bulletin, $\underline{XII}$, 24, (1987).

153. J. G. Bednorz, K.A. Müller, Z. Phys. B$\underline{64}$, 189 (1986).

154. M. K. Wu, et al., Phys. Rev. Lett. $\underline{58}$, 908 (1987).

155. J. D. Jorgensen, et al., Phys. Rev. Lett. $\underline{58}$, 1024 (1987).

156. M. A. Beno, L. Soderholm, D. W. Capone II, D. G. Hinks, J. D. Jorgensen, Ivan K. Schuller, C. U. Segre and J. D. Grace, Appl. Phys. Lett. (in press).

157. V. L. Ginzburg, Sov. Phys. Uspekhi $\underline{13}$, 335 (1970).

158. M. H. Cohen and D. H. Douglas Jr., Phys. Rev. Lett. $\underline{19}$, 118 (1967).

159. D. L. Miller, M. Strongin and O. F. Kammerer, Phys. Rev. B$\underline{13}$, 4834 (1986).

160. M. Strongin, et al., Phys. Rev. Lett. $\underline{21}$, 1320 (1968).

161. D. Allender, J. Bray and J. Bardeen, Phys. Rev. B$\underline{7}$, 1020 (1973).

162. C. Uher, R. Clarke, G. G. Zheng, and I. K. Schuller, Phys. Rev. B$\underline{30}$, 453 (1984).

163. C. Uher, J. L. Cohn, and I. K. Schuler, Phys. Rev. B$\underline{34}$, 4906 (1986).

164. Y. Asada and K. Ogawa, Sol. St. Commun. $\underline{60}$, 161 (1986).

165. V. Matijasevic and M. R. Beasley, Phys. Rev. B$\underline{35}$, 3175 (1987).

166. A. M.Kadin, et al., Proc. 1984 MRS Symposium on Layered Structures, Epitaxy and Interfaces.

167. A. M. Kadin, et al., LT-17, U. Eckern, A. Schmidt, W. Weber, and H. Wühl (eds.) Elsevier Science Publishers, B.V. (1984).

168. J. Wood, et al., IEEE Trans. Mag. $\underline{MAG-21}$, 842 (1985).

169. T. W. Haywood and D. G. Ast, Phys. Rev. B$\underline{18}$, 2225 (1978).

170. G. C. Granqvist and T. Claeson, Sol. St. Commun. $\underline{32}$, 531 (1979).

171. H. Raffy and E. Guyon, Physica $\underline{108}$B, 947 (1981).

172. H. Raffy, J. C. Renard, and E. Guyon, Sol. St. Commun. $\underline{11}$, 1679 (1972).

173. H. Raffy, E. Guyon, and J. C. Renard, Sol. St. Commun. $\underline{14}$, 427 431 (1974).

174. T. Claeson, Thin Solid Films $\underline{66}$, 151 (1980).

175. P. Ceuterickx, Licentiaat Thesis, Katholieke Universiteit Leuven, (1979) (unpublished).

176. J.-P. Locquet, et al., IEEE Trans. Mag. Proc. of the 1987 Applied Superconductivity Conference (IN PRESS).

177. W. E. Yetter, E. J. Kramer, and D. G. Ast, J. Low Temp. Phys. $\underline{49}$, 2271 (1982).

178. H. Homma, C. S. L. Chun, G.-G. Zheng, and I. K. Schuller, Phys. Rev. B$\underline{33}$, 3562 (1986).

179. H. Homma, C. S. L. Chun, G.-G. Zheng, and I. K. Schuller, Proc. Int. Conf. on Material and Mechanisms of Superconductivity (K. A. Gschneider, Jr. and E. L. Wofl, eds.) North-Holland, Amsterdam, (1985), p. 173.

180. H. K. Wong, et al., Superlattices and Microstructures $\underline{1}$, 259 (1985).

181   K. Kanoda, et al., Phys. Rev. B$\underline{33}$, 2052 (1986).

182. K. Kanoda, et al., Phys. Rev. B$\underline{35}$, 415 (1987).

183. M. G. Karkut, D. Ariosa, J.-M. Triscone, and Ø. Fischer, Phys. Rev. B$\underline{32}$, 4800 (1985).

184. M. G. Karkut, J.-M. Triscone, D. Ariosa, and Ø. Fischer, Phys. Rev. B$\underline{34}$, 4390 (1986).

185. J. Geerk, M. Gurvitch, D. B. McWhan and J. M. Rowell, Physica $\underline{109}$, 1775 (1982).

186. D. B. McWhan, M. Gurvitch, J. M. Rowell, and L. R. Rowell, J. Appl. Phys. $\underline{54}$, 3886 (1984).

187. J. C. Villegier, B. Blanchard, and O. Laborde, in"Proceedings of LT-17", U. Eckern, A. Schmid, W. Weber, and H. Wühl (eds.) Elsevier Publishers, B.V. (1984), pg. BD1.

188. J. Guimpel, et al., J. Low Temp. Phys. $\underline{63}$, 151 (1986).

189. I. K. Schuller and C. M. Falco, in AIP Conf. Proc. $\underline{58}$, 197 (1979).

190. I. K. Schuller, Phys. Rev. Lett. $\underline{44}$, 1597 (1980).

191. W. P. Lowe, T. W. Barbee, T. H. Geballe, and D. B. McWhan, Phys. Rev. B$\underline{24}$, 6193 (1981).

192. I. Banerjee, Q. S. Yang, C. M. Falco, and I. K. Schuller, Sol. St. Comm. $\underline{41}$, 805 (1982).

193. I. Banerjee, Q. S. Yang, C. M. Falco, and I. K. Schuller, Phys. Rev. B$\underline{28}$, 5037 (1983).

194. I. Banerjee, Ph.D. Thesis, Northwestern University (1982) (unpublished).

195. Q. S. Yang, C. M. Falco, I. K. Schuller, Phys. Rev. B$\underline{27}$, 3867 (1983).

196. C. S. L. Chun, G. Zheng, J. L. Vicent, and I. K. Schuller, Phys. Rev. B$\underline{29}$, 4915 (1984).

197. I. Banerjee and I. K. Schuller, J. Low Temp. Phys. $\underline{54}$, 501 (1984).

198. C. M. Falco and I. K. Schuller, in "Superconductivity in d- and f-band Metals - 1982", edited by W. Buckel and W. Weber, Kernforschungzentrum, Karlsruhe (1982), p. 283.

199. W. Sevenhans, J.-P. Locquet, A. Gilabert, and Y. Bruynseraede, LT-17, U. Eckern, A. Schmidt, W. Weber, and H. Wühl (eds.) Elsevier Publishers, B.V. (1984) pg. BD2

200. J. Guimpel, F. de la Cruz, J. Murduck, and I. K. Schuller, Phys. Rev. B35, 3655 (1987).

201. R. Vaglio, A. Cucolo and C. M. Falco, Phys. Rev. B35, 1721 (1987).

202. S. T. Ruggiero, T. W. Barbee, Jr., and M. R. Beasley, Phys. Rev. Lett. 45, 1299 (1980).

203. S. T. Ruggiero, T. W. Barbee, Jr., and M. R. Beasley, Phys. Rev. B26, 4894 (1982).

204. S. T. Ruggiero, Ph.D. Thesis, Stanford University, (1982) (unpublished).

205. S. M. Durbin, J. E. Cunningham, M. E. Mochel, and C. P. Flynn, J. Phys. F11, L223 (1981).

206. P. R. Broussard, Ph.D. Thesis, Stanford University, (1986) (unpublished).

207. S. M. Durbin, Ph.D. Thesis, University of Illinois (1983) (unpublished).

208. G. Hertel, D. B. McWhan, and J. M. Rowell, in "Superconductivity in d- and f-band Metals - 1982", edited by W. Buckel and W. Weber, Kernforschungzentrum Karlsruhe (1982), p. 299.

209. J. Q. Zheng, J. B. Ketterson, C. M. Falco, and I. K. Schuller, Physica 108B, 945 (1981).

210. Y. J. Qian, et al., J. Low Temp. Phys. 49, 279 (1982).

211. W. P. Lowe and T. H. Geballe, Phys. Rev. B29, 4961 (1984).

212. P. R. Broussard, D. Mael, and T. H. Geballe, Phys. Rev. B30, 4055 (1984).

213. L. H. Greene, W. L. Feldman, and J. M. Rowell, Physica B&C 135, 77 (1985).

214. H. Yamamoto, M. Idead, and M. Tanaka, Jpn. J. Appl. Phys. 24, L314 (1985).

215. P. H. Schmidt, J. M. Vandenberg, R. Hamm, J. M. Rowell, in "Superconductivity in d- and f-band Metals" (H. Suhl and M. B. Maple, eds.), Academic Press, NY, (1980) p. 57.

216. R. E. Howard, et al., IEEE Trans. Magn. MAG-13, 138 (1977).

217. B. Y. Jin, et al., Jour. Appl. Phys. 57, 2543 (1985).

218. J. Murduck, J. Vicent, I. K. Schuler, and J. B. Ketterson, J. Appl. Phys. (IN PRESS), Appl. Phys. Lett. (IN PRESS).

219. See for example, "Radiation Effects on Superconductivity", eds. B. S. Brown, H. C. Freyhardt, and T. H. Blewitt, North Holland Publishing Co. (1978).

220. C. M. Varma and R. C. Dynes, in "Superconductivity in d- and f-band Metals--Second Rochester Conference", edited by D. H. Douglass (Plenum, NY, 1976), pg. 507.

221. S. A. Wolf, J. J. Kennedy, and M. Nisenoff, J. Vac. Scie. Technol. 13, 145 (1976).

222. E. L. Wolf, et al., J. Low Temp. Phys. 40, 19 (1980).

223. Q. S. Yang, E. L. Wolf, and I. K. Schuller (unpublished).

224. See for instance, S. T. Ruggiero and M. B. Beasley in Ref. 1, pg. 365.

225. W. E. Lawrence and S. Doniach "Proc. 12th Int. Conf. on Low Temperaure Physics", E. Kanda ed., Academic Press, Kyoto, Japan (1970), pg. 361.

226. R. A. Klemm, A. Luther, and M. R. Beasley, Phys. Rev. B12, 877 (1975).

227. M. Tachiki and S. Takahashi, Physica 135B, 178 (1985).

228. S. Takahashi and M. Tachiki, Phys. Rev. B33, 4620 (1986).

SEMICONDUCTOR INTERFACES

R.H.Williams

Department of Physics
University College, Cardiff, CF1 1XL
United Kingdom

1. INTRODUCTION

The bulk properties of solids are relatively well understood and a
large number of experimental techniques have been developed to investigate
and establish different aspects of these. Theoretical understanding of
bulk solids is also at an advanced stage. The situation for interfaces
between solids however, is in relative terms entirely different, and the
experimental methods available to study interfaces are at a more primitive
stage. Yet many of the properties of the semiconductor multilayer
structures currently being produced are governed by the nature of
interfaces between solids. To gain a full understanding of the behaviour
of superlattices and multi-quantum well structures it is essential to
develop suitable models to describe semiconductor interfaces. In this
article, therefore, we will concentrate on some elementary aspects of the
semiconductor-vacuum,   semiconductor-metal,   semiconductor-semiconductor,
and semiconductor-insulator boundaries. These interfaces are not only of
fundamental interest but are also of the utmost importance in solid state
devices such as ultra-fast transistors and lasers. In the article we will
deal with single interfaces, but the treatment may readily be extended to
solids containing multiple interfaces.

A large number of experimental techniques have been developed to
probe semiconductor interfaces. In this article we will concentrate in
particular on the use of photoelectron spectroscopy to study interfaces;
the application of other techniques such as RHEED and TEM are considered
more thoroughly elsewhere in this volume.

2. CLEAN AND REAL SURFACES

We start by considering the semiconductor-vacuum boundary, and
emphasize that all free surfaces under normal atmospheric conditions are
covered by a layer of contaminant atoms from the environment. In order to
properly investigate surfaces it is necessary to generate them in an
atomically clean condition and this may be done in one of several ways:

a) By cleaving in ultra high vacuum. This method yields (111) surfaces of
silicon and germanium (corresponding to the fracture of the minimum number
of bonds) and gives clean (110) surfaces of the III-V semiconductors. The

disadvantage of the method is that it is limited to just one plane for most solids.

(b) By sputtering and annealing. Usually inert gas ions of energy 500 eV – 3 KeV are used to bombard the surface, and the surface damage produced is removed by annealing. The technique may be applied to any plane but often leads to non-stoichiometric surfaces for compound semiconductors.

(c) By heating in ultra high vacuum. The solid is first chemically etched in order to produce a thin surface layer of a volatile oxide. It is then heated in high vacuum to remove this oxide and to generate a clean surface. The Si (111) 7 x 7 reconstructed surface may be prepared in this way and ways of lowering the cleaning temperature are always being sought.

(d) By molecular beam epitaxial growth. The growth of layers of MBE may leave the surface of the layer atomically clean. However, in order to grow the layer in the first instance it is necessary to clean the substrate by methods (a), (b) and (c).

2.1 Surface Structure

The structure of a clean surface generally differs from that associated with the underlying bulk. If the surface structure is different from the bulk whilst retaining the bulk unit mesh parallel to the surface, that surface is said to be relaxed. The clean cleaved (110) surfaces of the III-V semiconductor fall into this category. This structure, for the case of GaAs (110) surfaces is shown in Fig. 1. The structure was derived following extensive studies by several techniques, in particular low energy electron diffraction[1], and it may be seen that the surface relaxation involves the movement of the surface cations and anions towards and away from the bulk, respectively. The surface

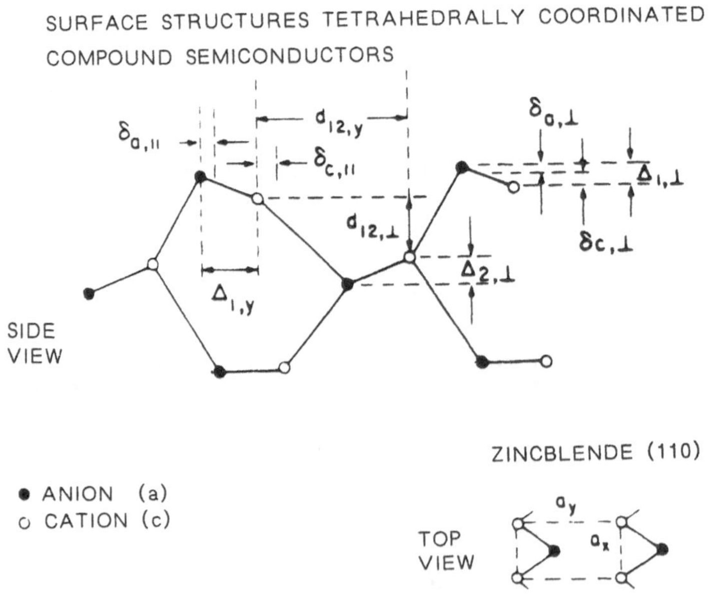

Figure 1   Illustration of the relaxation of the GaAs (110) surface[1]

TABLE 1

Relative As coverage for various GaAs(100) reconstructed surfaces

Surface Structure	As Coverage Ref 2	As Coverage Ref 3.
C(4x4)	1.0	0.86
C(2x8)	0.89	0.61
C(8x2)	0.52	0.22
1x6	0.42	0.52
4x6	0.31	0.27

structures associated with the (100) and (111) planes of GaAs are more complicated and these surfaces are reconstructed, i.e., the unit mesh at the surface is quite different to that anticipated based on the bulk structure. The reconstructed form of the GaAs (100) surface appears to be dependent on the surface composition and a number of forms are presented in Table 1[2,3]. The column labelled 'As coverage' refers to the fraction of the outermost layer composed of As atoms. For all but the (110) non polar surfaces of III-V semiconductors the detailed form of the surface structures are not well understood.

2.2 Surface Electronic Structure and Band Bending

The periodic potential which leads to a band gap in the semiconductor bulk no longer prevails at the surface. As a result, surface states may occur and may have energies corresponding to the bulk band gaps. Consider the situation illustrated in Fig. 2(a) where a band of acceptor like surface states is present on the surface of an n-type semiconductor. Electrons from the bulk may be captured in these states leading to a depletion layer of width W below the surface, as illustrated in Fig. 2(b). For a solid of permittivity ϵ and donor density N_d cm^{-3} the space charge width is given by

$$W = \left[\frac{2\epsilon\epsilon_0 V_s}{e N_d} \right]^{1/2}$$

where V_s is the band bending. For a value of $V_s = 0.53$ eV, $\epsilon_s = 12$ and $N_d = 10^{15}$cm^{-3} it may be seen that $W \sim 2.6 \times 10^{-5}$ cm. Clearly, if one has a thin film or a small device with dimension $\sim 10^{-5}$cm the depletion of electrons from the bulk will significantly influence the bulk electrical conductivity. Surface states may therefore influence the bulk properties significantly for thin layers or very small structures. It is of interest to note that the numbers of electrons in the acceptor states in the above example is $N_d \times W = 2.6 \times 10^{11}$ electrons cm^{-2}. As we shall discuss shortly the distribution of surface states on the clean cleaved Si and Ge (111) surfaces have the precise form illustrated in Fig.2.

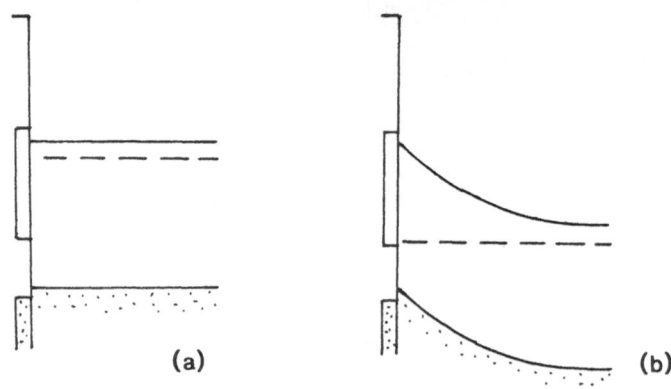

(a)   (b)

Figure 2   Illustration of the way acceptor sur-
face states lead to band bending

Space charge layers of the kind considered here are of great
importance in semiconductor multilayer structures. Accumulation layers,
of course, are just as important as depletion layers.

2.3 Mapping Surface States

Many methods have been developed over the years to study surface
states on semiconductors; most of them are indirect methods such as the
Kelvin probe to measure contact potential differences. During the last
ten years or so angle resolved photoelectron spectroscopy has made
possible the determination of surface state distribution with
unprecedented accuracy. In the photoemission process electrons are excited
from occupied initial states to empty states, and if the excitation is
close enough to the surface these electrons may escape without degredation
of energy, if their energies are sufficient, i.e., in excess of the vacuum
level. The photoemission process is illustrated in Fig.3. Maximum
surface sensitivity is achieved by ensuring that the emerging electrons
have energies in the range 20 to 500 eV (see Fig.4) which necessitates
photon energies in this same range for emission from valence band states
and surface states. The only tunable photon source covering this range
involves the use of synchrotron radiation[4]. Photoemission from the
atomic core levels enables the chemical composition of the surface to be
monitored, an example of this will be presented in Section 3.2.

Energy distribution curves for emission from the clean cleaved (111)
of germanium are presented in Fig.5[5]. If E_K corresponds to the measured
kinetic energy of electrons forming a peak, then the energy of the initial
state, E_i, with respect to the Fermi level is given by

$$E_i = \hbar\omega - \phi - E_K \quad ,$$

where $h\nu$ is the incident photon energy and ϕ is the work function. In
angle integrated photoemission the energy distribution of electrons
emitted over a large solid angle is measured. In the more powerful angle
resolved version of the method electrons emitted into a narrow cone $\Delta\theta$ are
studied, for various angles of emission θ. For an emitted electron the
momentum parallel to the surface is given by

$$k_{\parallel} = \left[\frac{2mE_K}{\hbar^2} \right]^{1/2} \sin \theta$$

174

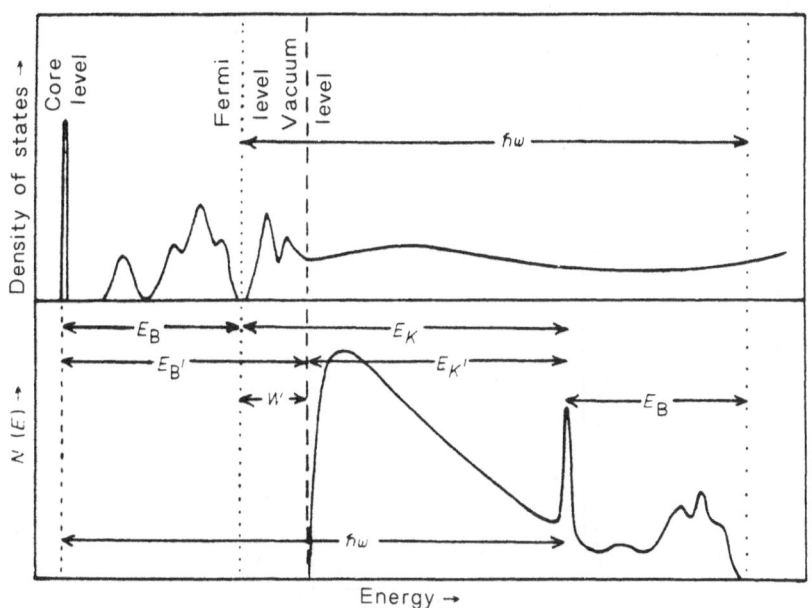

Figure 3   Illustration of the photoemission mechanism

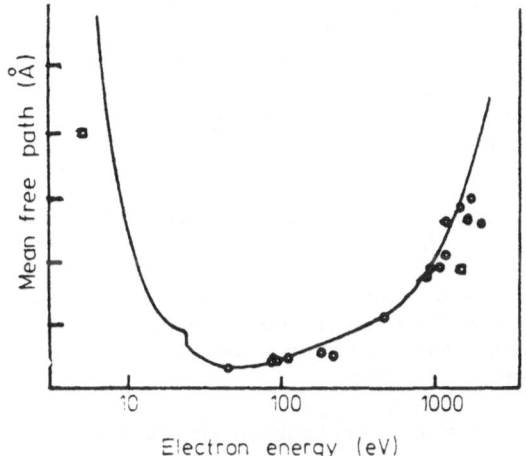

Figure 4   Illustration of the mean free
path of an electron in a solid as
a function of its energy

and can be determined if E_K and θ are measured.   For emission from surface
states, which of course are two dimensional in nature, a plot of E_i vs k_{11}
yields the dispersion of the surface state bands.

Figure 6 shows the dispersion of peak A in Fig. 5, i.e., shows E_i vs
k_{11}. This emission peak is known to correspond to a surface state[5]. The
hatched region in Fig. 6 corresponds to the bulk band structure, so that
this occupied surface state is below the valence band edge in energy, as
illustrated in Fig. 2.   Nicholls et al[5], in their photoemission
investigations, used highly n-type crystals of germanium, such that the

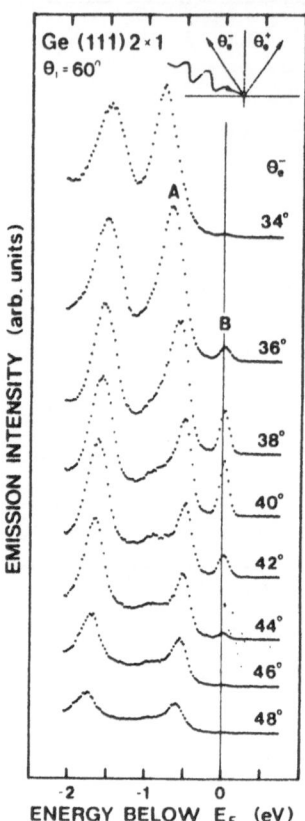

Figure 5   Angle resolved photoemission spectra for a clean cleaved
            n-type Ge(111) 2 x 1 surface

(normally empty) acceptor like surface states in the band gap were
partially occupied, as in Fig. 2(b).   Peak B in Fig. 5 shows the
photoemission from this state, and again by monitoring its position with
changing escape angle the dispersion could be obtained and is shown in Fig.
6.   The energy difference between A and B in Fig. 6 is just the germanium
surface state band gap.   The result for the Si (111) 2 x 1 reconstructed
surface is very similar.

        Attempts to calculate a distribution of surface states in agreement
with angle resolved photoemission were quite unsuccessful until Pandey[6]
suggested a new model to account for the 2 x 1 reconstruction of the Si
and Ge (111) surfaces.   This model is referred to as the π-bonded chain
model, and is shown in Fig.7(b), compared with the unreconstructed form of
Fig.7(a).   The surface state energies and dispersion calculated for the
π-bonded structure are shown in Fig.6; it may be seen that the dispersion
agrees   well   with   experiment   though   not   the   absolute   energies.
Nevertheless, the π-bonded chain model is now believed to represent the
correct structure for Si and Ge (111) 2 x 1 surfaces.

        It is also worth remarking that the band gap between the occupied and
unoccupied surface states for Si (111) reconstructed surfaces has been
established by other techniques, notably optical reflectivity[7] and low
energy electron loss spectroscopy[8].   The example described above simply
illustrates how surface states may be probed, and how the information
obtained may in turn lead to an understanding of the surface structure.
These techniques have been applied to study a wide range of semiconductor
surfaces.

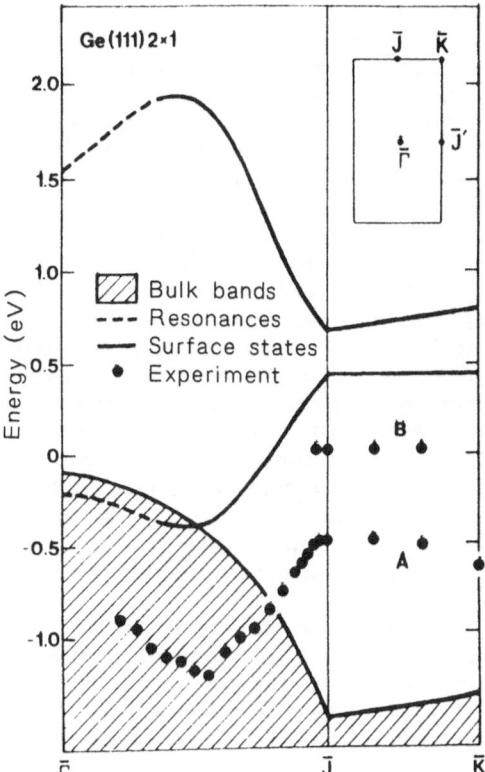

Figure 6 Dispersion of surface states on a Ge (111) 2 x 1 surface[5].
The hatched region represents the occupied bulk energy bands

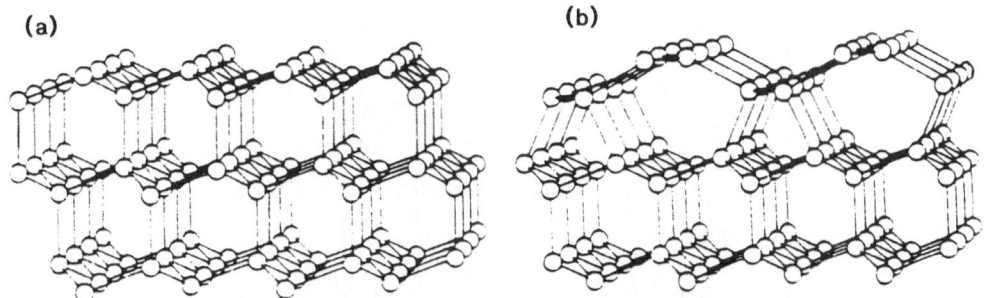

Figure 7 The x-bonded chain model of the Si and Ge (111) 2 x 1
surface (b), compared with the unrelaxed form (a)

2.4 Overlayers on Surfaces

Atomically clean semiconductor surfaces are generally highly reactive
and the adsorption of gases such as oxygen and hydrogen often leads to a
modification of the surface structure and a drastic modification in the
form and energies of surface states. For example, the adsorption of
hydrogen on the Si (111) 2 x 1 surface leads to a removal of the
reconstruction (i.e., to a 1 x 1 structure).

In order to illustrate the earliest formation of interfaces in multilayer structures we consider the case of Sb adsorption on InP (110) cleaved surfaces. The (110) surface of InP is relaxed· in a similar manner to that of GaAs shown in Fig.1· and the surface state distribution is well understood. Photoemission spectra for the clean surface for two angles of emission, corresponding to the Γ and M points in the surface Brillouin Zone, are shown in Fig. 8(a). ES_1 and ES_2 correspond to surface state emission from A5 and A3, at the M point in Fig.8(b). The deposition of one monolayer of Sb leads to an ordered 1 x 1 overlayer and two new surface states labelled ES_3 and ES_4[2]. At the same time core level photoemission from the Sb 4d core indicates that Sb must be bonded on two different surface sites for one monolayer coverage. The model that is found to be constant with these studies as well as with LEED studies[10] is shown in Fig.8(c); Sb atoms are bound to P and In surface atoms as well as to each other, forming zig zag chains along the surface. The corresponding relaxation is reversed and the surface state distribution calculated[10] for this situation is illustrated in Fig. 8(d). Photoemission peaks ES_3 and ES_4 correspond to new surface states at the M point. Here it is of great interest to note the empty acceptor-like state S_7 that has appeared in the band gap. This state accounts for the band bending of around 0.3 eV observed when one monolayer of Sb is adsorbed.

This example illustrates how a combination of surface science techniques have been linked with calculations of surface state energies to deduce the precise form of an admittedly relatively simple interface. This approach is now fairly standard and has yielded considerable insight regarding metal-semiconductor, and semiconductor-heterojunction interfaces.

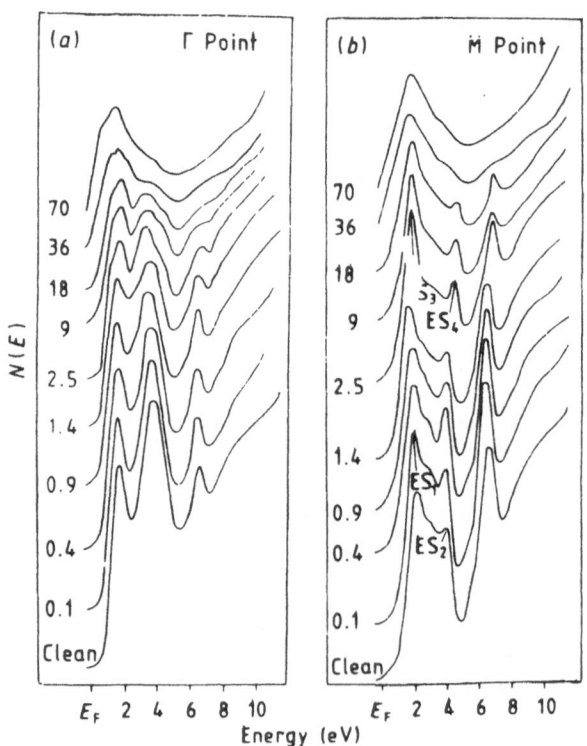

Figure 8(a)   Angle resolved photoelectron spectra for a clean InSb (110) surface and for various amounts of Sb, in Å, deposited on the surface

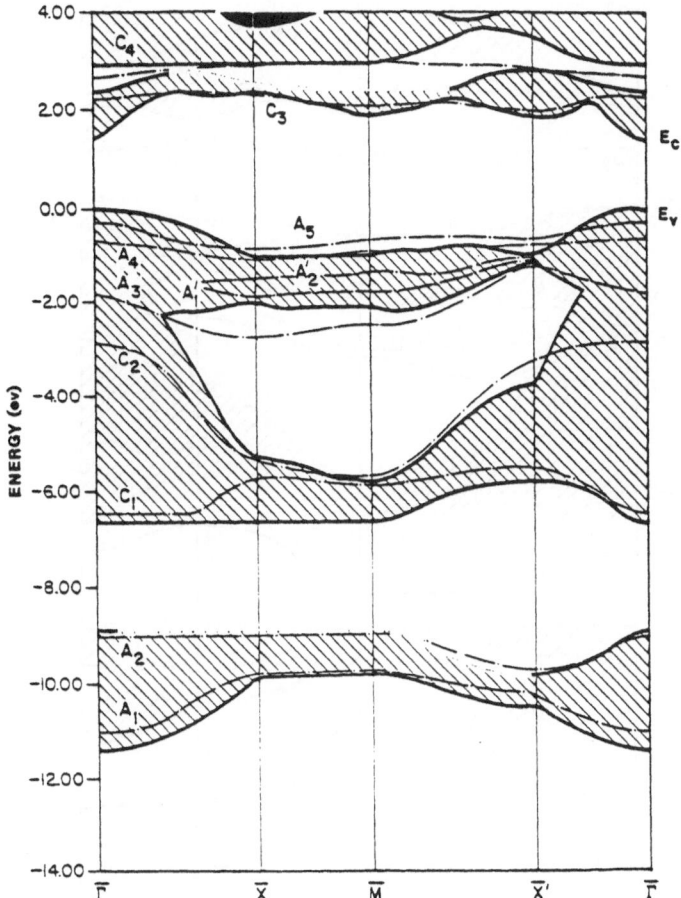

Figure 8(b)   Band structure of InP bulk (hatched) and
(110) surface (dot-dashed line)[8]

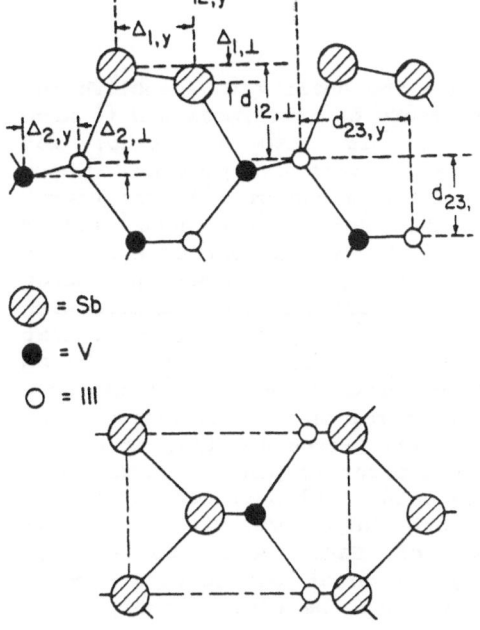

⊘ = Sb

● = V

○ = III

Figure 8(c)   The proposed structure
for a monolayer of Sb
on an InP (110) sur-
face[9]

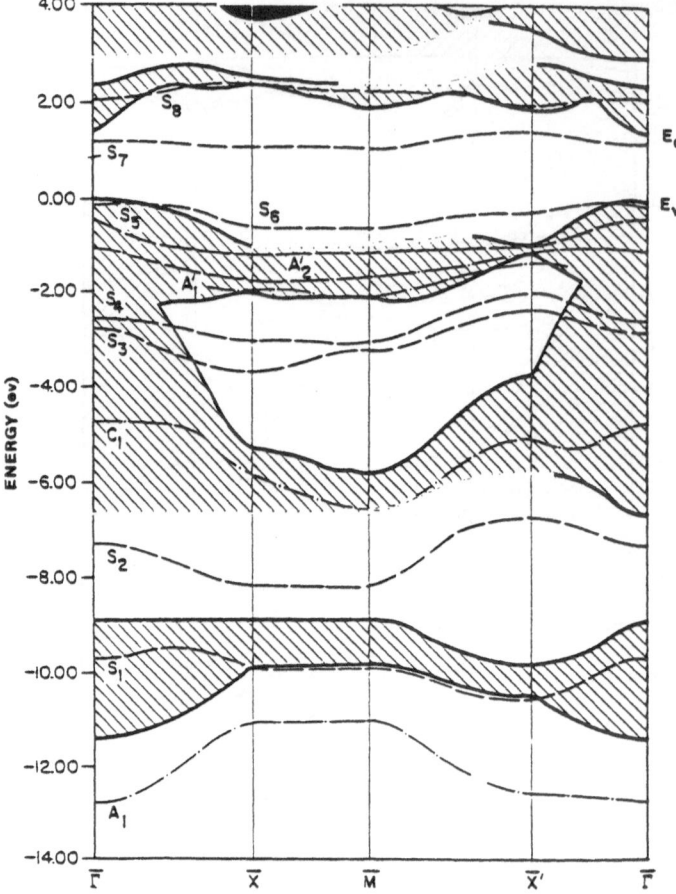

Figure 8(d)   Band calculation for a monolayer of
Sb on an InP (110) surface with the
structure shown in Figure 8(c)

2.5 Band Bending Due to Overlayers

For multilayers it is important to have the ability to establish the crystalline order and crystallographic structure at interfaces and to know if new chemical phases are formed. It is also important, for semiconductors, to know whether space charge layers are present. There are many techniques available for this, and again one of the most useful is photoemission. Figure 9 illustrates this point; here photoelectron spectra are sketched for n- and p-type crystals of the same material for the case where there is no band bending, (a) and (b). The binding energy E_1 of features such as the strong core level emission are measured with respect to the Fermi level at the surface in most spectrometers so that the <u>apparent</u> binding energy of the core level emission will differ by an amount nearly equal to the band gap for the cases shown in (a) and (b). Conversely, if the core level emission does differ in binding energy by an amount nearly equal to the band gap for n- and p-type material it may be stated that there are no surface states sufficient to pin the Fermi level in the band gap. This is the case for the GaAs (110) surface, and the Ga 3d and As 3d core level photoemission are in accord with the above discussion. However, if the cleavage of GaAs leads to a high concentration of steps, the Fermi level position at the surface becomes pinned close to mid gap and one then finds that the binding energies of

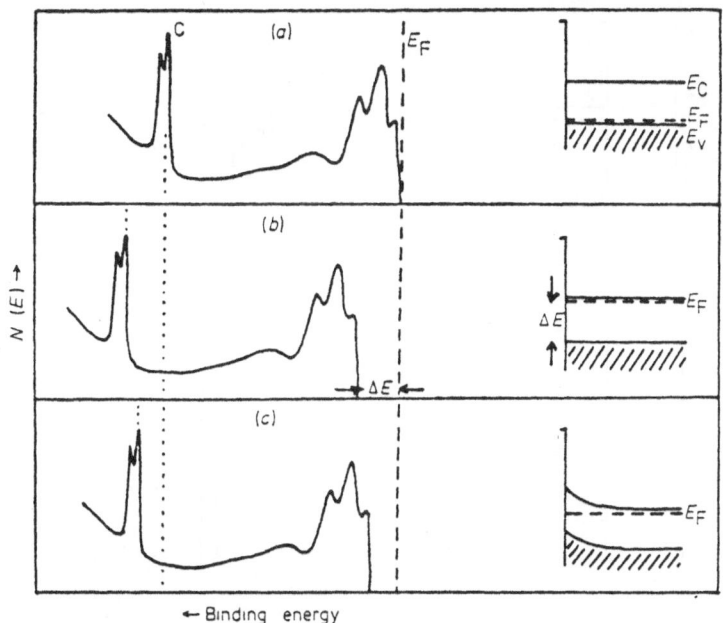

Figure 9 Schematic photoelectron spectra for an n- and a p-type
semiconductor with flat bands and with band bending (c)

the Ga 3d and AS 3d photoemission peaks are in between those appropriate
for the unpinned n- and p-type situations. This is illustrated in Fig.
9(c). This method is a particularly valuable one to monitor the shifts of
the Fermi level at the surface, e.g., when adsorbed atoms are incident on
the surface, and is widely used. Great care must be taken, however, to
distinguish such Fermi level shifts from 'chemical shifts' associated with
chemical reactions at interfaces; the latter are discussed in Section 3.2.

3. METAL-SEMICONDUCTOR INTERFACES

3.1 Schottky Barriers and Ohmic Contacts

Practically every semiconductor device requires a stable and reliable
electrical contact to it and indeed contact technology is a large field in
its own right. Many applications need low-resistance ohmic contacts,
whereas others need contacts with electrical barriers, so-called Schottky
barriers, at the interface. The simplest model of Schottky barrier
formation[11] is illustrated in Fig. 10(a), where a metal of large work
function, ϕ_m, makes contact with an n-type semiconductor with no surface
states. The Schottky barrier height, ϕ_b, is then given by

$$\phi_b = \phi_m - \chi_s \quad .$$

However, it has long been known that ϕ_b does not show such a large
dependence on ϕ_m and an explanation in terms of surface states was offered
by Bardeen[12]. As shown in Fig.10(b), charge transfer can take place
between the surface states and the metal, leading to a very thin dipole
layer at the interface through which electrons can easily tunnel. The
interior of the semiconductor is thus screened from the metal so that ϕ_b
is independent of ϕ_m in the so-called Bardeen limit. In between the
Schottky and Bardeen limits it is often assumed that a linear model can be
applied and that

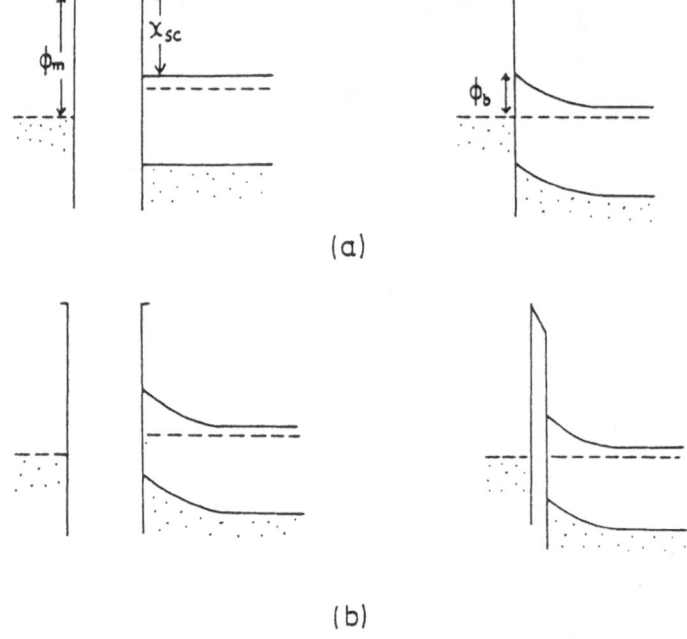

(a)

(b)

Figure 10   A metal in contact with a semiconductor. Schematic illustration of the energy bands in (a) the Schottky
limit, and (b) the Bardeen limit

$$\phi_b = \gamma(\phi_m - \chi_s) + (1-\gamma)(E_g - \phi_o) \quad ,$$

where ϕ_o is a constant referred to as the neutral level. Here it is
assumed that the metal and semiconductor are separated by a layer of
permittivity ϵ and with δ, and that a uniform density of surface
(interface) states, D_s cm^{-2}eV^{-1}, exists on the semiconductor. Then it may
be readily shown that

$$\gamma = \frac{\epsilon \epsilon_o}{\epsilon \epsilon_o + \epsilon \delta D_s} \quad .$$

For the situation where $D_s = 0$ it is seen that $\gamma = 1$ and the Schottky
model is recovered. For $D_s \sim 10^{14}$cm^{-2}eV^{-1} and a value of δ of a few Å,
one sees that γ is small and ϕ_b is relatively independent of the metal.
The Fermi level at the interface is then pinned close to ϕ_o, which is the
level to which the interface states are filled so as to leave the surface
layer neutrally charged. Clearly, therefore, interface densities of $\sim$
10^{14}cm^{-2}eV^{-1} lead to strong Fermi level pinning at metal semiconductor
interfaces.

So far we have assumed an interface density D_s cm^{-2}eV^{-1} but without
making any reference to the detailed origin of these states. This has
been, and indeed still is, the subject of considerable debate and
discussion in the literature. Clearly, interface states at a
metal–semiconductor boundary cannot be identical to the surface states on
the free semiconductor, because as pointed out already, adsorbed atoms
change the form of these states. Heine (1966) was the first to point out
that for a metal on a semiconductor the wavefunctions can tunnel a few Å
from the metal into the band gap of the semiconductor, leading to an

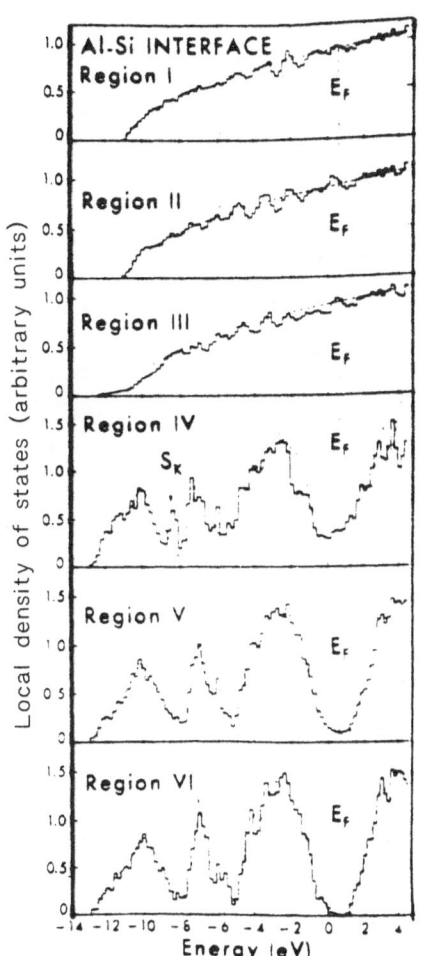

Figure 11

Density of states for Al on an Si surface, for different atomic layers on either side of the interface[14]

interface dipole layer. These wave functions are often referred to as metal induced gap states, or MIGS, and their origin has been considered in detail by Louie, Chelikowsky and Cohen[14]. Figure 11 shows the data for Al on Si. The density of states has been calculated for the equivalent of three atomic layers in the Al (assumed to be a jellium) and a similar number of layers in the Si. Well away from the interface the band gap of Si is clear, but near the interface some states are pulled into the gap from the valence and conduction bands, so that the band gap in the semiconductor closes near the interface. The neutral level ϕ_O, associated with these states has been considered in detail by Tejedor et al.[16,17] and more recently by Tersoff[17], and it has been assumed that in the limit of a high density of MIGS, the Fermi level will be pinned close to ϕ_O. Schottky barrier heights calculated for a number of compounds by Tersoff are shown in Table 2.

In general, the models discussed so far do not take into account the possibility of imperfections at the metal-semiconductor interface. The question then arises as to how perfect these interfaces really are and whether or not D_S could be accounted for in terms of imperfections. In the next sub-section we therefore consider chemical reactions and intermixing at metal-semiconductor interfaces.

TABLE 2

Band gaps, Schottky barriers for gold contacts on p-type material, and calculated 'neutral level' energy, E_B. After Tersoff (Ref 17)

	E_g	Φ_{bp}	E_B
Si	1.11	0.32	0.36
Ge	0.66	0.07	0.18
GaP	2.27	0.94	0.81
InP	1.34	0.77	0.76
AlAs	2.15	0.96	1.05
GaAs	1.43	0.52	0.70
InAs	0.36	0.47	0.50
GaSb	0.70	0.07	0.07

3.2 Chemical Reactions at Interfaces

On thermodynamic grounds alone one would expect certain metal-semiconductor combinations to react chemically while other combinations should be inert, and this indeed turns out to be the case. We illustrate the case of a reactive interface by considering aluminium contacts deposited on InP (110) surfaces held at room temperature. Figure 12 shows core level spectra for the In 4d and P 2p electrons from an atomically clean cleaved (110) InP surface, and following the deposition of increasing amounts of Al on the surface. It is immediately obvious that the interface is not abrupt and simple; following the deposition of 20 Å of Al the P 2p emission is fully attenuated as expected, but the In 4d emission suffers relatively little attenuation. Furthermore, the In 4d emission is chemically shifted, to lower binding energies, by a substantial amount. The reason is that a phosphide of Al is formed at the interface leading to the release of phase-segregated indium. The indium released is incorporated in the growing Al layer. Studies using LEED demonstrate that the interface is highly disordered.

The interaction of a range of metals with a large number of clean (and oxidised) semiconductors has been studied in this way. Some systems, e.g., Ag on InP (110) and GaAs (110) surfaces lead to abrupt interfaces unless the contact is heated. In other situations strong interdiffusion of the metal into the semiconductor and elements from the semiconductor into the metal are observed. By and large, the reactions are reasonably understood on the basis of bulk thermodynamic data, provided that the formation of alloys and compounds are taken into account. Brillson[18], McGilp[19], and Pugh and Williams [20] have considered various aspects of these reactions.

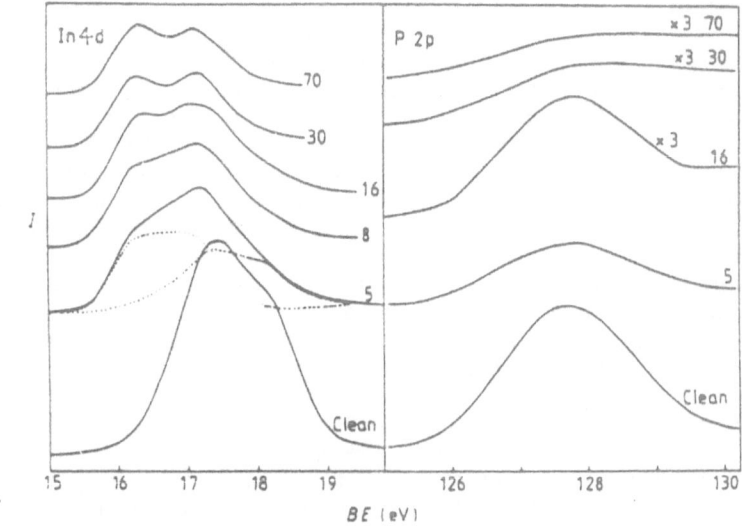

Figure 12 Core level photoemission spectra for the In 4d and
P 2p levels in InP for the clean (110) surface and
following the deposition of Al (in Å)

In order for exchange reactions to occur at metal-semiconductor
interfaces many bonds have to be broken and the reactions require
activation. Zunger[21] suggested that Al adatoms were likely to be mobile
on non polar cleaved surfaces, such as the (110) GaAs surface, and that
metallic nuclei would most likely be formed. He further suggested that
chemical reactions could somehow be activated by the energy released in
the formation of such nuclei. Ihm and Joanopoulos have investigated
theoretically the behaviour of Al adatoms on (110) GaAs surfaces, and
demonstrated that well-defined paths for surface diffusion are available
and that the formation of nuclei is likely on room temperature substrates.
There is a substantial body of experimental evidence to support this
assertion. Ludeke and Landgren[23] have carried out extensive
investigations of Al adsorption and layer growth on GaAs (100) surfaces
and show that the surface diffusion and nucleation is influenced by the
surface composition. Interestingly, exchange Al:Ga reactions were always
observed to accompany the formation of nuclei.

Many metal layers prepared by vacuum deposition on (110) surfaces of
III-V semiconductors are polycrystalline in nature. However, good single
crystal layers may often be grown on the polar surfaces by molecular beam
epitaxial growth. These growth processes are discussed elsewhere in this
Volume.

3.3 Models of Interfaces

A large body of experimental data has been assembled for metal
contacts to III-V semiconductors, particularly GaAs, and in this section
we summarise some of the most important models put forward to describe
barrier formation at interfaces, and the factors leading to these models.
There has been considerable discussion in the literature regarding the
relative merits of these models.

The first model is that put forward by Woodall and Freeouf[24] which
makes the assumption that the Schottky barriers are governed by anion rich
clusters at the metal-semiconductor interface, i.e., phosphorus rich

clusters for metals on InP. To date there is no conclusive support in favour of the model.

Most metals on clean n-GaAs lead to values of Φ_b in the range of 0.75 eV to 0.9 eV, with little dependence on the work function of the metal. In order to account for the data Spicer et al. put forward the unified defect model, whereby the Fermi level at the interface is pinned by point defects in the semiconductor, near the interface. The point defects are generated, it is suggested, when the metal is incident on the semiconductor surface. Charge transfer, and the consequent dipole layer, occurs between the metals and the defects, i.e., the defect levels form energy states in the gap and act like surface states. This model has been considered in detail by a number of theoretical groups and the results of Allen et al[25] are shown in Fig. 13. Here it is assumed that the defects which lead to Fermi level pinning are antisite defects, in this case cations (e.g., Ga) on anion sites. The barrier heights calculated on this basis for a series of III-V compounds and their alloys are shown by the full line in Fig. 13. The full points are experimental values of barrier heights for Au on the same materials and it may be seen that the general trends are well accounted for. However, care must be taken not to take this too far since the trends are also accounted for on the basis of Fermi level pinning by vacancies[27] rather than antisite defects.

Even more important is the fact that models based on metal induced gap states (MIGS) also account very well for the data for Au contacts on III-V semiconductors and their alloys shown in Fig. 13. The reasons why it is often difficult to distinguish between MIGS and the defect model have been discussed recently by Tejedor and Flores[28]. The antiside defect energy tends to follow the mid point energy of anion and cation dangling bonds; this point is just the neutral level Φ_o in the MIGS model.

Investigations have also concentrated on identifying the importance of chemical reactions at metal-semiconductor interfaces. For metals on III-V semiconductors their importance is somewhat unclear, but it is certain that chemical reactions have a pronounced effect on Schottky barrier heights for metals on the III-VI semiconductor CdTe[29]. It is also relevant to note that a great variation of barrier heights is

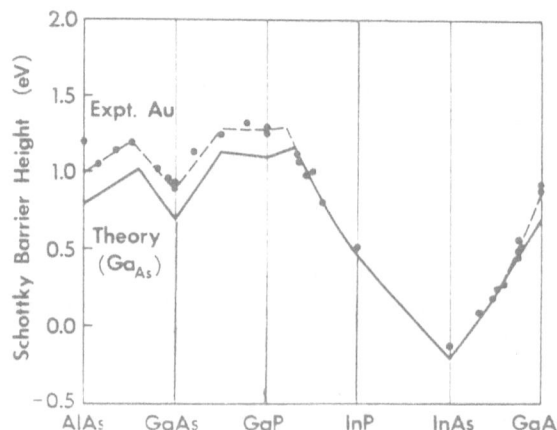

Figure 13   Theoretically predicted Schottky barrier heights (solid line) in III-V semiconductors based on the assumption of Fermi level pinning by antisite defects. The points are experimental data[26]

observed for metals on the more ionic large gap semiconductors, such as CdS, than on the covalent solids. In terms of MIGS this is explained in terms of a smaller penetration depth of the metal wave function into the larger gap semiconductor, and a consequently smaller value of D_s. The combination of MIGS and defects has been recently considered by Monch[30].

3.4 Case Study - 'Ideal' Interface

In view of the possible effects of imperfections at interfaces a great deal of effort has been expended in generating 'ideal' interfaces which are as free of defects as possible. Probably the most perfect interface yet made for metals on semiconductors is that associated with $NiSi_z$ on Si (111) surfaces. The silicide is metallic, with the CaF_z structure, and the lattice match to silicon is excellent. It appears that by using slightly different preparation procedures it is possible to fabricate what are called type A and type B contacts. These are illustrated in Fig. 14 where it may be seen that the type A silicide has the lattice rotated through 60° with respect to type B, i.e., one is a mirror image of the other. Tung[31] has fabricated Schottky diodes using both types of contacts and demonstrated that I-V yield barrier heights differing by around 0.15 eV. The I-V characteristics for these diodes are shown in Fig. 15. There has been some debate as to whether or not these differences are due to defects at the interface but all indications are that the interfaces are highly perfect and the variation in ϕ_b is associated with the different interface structure for type A and type B contacts.

The above result is of considerable importance for it is the first to demonstrate clearly the influence of interface structure on Schottky barrier height. Most nickel contacts to silicon consist of mixtures of types A and B, and possibly other phases as well, and the relative amounts probably depend on the fabrication conditions of the contact. One might then expect quite a large spread in the values of ϕ_b measured for Ni on Si. A similar variation may well be possible with other metals also.

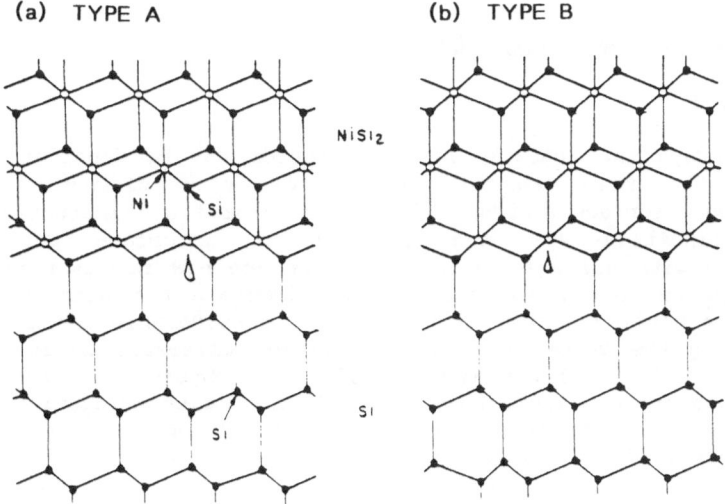

Figure 14   The structure of Type A and Type B $NiSi_2$ on Si (111) surfaces[31]

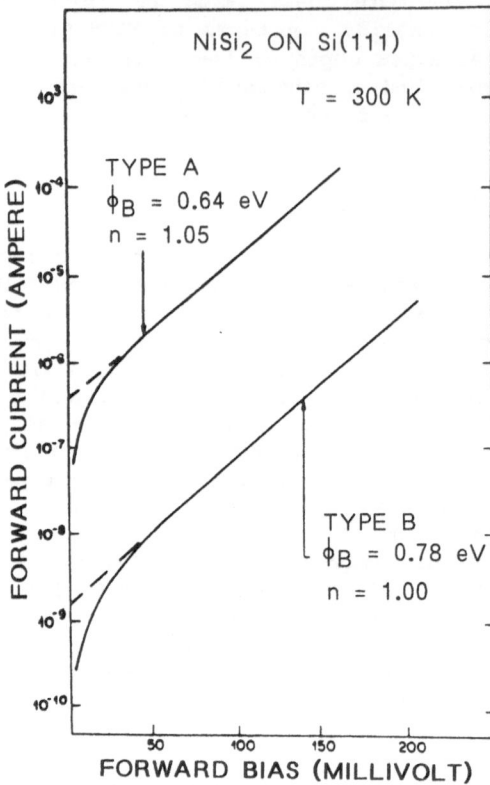

Figure 15 Forward current characteristics for Type A
and Type B contacts in Si Schottky diodes[31]

At this stage it is difficult to state conclusively which mechanisms dominate Schottky barrier formation at metal semiconductor interfaces. It is likely that several mechanisms play a role with relative importance of each depending on the particular conditions prevailing when the contact is made. It is clear that more studies of the kind carried out for $NiSi_2$ on Si (111) are urgently needed before substantial further progress can be made.

4.   SEMICONDUCTOR HETEROJUNCTIONS

4.1 Background

In many ways the Schottky barrier problem may be considered as one extreme of the more general problem of band discontinuities at heterojunction interfaces. Figure 16 shows a schematic illustration of the band discontinuities ΔE_v and ΔE_c at an interface between two semiconductors with different band gaps. If one now imagines one of the semiconductors to have zero band gap one recovers the Schottky barrier situation. For heterojunctions also there is considerable research effort in establishing the values of ΔE_v and ΔE_c for different combinations of semiconductors and, equally important, the basic physics underlying these. Several methods are available to measure the band discontinuities; here we will only consider their investigation via surface science techniques, in particular photoelectron spectroscopy.

4.2 Chemical Reactions at Heterojunctions

Before considering band discontinuities at heterojunctions we first briefly look at the chemical nature, in the same way as we considered reactions at metal-semiconductor interfaces. Figure 17 shows the usual band gap and lattice parameter for a wide range of semiconductors. Probably the most widely studied system is GaAlAs on GaAs and TEM studies indicate interfaces of remarkably high quality. It also seems that smoother interfaces are obtained when GaAlAs is deposited on GaAs than when GaAs is deposited on GaAlAs.

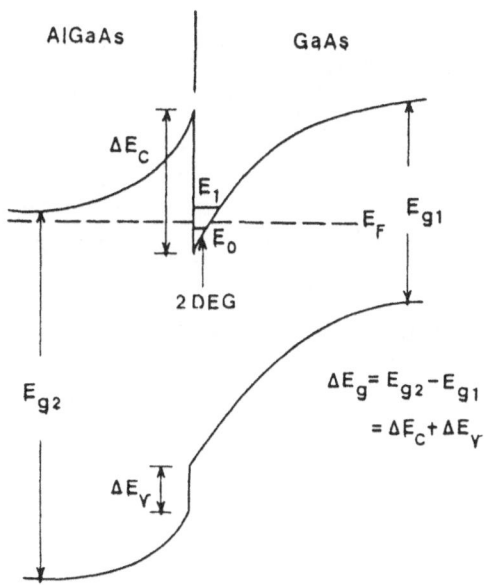

Figure 16   Energy bands at a semiconductor heterojunction

All interfaces grown by molecular beam epitaxy, however, are not perfect. Consider the case of CdTe deposited on InSb clean surfaces. Figure 17 demonstrates that the lattice matching condition is well satisfied. The large gap – small gap combination is also an interesting one from the point of view of application. Deposition of CdTe onto InSb (100) surfaces held at room temperature leads to a relatively abrupt interface. But the CdTe layer is disordered. However, if the substrate temperature is in the region of 200°C, a typical temperature used in MBE growth, photoemission investigations show that the interface is far from simple[32]. Figure 18 illustrates typical spectra. Here, the photon energy used was 100 eV and the spectrum for the clean InSb surface clearly shows the In 4d and Sb 4d emission lines. The difference in amplitude between these is simply due to variations in photoionisation cross section with energy. Upon deposition of CdTe, from a single source, it may be seen that the Sb 4d emission is attenuated and the Te 4d emission gradually increases, as expected. However, there is little evidence of any Cd 3d emission and the In 4d line is attenuated much more slowly than expected. Detailed analysis of the spectra leads to the conclusion that a telluride of indium is formed, quite likely In_xTe_y[33]. Studies of the same interface by Raman spectroscopy[33] supports this view.

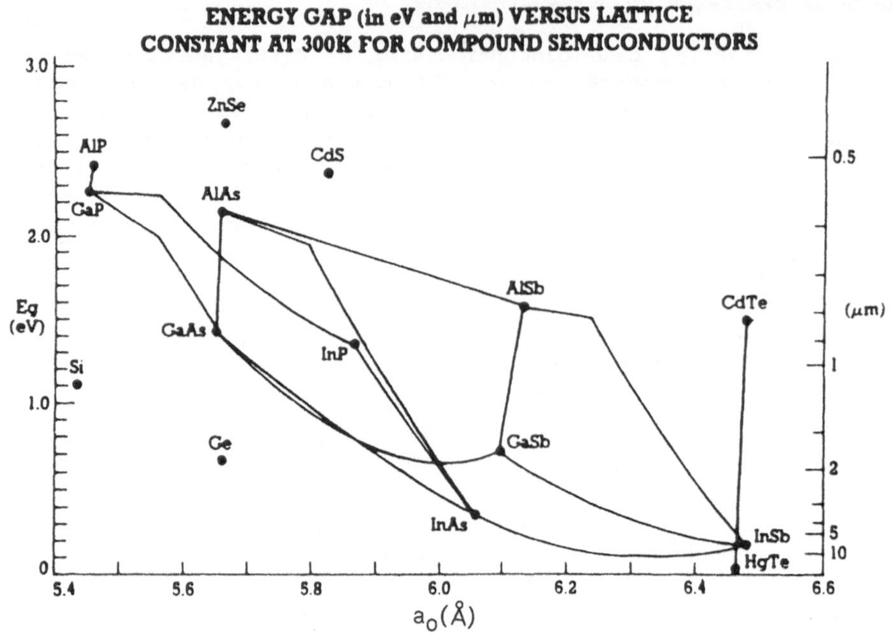

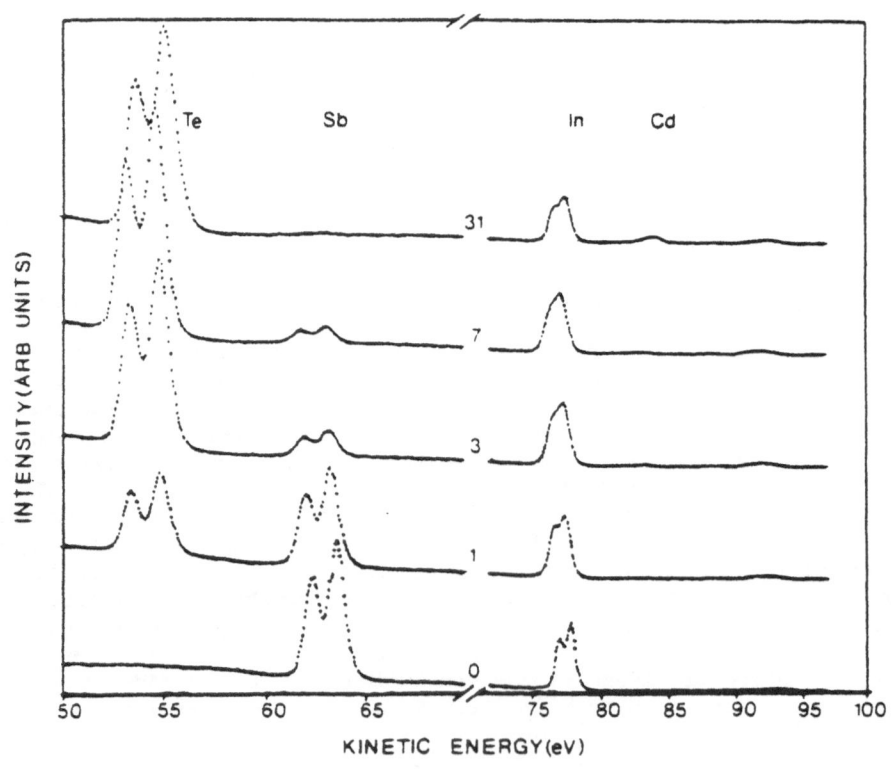

ENERGY GAP (in eV and μm) VERSUS LATTICE CONSTANT AT 300K FOR COMPOUND SEMICONDUCTORS

Figure 17   Band gaps and lattice spacings for selected semiconductors

Figure 18   Photoelectron spectra for a clean InSb surface and following the deposition of CdTe under MBE[32] conditions

The thermodynamics of the above system has recently been considered[33], and it appears that CdTe and InSb in a closed system should not chemically react. However, InSb is quite unstable in the prescence of tellurium and will readily react. The sticking coefficient of Cd on InSb at 200°C is substantially smaller than that of Te, and this in turn leads to the reaction to form tellurides. However, the deposition of InSb on CdTe may well lead to abrupt interfaces with no intermediate indium telluride phase according to the above arguments. Indeed TEM studies show that interfaces formed by depositing InSb on CdTe are much smoother than those formed when the sequence is reversed. A major problem in the growth of CdTe on InSb, therefore, is to find a temperature window which is high enough to enable the growth of high quality crystals but at the same time low enough to avoid strong interface chemical reactions. This is a common problem in the growth of semiconductors by techniques such as molecular beam epitaxy.

4.3 Band Discontinuities

It is important to understand the physics determining band discontinuities not only for academic reasons, but also because the rules governing them are required by device engineers. One of the earliest approximate models applied was the so-called Anderson electron affinity rule which states that the conduction band offset for two semiconductors in contact is simply the difference in electron affinity for the two materials. However, the electron affinities are composed of bulk components and surface dipole contributions and the latter change in an unknown way when the two solids make contact. One cannot, therefore, expect the 'electron affinity rule' to account accurately for band discontinuities. Other models have been put forward by Harrison[34] and by Tersoff[17]. The latter model is similar to that discussed earlier for Schottky barriers, and it is suggested that it is the neutral levels ϕ_0 of the two semiconductors in contact that should be aligned. Band discontinuities calculated on this basis for several pairs of semiconductors are presented in Table 3.

A number of methods have been used to measure band discontinuities; here we consider the application of the photoemission technique. The

TABLE 3

Experimental valence band offsets at selected heterojunctions, compared with theoretically derived values by Tersoff

	Experimental	Theory
AlAa/GaAs	0.19–0.5 eV	0.35
InAs/GaSb	0.51	0.43
GaAs/InAs	0.17	0.20
Si/Ge	0.2	0.18
GaAs/Ge	0.53	0.52

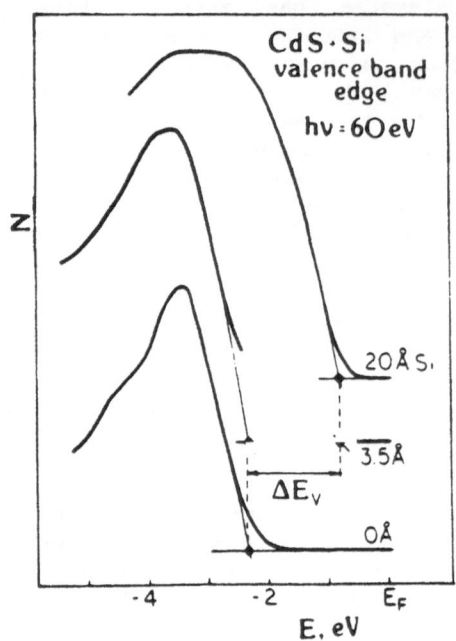

Figure 19

Valence band photoemission for clean CdS and following the depositons of Si[35]

method has been used in two ways; the first of these is illustrated in Fig. 19 for the case of silicon deposited on CdS[35]. The valence band edge of clean CdS is first established from the photoemission spectrum. A thin layer of silicon is then deposited, and for small coverages the emission from the valence band edges of Si and CdS can be observed simultaneously in a spectrum; the differene between these gives ΔEv. This method, however, is not always straightforward to apply, particularly if the emission from the upper edge of the valence band is masked by emission from surface states, making the precise determination of the band edge energy difficult.

A second method makes use of photoemission from core levels, and an example of its application is presented in Fig. 20, for HgTe–CdTe interfaces[35]. First of all the location of the core levels E_{Cd4d}^{CdTe} and E_{Hg5d}^{HgTe} are determined accurately with respect to the valence band edge for the free surfaces. The energies of these levels are also monitored when a thin layer of HgTe is deposited on CdTe. Then the valence band discontinuity may be obtained from

$$\Delta Ev = \left[E_{Hg5d5/2}^{HgTe} - Ev^{HgTe} \right] - \left[Ev_{Cd4d}^{CdTe} - Ev^{CdTe} \right] + \Delta E_{CL} \quad ,$$

where ΔE_{CL} is the difference of core level binding energies shown in Fig. 20. In Fig. 21 typical experimental spectra are shown. To obtain an accurate determination of the valence band edge the calculated valence band density of states was covoluted with the spectrometer response function giving the solid line shown. The agreement with experiment (dots) is excellent. In this way it was concluded that ΔEv = 0.35 ± 0.06 eV for the HgTe–CdTe interface.

The above example is an important one in view of the so-called 'common anion rule'. It has been suggested that ΔEv for a pair of semiconductors with the same anion (Te in this case) should be close to zero but clearly this is not the case for HgTe–CdTe.

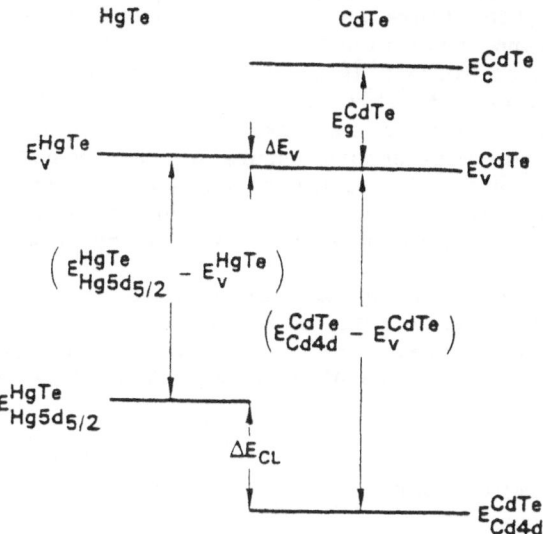

Figure 20   Energy level diagram for HgTe-CdTe interface

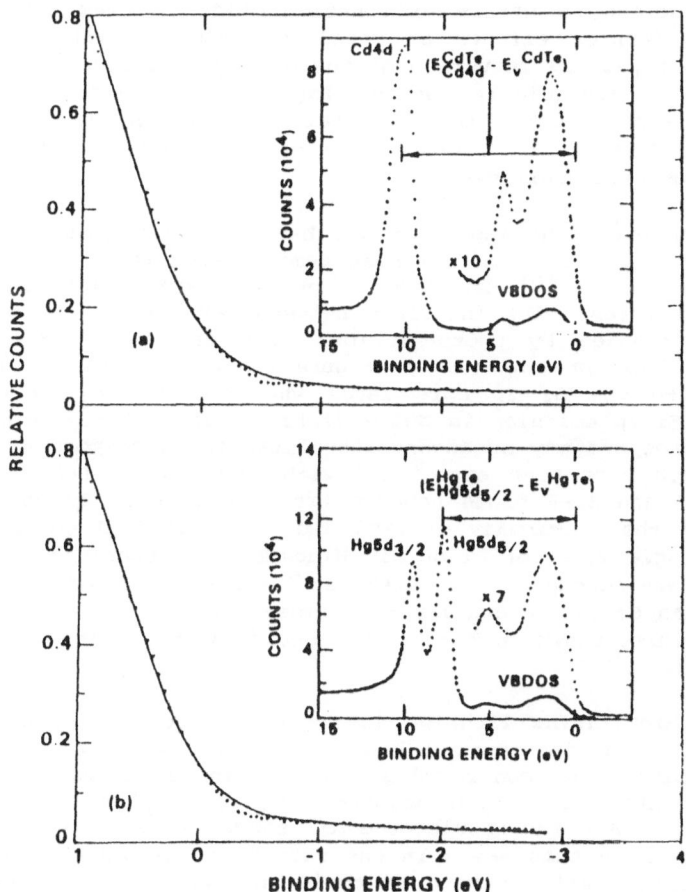

Figure 21   Photoemission spectra for HgTe on CdTe[36]

At the present time there is no single theory which can be said to fully account for band discontinuities at semiconductor heterojunctions. Certainly, the model of Tersoff is as good as any to date and is superior to either the 'electron affinity' or the 'common anion' rules. However, further accurate experiments are required and in particular the dependence of ΔE_v and ΔE_c on the microscopic nature of the interface needs to be fully evaluated. The studies of Bauer and Sang[37] indicate that ΔE_v is relatively independent of the reconstructed form of the starting surface and does not appear to be influenced significantly by interface defects.

The methods of determining ΔE_v illustrated in Figs. 20 and 21 is useful provided the valence band edge photoemission can be accurately established and provided the core level emission is not too severely influenced by the surface. Separating bulk effects from surface ones can often be a problem.

5. INSULATORS ON SEMICONDUCTORS

Interfaces between semiconductors and insulators are of great importance in solid state electronics and indeed the enormous success of silicon technology owes much to the fact that the SiO_2-Si interfaces can be of a remarkably high quality. Interface states at boundaries between semiconductors and insulators have been widely probed by a large range of methods[38].

The interaction of metals with native oxides on semiconductors can often lead to complex effects at metal-semiconductor interfaces. Some metals lead to a partial reduction of the oxide on silicon, and this kind of effect is particularly acute for oxide layers on compound semiconductors such as CdTe. Here, we shall not discuss these interactions but we will concentrate on a case study of one insulating layer, namely CaF_2, on silicon (111) surfaces.

CaF_2 is an excellent insulator and has a lattice constant differing from that of Si by 0.6%. It has been demonstrated that it can be grown epitaxially on silicon[39] and efforts are under way to fabricate Si-CaF_2 multilayer structures. The interface between CaF_2 and Si is of interest and has been studied by photoemission. The data of Himpsel and his coworkers[40,41] are presented here. Figure 22 shows photoemission spectra corresponding to the Si 2p core level when 8 Å of CaF_2 is on the Si surface. The Si 2p emission is quite different from that appropriate for the clean surface and may be deconvoluted into three components as shown. These components have been associated with Si-Ca bonds (charge transfer from Ca to Si) and Si-F bonds (charge transfer to F). By comparing the intensities of the chemically shifted and the unshifted components with those for a single layer of Cl on Si, Himpsel et al. concluded that the Ca atoms in CaF_2 are bonded to the outermost layer of Si atoms whereas the F atoms are bonded to the second layer. Himpsel et al. considered a range of possible structures shown in Fig. 23 and concluded that model 4 is the most likely structure.

Epitaxial insulating fluoride layers have also been grown on III-V semiconductors. An interesting point is that the lattice matching condition in many cases can be obtained by using mixtures of CaF_2, SrF_2 and BaF_2 with the composition adjusted in order to yield the correct lattice match. A GaAs-insulator-GaAs sandwich structure has been successfully made in this way. In the future, it is likely that a great deal more attention will be paid to these epitaxial insulating layers on semiconductors.

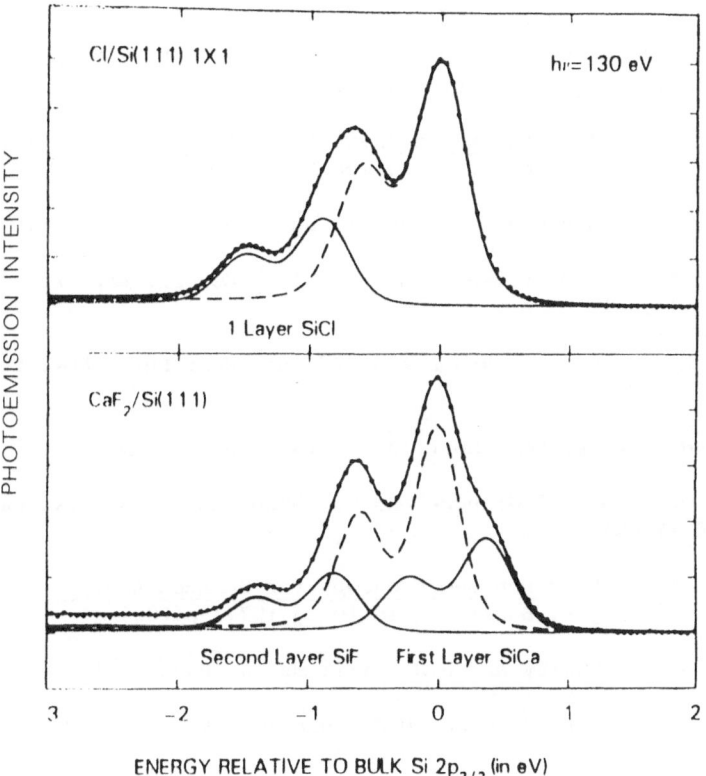

Figure 22   Si 2p core level photoemission spectra with a monolayer
of CaF_2 (lower) and Cl (upper)[40]

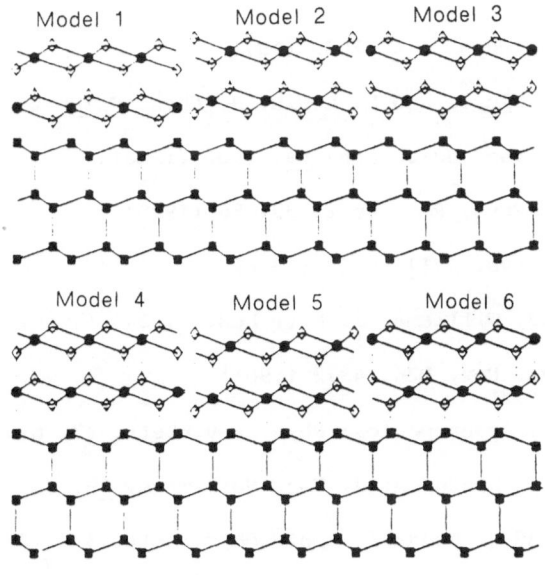

Figure 23   Various possible models of CaF_2 on an Si (111) surface

REFERENCES

1. R.J. Meyer, C.B. Duke, A. Paton, J.C. Tsang, J.L. Yeh, A. Kahn and P. Mark: J. Vac. Sci. Tech. 17, 971 (1980)

2. R.Z. Bachrach: In Progress in Crystal Growth and Characterisation, ed. by P. Pamplin, Vol. 2 (Pergamon Press, Oxford 1979) p.115

3. P. Drathen, E. Ranke and K. Jacobi: Surf. Sci. 77, L162 (1978)

4. R.H. Williams, G.P. Srivastava and I.T. McGovern: Rep. Prog. Phys. 43, 1357 (1980)

5. J.M. Nicholls, P. Martensson and G.V. Hansson: Phys. Rev. Lett. 54, 2363 (1986)

6. K.C. Pandey: Phys. Rev. Lett. 47, 1913 (1981)

7. G. Chiarotti, G. Del Signore and S. Nannarone: Phys. Rev. Lett. 21, 1170 (1968)

8. H. Froitzheim: Electron Spectroscopy for Surface Analysis, ed. by H. Ibach (Springer, Berlin, Heidelberg 1977) p.205

9. C. Maani, A. McKinley and R.H. Williams: J. Phys. C 18, 4795 (1985)

10. C.B. Mailhiot, C.B. Duke and D.J. Chadi: J. Vac. Sci. Tech. A3, 915 (1985)

11. W. Schottky: Z. Phys. 113, 367 (1939)

12. J. Bardeen: Phys. Rev. 71, 717 (1947)

13. V. Heine: Phys. Rev. A138, 1689 (1965)

14. L. Louie, J.R. Chelikowsky and M.C. Cohen: Phys. Rev. B15, 2154 (1977)

15. F. Yndurain: J. Phys. C 4, 2849 (1971)

16. C. Tejedor, F. Flores and L. Louis: J. Phys. C 10, 2163 (1977)

17. J. Tersoff: J. Vac. Sci. Tech. B4, 1066 (1986)

18. L.J. Brillson: Phys. Rev. Lett. 35, 56 (1975)

19. J. McGilp: J. Phys. C 17, 2249 (1984)

20. B. Pugh and R.S. Williams: J. Mat. Res. 1, 343 (1986)

21. A. Zunger: Phys. Rev. B24, 4372 (1981)

22. J. Ihm and J.D. Joannopoulos: Phys. Rev. Lett. 47, 679 (1981)

23. R. Ludeke and G. Landgren: J. Vac. Sci. Tech. 19, 674 (1981)

24. J.L. Freeouf and J. Woodall: Appl. Phys. Lett. 39, 727 (1981)

25. W.E. Spicer, I. Lindau, P.R. Skeath, C.Y. Su and P.W. Chye: Phys. Rev. Lett. 44, 420 (1980)

26. R.E. Allen, O.F. Sankey and J.D. Dow: Surf. Sci. **168**, 376 (1986)

27. M.S. Daw and D.L. Smith: Phys. Rev. **B20**, 5150 (1980)

28. F. Flores and T. Tejedor: J. Phys. C **20**, 145 (1987)

29. I.M. Dharmadasa, W.G. Herrenden-Harker and R.H. Williams: Appl. Phys. Lett. **48**, 1082 (1986)

30. W. Monch: J. Vac. Sci. Tech. **B4**, 1085 (1986)

31. R. Tung: Phys. Rev. Lett. **52**, 461 (1984)

32. K.J. Mackey, P. Allen, W.G. Herrenden-Harker, R.H. Williams, G.M. Williams and C. Whitehouse: Appl. Phys. Lett. **49**, 354 (1986)

33. K.J. Mackey, D.T. Zahn, P. Allen, W. Richter and R.H. Williams: J. Vac. Sci. Tech. (in press)

34. W.A. Harrison: J. Vac. Sci. Tech. **14**, 1016 (1977)

35. G. Margaritondo: Sol. St. El. **26**, 499 (1983)

36. S.P. Kowalczyk, J.T. Cheng, E.A. Kraut and R.W. Grant: Phys. Rev. Lett. **56**, 1605 (1986)

37. R.S. Bauer and H.W. Sang: Surf. Sci. **132**, 465 (1983)

38. A.M. Sze: Physics of Semiconductor Devices (Wiley, New York 1981)

39. R.F.C. Farrow, P.W. Sullivan, G.M. Williams, G.R. Jones and . C. Cameron: J. Vac. Sci Tech. **19**, 415 (1981)

40. F.J. Himpsel, F.U. Hillebrecht, G. Hughes, J.L. Jordan, U.O. Karlsson, F.R. McFeely, J.F. Morar and R. Dieger: Appl. Phys. Lett. **48**, 596 (1986)

41. U.O. Karlsson, F.J. Himpsel, J.F. Morar, D. Rieger and J. Yarmoff: J. Vac. Sci. Tech. **B4**, 1117 (1986)

STABILITY OF ARTIFICIAL MULTILAYERS

Frans Spaepen

Division of Applied Sciences
Harvard University
Cambridge, Massachusetts 02138

INTRODUCTION

This paper is a tutorial overview of the factors that affect the stability of artificial multilayers. First, the basic thermodynamics of solutions will be briefly reviewed. Next, the stability of multilayers with continuous structure will be considered. In these multilayers the structure is the same throughout the film and the layering corresponds to a modulation of the composition. Examples are the face-centered cubic films with a [111] texture studied first by Hilliard and co-workers at Northwestern, and amorphous metal or amorphous semiconductor films. Depending on the thermodynamics of the system, the composition modulation can either disappear or sharpen by diffusion. For short modulation repeat lengths, this process must be analyzed with the Cahn-Hilliard theory for the free energy of inhomogeneous systems, which takes into account the contributions from composition gradients. The effect of elastic strain, first formulated by Cahn, needs to be taken into account as well. For large amplitude modulations, the diffusion equations may also become non-linear.

In the next section, the stability of non-continuous structures, such as face-centered cubic/body-centered cubic or crystalline/amorphous multilayers will be discussed under the constraint that the layer interfaces remain parallel. The breakdown of the layer structure is driven by the decrease in interfacial free energy. A section is therefore devoted to a brief review of the coarsening mechanism in eutectic lamellar structures, which is well understood, and may be the most important factor in the coarsening of artificial multilayers as well.

In the last section, the use of artificial multilayers for the study of interdiffusion in amorphous metals and semiconductors is briefly discussed. Not only is this technique extremely sensitive, it also allows time-dependent measurements, which is essential for the characterization of the effects of structural relaxation.

More advanced treatments and more detailed discussion of applications can be found in a number of earlier review articles [1-4].

THERMODYNAMICS OF SOLUTIONS

The stability of a system at constant pressure is determined by its Gibbs free energy, $G=U+pV-TS$. For a condensed system at ambient pressure, the pV-term can be neglected, and the Helmholtz free energy $F=U-TS$ can be used. The energy of a random mixture of N atoms in a binary alloy, consisting of Nc A-atoms and $N(1-c)$ B-atoms, can be expressed as

$$U = \frac{1}{2} N Z [c^2 V_{AA} + (1-c)^2 V_{BB} + 2c(1-c) V_{AB}] \qquad (1)$$

where Z is the number of nearest neighbors and V_{ij} is the energy of a bond between an i- and a j-atom. The energy of the pure components can be written as

$$U = \frac{1}{2} N Z [c V_{AA} + (1-c) V_{BB}] \qquad (2)$$

The energy of mixing is the difference between those two quantities:

$$\Delta U = N Z c (1-c) \varepsilon \qquad (3)$$

where

$$\varepsilon = V_{AB} - \frac{1}{2} (V_{AA} + V_{AB}) \qquad (4)$$

is called the interaction parameter.

In the regular solution model the entropy of mixing is taken to be the same as in an ideal solution (i.e., $\varepsilon=0$):

$$\Delta S = - k_B N [c \ln c + (1-c) \ln(1-c)] \qquad (5)$$

Finally, the free energy of the mixture, using the pure systems as the reference states is:

$$F = \Delta U - T\Delta S \qquad (6)$$

For ordering solutions (i.e., pairs of unlike atoms are energetically preferred, or $\varepsilon<0$) F(c) has a concave shape (F">0), as shown in Figure 1. Also shown on that figure are the chemical potentials μ_A and μ_B of the components in a particular solution. Note that, quite generally,

$$\mu_A - \mu_B = \frac{dF}{dc} \qquad (7)$$

since c varies between 0 and 1 across the diagram.

For phase-separating solutions (i.e., pairs of like atoms are energetically preferred, or $\varepsilon>0$), F(c) still has the concave shape of Figure 1 at high temperatures, because the entropy term is relatively more important. At low temperatures, F(c) has the shape of the curve of Figure 2, with a convex part (F"<0) for concentrated solutions. As shown on that figure, the free energy, of a solution with composition c_0 between the two inflection points P' and P", is lowered by separation into two phases of adjacent compositions c' and c". This means that phase separation can occur by the growth of infinitesimal composition fluctuations, known as "spinodal decomposition". The locus of the inflection points on the temperature-composition phase diagram of Figure 3 is known as the "spinodal line", and the phase boundaries are determined by the locus of the common tangent points Q' and Q".

For a more extensive treatment of solution theory the reader is referred to several standard texts on the subject [5-7].

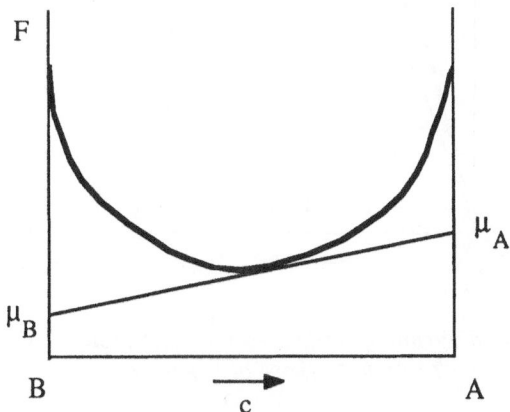

Figure 1. *Schematic diagram of the isothermal free energy of an ordering solution as a function of composition. μ_A and μ_B are the chemical potentials of A and B in a solution of composition corresponding to the tangent point.*

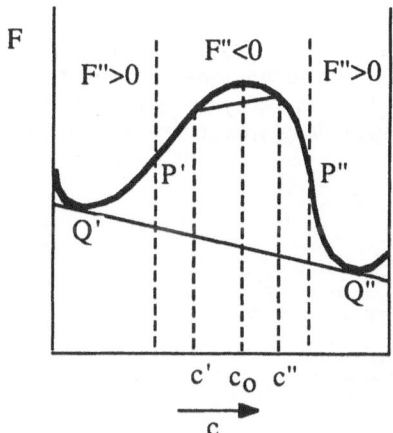

Figure 2. *Schematic diagram of the isothermal free energy of a phase-separating solution as a function of composition. The spinodal points P' and P'', and the equilibrium compositions Q' and Q'' are indicated.*

STABILITY OF INHOMOGENEOUS SYSTEMS

The free energy of an inhomogeneous system can be obtained from the theory of section 2 only if the composition varies sufficiently slowly with position, so that the free energy of an elementary volume is only a function of the local composition. For an inhomogeneity $c(x)$, for example, the total free energy of the system is then

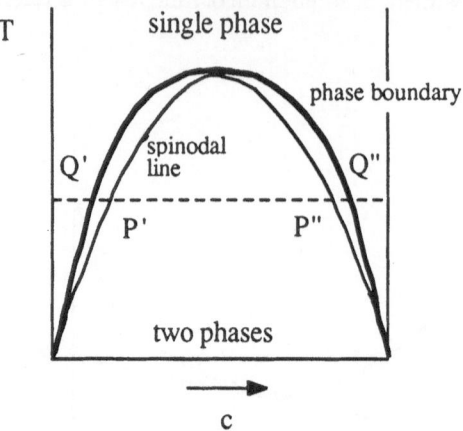

Figure 3. *Schematic phase diagram of a phase-separating system. The indicated points correspond to those of Figure 2.*

$$F = A \int f(c) \, dx \tag{8}$$

where A is the area perpendicular to the x direction, $f(c)$ is the bulk free energy of section 2 (per unit volume), and the integration is over the entire sample volume.

If the composition varies strongly with position, the free energy of an elementary volume also depends on the composition of the neighboring volumes. Cahn and Hilliard [8-11], in approach similar to the Ginzburg-Landau theory for phase transformations, first treated this problem analytically by expanding the local free energy per unit volume in a Taylor series form:

$$f(c, \frac{\partial c}{\partial x}, \frac{\partial^2 c}{\partial x^2}) = f(c)$$
$$+ \kappa_{11} \frac{\partial c}{\partial x} + \kappa_{12} (\frac{\partial c}{\partial x})^2 + \dots$$
$$+ \kappa_{21} \frac{\partial^2 c}{\partial x^2} + \dots$$
$$+ \dots$$

Since the free energy of the system does not change if it is pointed the other way along the x-axis, κ_{11} must be zero. By keeping only the first three remaining terms in the expansion, the total free energy of system can be wrtitten as

$$F = A \int [f(c) + \kappa_{12} (\frac{\partial c}{\partial x})^2 + \kappa_{12} (\frac{\partial^2 c}{\partial x^2})] \, dx \tag{10}$$

Integration by parts of the last term, taking into account that the contribution to the total free energy of the integrated term is negligible, gives

$$F = A \int [f(c) + \kappa_{12} (\frac{\partial c}{\partial x})^2 - \frac{\partial \kappa_{21}}{\partial c} (\frac{\partial c}{\partial x})^2] \, dx \qquad (11)$$

Combination of the two gradient terms gives

$$F = A \int [f(c) + \kappa (\frac{\partial c}{\partial x})^2] \, dx \qquad (12)$$

where κ is called the gradient energy coefficient, and is defined as

$$\kappa = \kappa_{12} - \frac{\partial \kappa_{21}}{\partial c} \qquad (13)$$

It is instructive to calculate κ in a simple nearest neighbor interaction model. Consider two atomic planes in a binary alloy, each containing a total of N atoms per unit area. There are y bonds per atom between the planes. One plane contains cN A-atoms, the other $(c+\Delta c)N$ A-atoms. The energy of the bonds between the planes can then be written as

$$U = N \, y \, [c \, (c+\Delta c) \, V_{AA} + (1-c)(1-c-\Delta c)V_{BB} + c(1-c-\Delta c)V_{AB} + (1-c)(c+\Delta c)V_{AB}] \qquad (14)$$

The bond energy between two planes of equal composition c is easily found by putting Δc equal to zero in equation (14). The *excess* bond energy associated with having a different composition on the second plane is then the difference between these two quantities:

$$\Delta U = \frac{1}{2} N \, y \, \Delta c \, [c V_{AA} - (1-c)V_{BB} + (1-2c)V_{AB}] \qquad (15)$$

Note that this excess bond energy is *linear* in Δc. This means that the total excess bond energy of a plane in a uniform composition gradient (i.e., a plane of composition c sandwiched between one of composition $c+\Delta c$ and one of composition $c-\Delta c$) is zero. According to equation (9), κ_{12} must therefore be zero in this model. Curvature in the composition profile is obtained by surrounding the plane of composition c with two planes of composition $c+\Delta c$. The total excess bond energy per unit area of the central plane is then $2\Delta U$. Considering that, for a plane spacing d, the excess bond energy per unit volume is $2\Delta U /d$. Since, according to equation (9), this quantity must also equal

$$\kappa_{21} \frac{\partial^2 c}{\partial x^2} = \kappa_{21} \frac{2 \, \Delta c}{d^2} \qquad (16)$$

the following expression for κ_{21} is obtained:

$$\kappa_{21} = \frac{1}{2} N \, y \, d \, [c V_{AA} - (1-c)V_{BB} + (1-2c)V_{AB}] \qquad (17)$$

The gradient energy coefficient can now be obtained from equation (13):

$$\kappa = N \, y \, d \, \varepsilon \qquad (18)$$

Note that in this simple nearest neighbor interaction model the sign of the gradient energy coefficient is the same as that of the solution interaction parameter.

The condition of stability of an inhomogeneous system is found by minimizing the free energy of equation (12) under the condition of conservation of mass:

$$\int c \, dx = c_o L \tag{19}$$

where c_o is the average concentration of A-atoms, and L the length of the sample. This is a standard variational calculus problem, which yields the stability condition that the following potential must be constant throughout the system:

$$\alpha = \frac{\partial f}{\partial c} - 2\kappa \frac{\partial^2 c}{\partial x^2} \tag{20}$$

Note that in the absence of a gradient energy term this potential reduces to $f' = (\mu_A - \mu_B)/\Omega$ (see equation (7); Ω is the atomic volume), which is the classical potential for interdiffusion.

INTERDIFFUSION

The classical equation for the interdiffusion flux, in a reference frame chosen so that the fluxes of A- and B- atoms are equal, can be written as:

$$-J = M \frac{\partial(\mu_A - \mu_B)}{\partial x} \tag{21}$$

where J is the flux of the A-atoms, and M is the mobility (always positive). In the presence of gradient energy contributions, the classical diffusion potential is replaced by that of equation (20):

$$-J = M\Omega \frac{\partial \alpha}{\partial x} = M\Omega \left[f'' \frac{\partial c}{\partial x} - 2\kappa \frac{\partial^3 c}{\partial x^3} \right] \tag{22}$$

The diffusion equation is now obtained by requiring conservation of mass:

$$\frac{\partial c}{\partial t} = - \text{div } J = M\Omega \left[f'' \frac{\partial^2 c}{\partial x^2} - 2\kappa \frac{\partial^4 c}{\partial x^4} \right] \tag{23}$$

Note that for this differential equation to be linear, it had to be assumed that M, f'' and κ were independent of composition. This is usually only true for small modulation amplitudes. Using the definition of the bulk interdiffusion coefficient:

$$D = M\Omega f'' \tag{24}$$

equation (23) can be rewritten as

$$\frac{\partial c}{\partial t} = D \left[\frac{\partial^2 c}{\partial x^2} - \frac{2\kappa}{f''} \frac{\partial^4 x}{\partial x^4} \right] \tag{25}$$

For the stability of aritificial multilayers solutions of the form

$$c = c_0 + A(t) \cos\beta x \tag{26}$$

are of interest, where $\beta = 2\pi/\lambda$ (λ: modulation repeat length). Insertion of equation (26) into (25) gives for the time dependence of the modulation amplitude

$$A(t) = \exp\left[-D \beta^2 (1 + \frac{2\kappa\beta^2}{f''}) t \right] \tag{27}$$

The amplitude, or the relative change in amplitude, is then:

$$R = \frac{d \ln A}{dt} = -D (1 + \frac{2\kappa\beta^2}{f''}) \beta^2 = -D_\lambda \beta^2 \tag{28}$$

where is D_λ is an effective, repeat length dependent, interdiffusion coefficient. Since in the kinematic diffraction regime the intensity, I, of the (000) satellites is proportional to the square of the amplitude of the corresponding modulation wave, the slope of a plot of lnA vs. t equals $-2 D_\lambda \beta^2$. The same is true for the satellites of the Bragg peaks if there is no modulation of the lattice spacing. By comparing satellites on both sides of the Bragg peaks, the compositional and positional modulations can be determined separately.

Figure 4 is a schematic summary of the behavior of thermodynamically different modulated films, as a function of modulation repeat length. Figures 4(a) and 4(b) correspond to a phase separating system ($\varepsilon > 0$, and hence, according to equation (18), $\kappa > 0$). Figure 4(a) applies to the composition range inside the spinodal. In this range $f '' < 0$, and equation (24) shows that the bulk interdiffusivity is then negative (i.e., uphill diffusion). In this regime a composition modulation is expected to *grow* with annealing. The interdiffusivity increases with decreasing repeat length, and becomes positive for repeat lengths less than a critical value

$$\lambda_c = 2\pi \sqrt{\frac{2\kappa}{|f''|}} \tag{29}$$

This is the result of the positive contribution ($\kappa > 0$) of the interfaces to the free energy, which favors homogenization. The same tendency is seen for the case of Figure 4(b), which applies outside the spinodal ($f '' > 0$), although here the interdiffusivity is positive (i.e., homogenization, or the familiar downhill diffusion) for all repeat lengths.

For ordering systems ($\varepsilon < 0$, and hence according to equation (18), $\kappa < 0$), only the case of $f '' > 0$ needs to be considered under the regular solution approximation used here. The bulk interdiffusivity, according to equation (24), is positive. Figure 4(c) shows how the interdiffusivity decreases with decreasing repeat length due to the negative contribution ($\kappa < 0$) of the interfaces to the free energy, and becomes negative for repeat lengths less than λ_c. The latter behavior can be understood by considering the formation of an ordered crystalline compound as uphill diffusion on an atomic scale.

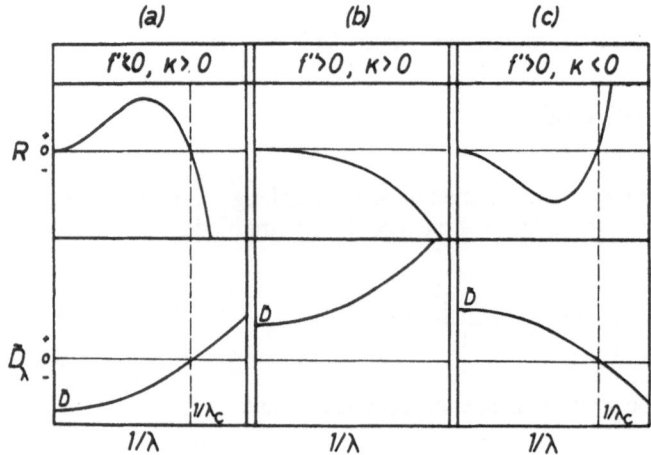

Figure 4. *Dependence of the amplification factor, R, and the interdiffusivity, Dλ, on the modulation repeat length for a phase separating system inside the spinodal (a), and outside the spinodal (b), and for an ordering system (c).*

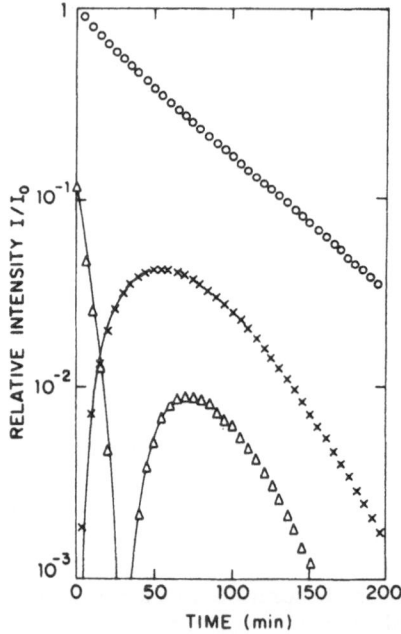

Figure 5. *Relative intensity versus annealing time for the first (circles), second (crosses) and third (triangles) Fourier components of the composition profile of a Ag/Au multilayer with an initial square symmetrical profile, calculated by de Fontaine [14].*

It should be kept in mind, of course, that the continuum theory discussed so far is expected to break down for repeat lengths on the order of the interatomic distance. Cook, de Fontaine and Hilliard [12] have formulated a discrete version of the theory that can be used in this regime as well.

NON-LINEAR DIFFUSION

If the interdiffusivity is a function of composition, the diffusion equation becomes non-linear. In linear diffusion , the amplitudes of the various Fourier components of the composition modulation decay independently according to equation (27). In the non-linear case, the Fourier modes are coupled, and there can be strong deviations from the exponential decay of equation (27), in particular for the higher harmonics. This is illustrated by Figure 5, which shows the decay of the intensity of the first three harmonics of a Ag/Au multilayer with an initial symmetrical square composition profile [14]. Note that the first harmonic still decays nearly exponentially. This is very different for the higher harmonics; the second harmonic, for example, would remain absent in a linear diffusion process.

For the diffusion equation (25) to be linear, it had to be assumed that the quantities M, f " and κ were independent of composition. This is expected to hold if the amplitude of the composition modulation is sufficiently small. A composition dependence of the mobility M, for example, may result from a strong dependence of the vacancy concentration on the composition. This is the case for interdiffusion of Si/Ge multilayers, where the vacancy concentration in Si is many orders of magnitude lower than in Ge at the same temperature [13].

The parabolic composition dependence of the free energy on the composition is also expected to break down at sufficiently large modulation amplitudes, especially in phase-separating systems. The simplest correction that gives the phase separation is

$$f = f_0 - \frac{1}{2} A w^2 + \frac{1}{4} B w^4 \qquad (30)$$

where f_0 is the free energy at the avarage composition c_0, and $w=c-c_0$. Figure 6 is a schematic illustration of f(c). The compositions of the phases in equilibrium with each other are given by the common tangent construction, or, in this case, f'=0. This gives: $w_{eq}= \pm \sqrt{(A/B)}$. The spinodal points are given by f''=0, which gives $w_s= \pm\sqrt{(A/3B)}$. Note that $f''(c_0)=-A$ and $f^{iv}(c_0)=6B$.

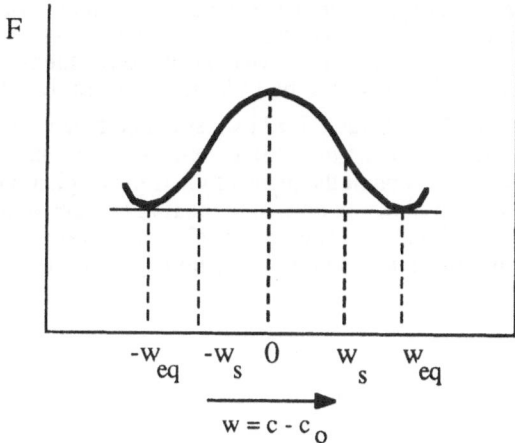

Figure 6. *Schematic diagram of the non-parabolic free energy used in the analysis of non-linear diffusion.*

To solve the non-linear diffusion equation, it is customary to expand the diffusion coefficient as a Taylor series of the composition [14,15]. Ignoring the gradient energy effects for the moment, the non-linear diffusion equation then becomes

$$\frac{\partial w}{\partial t} = \frac{\partial}{\partial x} [(D(c_0) + D'(c_0) w + \frac{1}{2} D''(c_0) w^2 + ...) \frac{\partial w}{\partial x}]$$ (31)

The coefficients of the expansion are proportional, respectively, to $f''(c_0)$, $f'''(c_0)$ and $f^{iv}(c_0)$ (see equation (24)). They can be determined from the analysis of non-linear diffusion data similar to the simulated ones of Figure 5, and can in turn be used to determine, for example, the width of the spinodal, $2w_s$, in systems such as Cu-Ni, where phase separation is difficult to observe due to slow kinetics [16].

Solving equation (31) gives for the the decay of the amplitude of the first harmonic, A_1, of the composition modulation

$$\frac{dA_1}{\partial t} = - D(c_0) \beta^2 A_1 + \frac{D''(c_0) \beta^3}{6\pi} \int w^3 \cos\beta x \ dx$$ (32)

If the modulation is sufficiently small and the dependence of the interdiffusivity on composition sufficiently smooth, so that

$$\frac{D''(c_0)}{3 D(c_0)} |w^3(x)|_{max} \ll 1$$ (33)

equation (31) gives exponential decay of the first harmonic, corresponding to an interdiffusivity $D(c_0)$. It is seen on the example of Figure 5 that the plot of ln I vs t is indeed straight for the first harmonic. It can be used to determine the interdiffusivity corrponding to the *average composition* of the film. By producing multilayers with different relative thicknesses of the two components, it is possible to determine D(c) over an entire composition range, as Cook and Hilliard first demonstrated for the Ag-Au system [17].

STRAIN EFFECTS

If the lattice parameter of the crystal that makes up the continuous structure of a multilayer depends on the composition, the preservation of lattice coherency throughout the film introduces coherency strains. These make a positive contribution to the free energy, and hence favor homogenization. This was first analyzed by Cahn for an isotropic solid [9]. The linear expansion with composition , a(c), is characterized by a strain parameter $\eta = d(\ln a)/dc$. For the composition modulation described by equation (26), the x-axis is perpendicular to the film. In a perfectly coherent crystal the distortion in the plane of the film (yz-plane) must be zero. On the other hand, the distortion in the direction normal to the film is unrestricted and there is no normal stress component in that direction. The elastic coherency strain is the difference between the total strain and the stress-free strain. Its components, for a modulation described by equation (26), are

$$\varepsilon_{xx} = A\eta \frac{2\nu}{1-\nu} \cos\beta x$$

$$\varepsilon_{yy} = \varepsilon_{zz} = - A\eta \cos\beta x$$

$$\varepsilon_{xy} = \varepsilon_{yz} = \varepsilon_{xz} = 0 \tag{34}$$

where ν is Poisson's ratio. The stress components are

$$\sigma_{xx} = 0$$

$$\sigma_{yy} = \sigma_{zz} = - A\eta \frac{E}{1-\nu} \cos\beta x$$

$$\sigma_{xy} = \sigma_{yz} = \sigma_{xz} = 0 \tag{35}$$

The total corresponding strain energy per unit volume is then

$$\frac{1}{2} \int \sigma_i \varepsilon_i \, dx = \frac{A^2 \eta^2 E}{2(1-\nu)} \tag{36}$$

For a general composition modulation around an average c_0 the coherency strain energy is found by adding the contributions from all the Fourier components. This leads to an additional, positive, term in the free energy of equation (12):

$$F = A \int [f(c) + \kappa \left(\frac{dc}{dx}\right)^2 + \eta^2 Y (c-c_0)^2] \, dx \tag{37}$$

where Y is an elastic modulus. For an isotropic solid $Y=E/(1-\nu)$. Expressions for Y for crystals have been derived as well [18]. If this equation (37) is used in the linear diffusion analysis of section 4, the relative change in amplitude becomes

$$R = \frac{\partial \ln A}{\partial t} = - D \left(1 + \frac{2\kappa\beta^2}{f''} + \frac{2Y\eta^2}{f''} \right) \tag{38}$$

It is clear that the coherency strain favors homogenization. Philofsky and Hilliard [19] first demonstrated this effect in the (ordering) Pd-Cu system. They observed a drop in the interdiffusivity at modulation repeat lengths greater than 28 Å, which corresponded to the loss of lattice coherence.

Stephenson [19,20] has recently extended the theory of strain effects on diffusion to take into account the possibility of strain relaxation by viscous flow, for example in glasses or by Nabarro-Herring creep in crystals. The effects he describes can be especially important at the small distance scales in the artificial multilayers.

STABILITY OF NON-CONTINUOUS STRUCTURES

The above sections apply to multilayers with continuous structure in which the free energy as a function of composition can be described by a single curve. If the structure is not

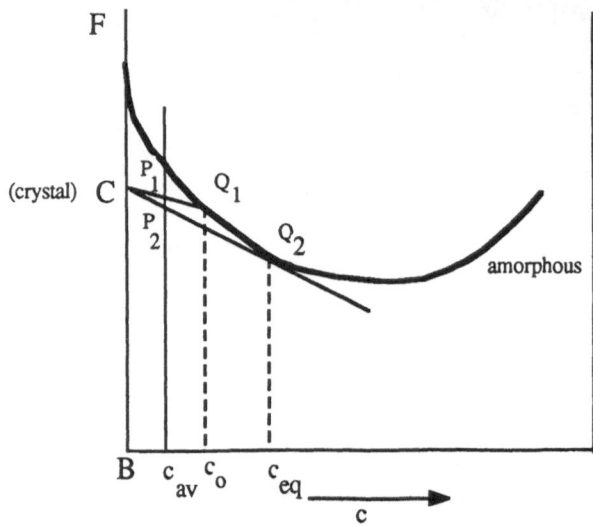

Figure 7: *Schematic free energy diagrams as a function of composition for the two phases in a crystalline/amorphous multilayer.*

continuous, free energy curves for both structures must be considered. As an example, we consider here the case of a pure B crystal (with zero solubility of A; c=0) layered with an amorphous A-B alloy of composition c_0. The average film composition is c_{av}. The free energies are represented on Figure 7. Because of the zero solubility, the free energy "curve" for the crystal is just a point (C). The average free energy per atom in the film is given by point P_1. It is clear that, if the interfaces remain parallel, the total free energy of the system can only be lowered if the amorphous phase becomes richer in A (increasing c). The lowest total free energy is reached at point P_2, when the amorphous phase has a composition c_{eq}, defined by the tangent drawn from the crystal point to the the amorphous free energy curve.

Since the average composition of the film obviously remains the same, this lowering of the free energy can only occur by a change in the relative thicknesses of the two types of layers. Originally the crystalline/amorphous thickness ratio is P_1Q_1/P_1C (line lengths); at equilibrium it is P_2Q_2/P_2C. (The atomic volumes in the two phases are assumed to be the same). The lowering of the free energy therefore happens by *crystallization*. Figure 8 shows schematically the evolution of one repeat length of an initially square composition profile. If the crystallization kinetics are much slower than the diffusion kinetics, the composition profile within the amorphous layer remains flat.

COARSENING

If the surface tension (interfacial free energy) between its layers is positive, the free energy of a multilayer is lowered by a decrease in the interfacial area. In non-continuous structures, the topological mismatch across the interfaces makes them similar to grain boundaries (or crystal/melt interfaces), which are known to have a positive surface tension. In continuous structures, a diffuse interface is formed, the surface tension of which arises from the gradient energy effects discussed in section 3. In phase separating systems, where the gradient energy coefficient, κ, is positive, the surface tension is positive as well. Cahn and Hilliard [8] showed that it equals

$$\sigma = \frac{f''^2}{f^{iv}} \sqrt{\frac{\kappa}{|f''|}} \tag{39}$$

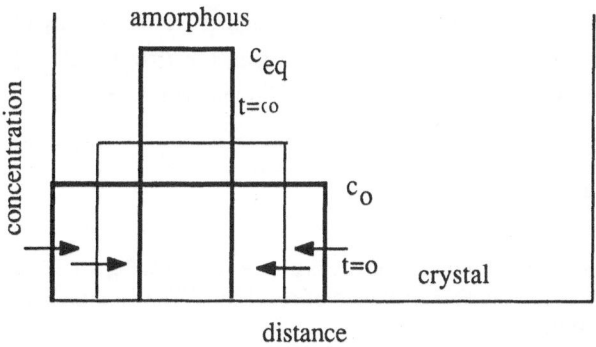

Figure 8. *Schematic diagram of the evolution of an initially square composition profile in the amorphous/crystalline multilayer corresponding to Figure 7. The arrows indicate the direction of interface motion (crystallization).*

Since in order to achieve stable two-phase equilibrium f^{iv} must be positive (see equation (30) and Figure 6), this is a positive quantity.

There are several possible mechanisms by which the decrease in the total interfacial area, or coarsening, can occur. One of the most likely ones is that observed in the coarsening of lamellar eutectics [22], which involves the motion of fault lines (i.e., edges of half-planes), as illustrated in Figure 9. The half plane recedes by diffusion of atoms from its tip to the adjacent layers, which thicken. Diffusion only occurs in the region indicated by the arrows, where there is a difference in local curvature between the receding and thickening lamellae. The coarsening rate can be calculated simply in the following way.

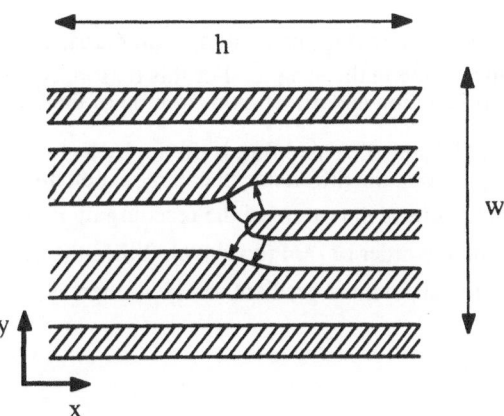

Figure 9. *Schematic diagram of coarsening in a multilayer structure by the motion of a receding fault line (the edge of a half-plane). The arrows show the direction of diffusion to the adjacent thickening layers.*

Consider a layered structure with an average repeat length λ and individual layer thicknesses of $\lambda/2$. If the tip of the fault line of Figure 5 recedes by dx, the total volume that must diffuse through the other layer is $(\lambda/2)$ L dx, where L is the length of the fault line (here taken along the z-direction). The total volume diffusing across the half-cylindrical interface at the receding tip in a time dt as a result of a diffusive flux J is $(\pi/4)L\lambda V_m$ Jdt, where V_m is the molar volume. The velocity of the moving fault line is then

$$v_f = \frac{\pi}{2} J V_m \tag{40}$$

The *average* repeat length, λ, in the multilayer is also equal to w/N, where w is the thickness of the multilayer and N is the number of layers of one kind. The change in average repeat length per unit time is then

$$\frac{d\lambda}{dt} = -\frac{\lambda^2}{w}\frac{dN}{dt} \tag{41}$$

The time it takes for one layer to disappear entirely is h/v_f, where h is the length of the sample along the direction of motion of the fault line. If we now imagine that we have N_f parallel fault lines, all moving at the same velocity, the total change in the number of layers per unit time is

$$\frac{dN}{dt} = -N_f \frac{v_f}{h} \tag{42}$$

Combining equations (41) and (42) then gives

$$\frac{d\lambda}{dt} = \lambda^2 \rho_f v_f \tag{43}$$

where $\rho_f = N_f/wh$ is the number of fault lines penetrating a unit area, or. alternatively, the total length of fault lines per unit volume in the sample. For this reason, equation (43) is generally valid, independently of the orientation and direction of motion of the lines.

As shown by equation (21), the diffusive flux is driven by a gradient in the chemical potentials. In a coarsening problem, the difference in chemical potentials arrives from a difference in curvature. The radius of curvature of the receding tip is on the order of $\lambda/4$; that of the receiving lamellae is on the order of $-\lambda/4$ (concave). Since the diffusion distance can be taken as $\lambda/2$, the driving chemical potential gradient can be written as

$$\nabla\mu = C \frac{2\sigma}{\lambda^2} \tag{44}$$

where C is a geometrical factor on the order of 30. The diffusion equation then becomes

$$J = M C \frac{\sigma}{\lambda^2} = C \frac{cD}{RT}\frac{\sigma}{\lambda^2} \tag{45}$$

where the standard relation between mobility and diffusivity for dilute solutions [10] is used. Combining equation (40), (43) and (45) then gives the final expression for the coarsening rate

$$\frac{d\lambda}{dt} = C' \frac{cD}{RT} \rho_f \sigma V_m \tag{46}$$

where C' is another geometrical factor on the order of 50. Note that the coarsening rate is independent of the repeat length. As expected, a high solubility, diffusivity, surface tension and fault line density are all unfavorable factors if one wants to prevent the layer structure from breaking down. Pinholes in the layers may be especially harmful, since they are sources of circular fault lines.

APPLICATION: INTERDIFFUSION IN AMORPHOUS MATERIALS

The artificial multilayer technique is especially well suited for study of diffusion in amorphous materials for two reasons. (i) It is the most sensitive method available (down to 10^{-27} m^2/s). Amorphous metals, for example, crystallize rapidly if their diffusivity exceeds 10^{-19} m^2/s, a sensitive technique is invaluable for obtaining data over a reasonable temperature range. (ii) It is non-destructive, and therefore allows measurements of the interdiffusivity as a function of time. Such measurements are crucial because structural relaxation continuously and strongly affects the atomic transport properties of amorphous materials.

The sensitivity of the artificial multilayer technique is illustrated on Figure 10. Conventional diffusion measurements are made on a single diffusion couple. The resolution of these techniques depends on the spatial resolution of the slicing technique that is used to analyze the composition profile. Mechanical slicing, such as microtoming, may be pushed to a 1μm resolution, but this only gives a lower limit of 10^{-17} m^2/s. Sputter profiling, Rutherford backscattering or nuclear reaction techniques can analyze profiles with a maximum resolution of about 10 Å, corresponding to a diffusivity of 10^{-23} m^2/s. The artificial multilayer technique can go four orders of magnitude further.

Since the first measurements of interdiffusion in an amorphous multilayer (Pd$_4$Si/Fe$_4$B) by Rosenblum et al. [23] in 1980, the technique has been exploited extensively to study amorphous materials: measurements in purely covalent amorphous multilayers (Si/Ge) [24]; measurements in binary metal-metal multilayers [25]; quantitative measurements of the effect of structural relaxation on the diffusivity [26]; determination of critical repeat lengths (see equation (29)) in amorphous metallic systems [25,27] and in a covalent amorphous system [28]; determination of the relation between diffusivity and viscosity, and the observation of the breakdown of the Stokes-Einstein relation [27, 29]. This last observation has led to improved insight into the atomic mechnisms that govern flow and atomic diffusion in amorphous metals.

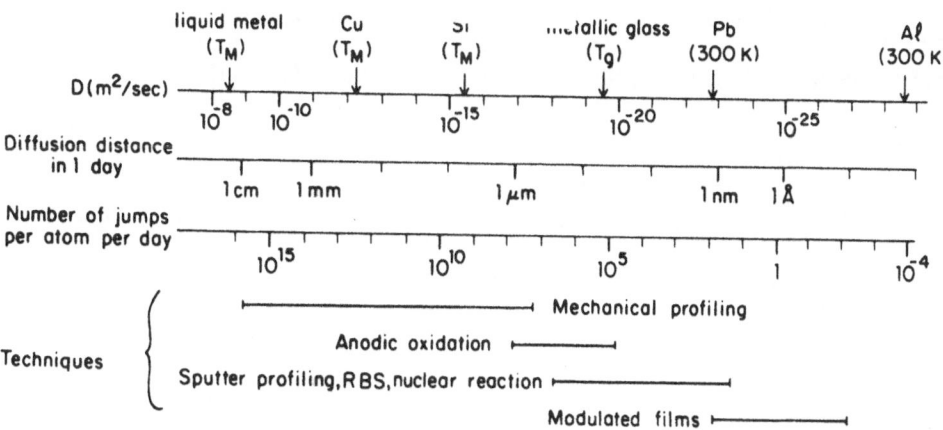

213

ACKNOWLEDGEMENTS

This paper is in many ways based on the more extensive treatment of reference [1], and it is a pleasure to acknowledge much discussion and collaboration with my co-author, Lindsay Greer. Our research in this area has been supported mostly by the Office of Naval Research, currently under contract number N00014-85-K-0023. Some support was also provided by the National Science Foundation through the Harvard Materials Research Laboratory, currently uner contract number DMR-83-16979.

REFERENCES

1. A.L. Greer and F. Spaepen, in "Synthetic Modulated Structure Materials", ed. by L.L. Chang and B.C. Giessen, Academic, NY (1985).
2. F. Spaepen, Mat. Res. Soc. Symp. Proc. **37**, 295 (1985).
3. A.L. Greer, Ann. Rev. Mat. Sci. **17**, 219 (1987).
4. Viewpoint Set on Artificially Layered Thin Films, Scripta Met. **20**, 441 (1986).
5. L.S. Darken and R.W. Gurry, "Physical Chemistry of Metals", McGraw-Hill, New York, (1953).
6. J.C. Slater, "Introduction to Chemical Physics", Dover, New York (1970)
7. A.H. Cottrell, "Theoretical Structural Metallurgy", Arnold, London (1965).
8. J.W. Cahn and J.E. Hilliard, J. Chem Phys. **28**, 258 (1958).
9 J.W. Cahn, Acta Met. **9**, 795 (1961).
10. J.E. Hilliard, in "Phase Transformations", ed. by H.I. Aaronson, ASM, Metals Park, Ohio (1970), p. 497.
11. J.W. Cahn, Trans. Met. Soc. AIME **242**, 166 (1968).
12. H.E. Cook, D. de Fontaine and J.E.Hilliard, Acta Met. **17**, 765 (1969).
13. S. M. Prokes and F. Spaepen, Mat Res. Soc. Symp. Proc. **77**, (1987) in press.
14. D. de Fontaine, Ph.D. thesis, Northwestern University, 1967.
15. T. Tsakalakos, Scripta Met. **20**, 470 (1986).
16. T. Tsakalakos, Scripta Met. **15**, 225 (1981).
17. H.E. Cook and J.E. Hilliard, J. Appl. Phys **40**, 2191 (1969).
18. J.W. Cahn, Acta Met. **10**, 179 (1962).
19. E.M. Philofsky and J.E. Hilliard, J. Appl. Phys. **40**, 2198 (1969).
20. G.B. Stephenson, J. Non-Cryst. Solids **66**, 393 (1984).
21. G.B. Stephenson, Scripta Met. **20**, 470 (1986).
22. L.D. Graham and R.W. Kraft, Trans. Met. Soc. AIME **236**, 94 (1966).
23. M.P. Rosenblum, F. Spaepen and D. Turnbull, Appl. Phys. Lett. **37**,184 (1980).
24. S.M. Prokes and F. Spaepen, Appl. Phys. Lett. **47**, 234 (1985).
25. M. Atzmon and F. Spaepen, Mat. Res. Soc. Symp. Proc. **80**, 55 (1987).
26. A.L. Greer, C.J. Lin and F. Spaepen, Proc. 4th Int. Conf. on Rapidly Quenched Metals (Japanese Institute of Metals, Sendai), eds, T. Masumoto and K. Suzuki, **I**, 567 (1981).
27. R.C. Cammarata and A.L. Greer, J. Non-Cryst. Solids **61/62**, 889 (1984).
28. S.M. Prokes and F. Spaepen, Mat. Res. Soc. Symp. Proc. **51**, 383 (1986).
29. E. Chason and T. Mizoguchi, Mat. Res. Soc. Symp. Proc. **80**, 61 (1987).

III-V SEMICONDUCTOR STRAINED-LAYER SUPERLATTICES

J-Y Marzin

Centre National d'Etudes des Telecommunications
Laboratoire de Bagneux*
196 av. Henri Ravera, 92220 Bagneux
France

ABSTRACT: The basic effects of the built-in strains on the band structure of strained-layer superlattices are described. We show how the strain state can be evaluated in the elastic deformation case, and discuss the strain relaxation. Optical properties on selected systems are presented to illustrate these strain related effects.

INTRODUCTION

For most of the III-V binary compound pairs, the difference in lattice parameters is well above 1% (see table 1). Close lattice-matching to the available substrates GaAs and InP thus strongly reduces the choice of epitaxial materials. Since the pioneering work of Frank and Van der Merwe [1-3], it is known that good quality mismatched layers, where the difference in lattice constant is accommodated elastically, can be grown. This requires that the thicknesses of the layers are small enough. For many device applications, and above all, for those based on quantum effects, the active layers are thin, so the restriction in the thicknesses is far less drastic than that on lattice-matching. This allows, for example, to use a smaller band-gap material ($In_xGa_{1-x}As$) on GaAs, and than $In_{.53}Ga_{.47}As$ on InP. This may be important in the optoelectronic field, because it enlarges the achievable operating wavelengths [4,5]. For microelectronics, larger band discontinuities lead to an increase of the channel carrier density in charge transfer devices. On the other hand, the built-in strains modify the properties of the materials: this leads to specific features of the strained-layer superlattices (SLS) or strained heterojunctions which can be turned into advantages. One can play with the relative effects of confinement and strain to taylor the valence band characteristics of the multilayered structure. To give again an example, the compressive strain experienced by a $In_{.2}Ga_{.8}As$ layer strained on GaAs entails a reduction by a factor of 3 of the in-plane effective mass [6], for the heavy holes, and so, an increase of their mobility.

In this paper, we first describe the strain state of the SLS and discuss the critical thickness. Then we show how the strain changes the material near Brillouin zone center electronic properties, for the bulk material and the heterostructures. We then present as illustrations the

TABLE 1. relative lattice mismatch (in %) for some III-V pairs

$a_2 \backslash a_1$	AlAs	InAs	GaSb	AlSb	InSb	InP
GaAs	-0.12	6.92	7.53	8.18	13.6	3.74
AlAs		6.8	7.4	8.06	13.4	3.62
InAs			0.62	1.27	6.71	-3.18
GaSb				0.65	6.09	-3.8
AlSb					5.45	-4.44
InSb						-9.9

band-edge optical properties of two such strained systems, and discuss shortly in the last part the case of highly strained systems.

II STRAIN STATE

In the first paragraph, the purely elastic situation is examined, whereas the limiting criteria for the occurrence of plastic relaxation are discussed in a second one.

II-1 ELASTICALLY STRAINED STRUCTURES

Consider a multilayered structure, schematized in Fig.1, in which each layer is homogeneous in the x,y plane. The constituents will be deformed so that, at each interface between two materials, the in-plane lattice parameters a_x and a_y are the same on both sides of the interface. We suppose that the total thickness (along Z) of the structure is much smaller than its other dimensions. The deformation in all the layers is quadratic of axis Z, so the only non-vanishing terms of the strain tensor ϵ are the diagonal ϵ_{ii} components. They are linked in each material n of unstrained lattice constant a_n to the lattice parameters $a_n^x = a_n^y$ and a_n^z of the quadratically strained unit cell by

$$\epsilon_{ii}^n = \left(\frac{a_n^i - a_n}{a_n}\right) \qquad (1)$$

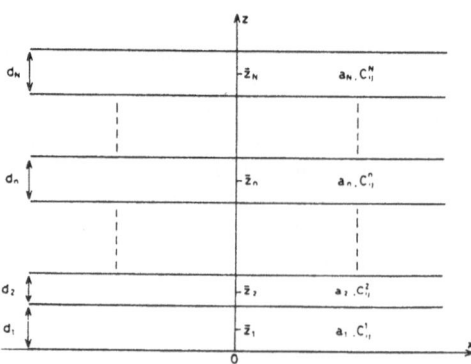

Figure 1

The stress tensor is then given by

$$\sigma_{ii} = C_{ij} \epsilon_{jj} \qquad (2)$$

where the C_{ij} are the elastic constants, which we will suppose to be the same in all materials. With the usual notations, (2) is written

$$\sigma_{xx} = C_{11} \epsilon_{xx} + C_{12} \epsilon_{yy} + C_{12} \epsilon_{zz} \quad \text{and circular}$$

permutation.

The density of elastic energy is

$$e = \frac{1}{2} \epsilon_{ii} \sigma_{ii} \qquad (3)$$

and the static equilibrium conditions read

$$\sigma_{ij,j} = 0 \qquad \text{in the volume} \qquad (4)$$

and

$$\sigma_{ij} n_j \, dS = F_i \qquad (5)$$

on the external surface dS where a force F is applied.

In our case, (4) and (5) implies that σ_{zz} is equal to zero, as no external forces are applied on the Z=cst external surfaces. Though the condition (5) is not realized in each individual layer, for the lateral edges (surfaces parallel to Z), the net resultant of the forces on these surfaces can be nulled. If we suppose that $a^{x,y}$ is constant in the whole structure, then ϵ and σ are constant in each layer and this condition gives:

$$a^{x,y} = \left[\frac{\sum\limits_{n=1}^{N} d_n a_n}{\sum\limits_{n=1}^{N} d_n} \right] = \bar{a} \qquad (6)$$

Although the resultant of the forces on the surfaces now vanishes, there is, in general, a torque which will tend to bend the sample. This can be taken into account by choosing $a^{x,y}$ of the form

$$a^{x,y} = \bar{a} + \alpha (z - z_0) \qquad (7)$$

The sample radius of curvature will then be $R = \bar{a}/\alpha$. In most epitaxial structures, this curvature radius is large enough, so that (6) applies, although bending may cause problems in the samples' processing. We neglect this effect in the following so that

$$\epsilon_{xx}^n = \epsilon_{yy}^n = \frac{\bar{a} - a_n}{a_n} \qquad (8)$$

and $\sigma_{zz} = 0$ gives

$$\epsilon^n_{zz} = - \frac{2\,C_{12}}{C_{11}}\,\epsilon^n_{xx} \qquad . \qquad (9)$$

Figure 2 shows two possible strain states for a S.L.S. In a), it is in self equilibrium: the in-plane parameter $a^{x,y}$ is the average lattice constant of the constituents, ponderated by the thicknesses. The material of larger lattice parameter is under biaxial compression and the other under biaxial tension. In b), it is strained as a whole on a thick substrate of material 2, which is one of the constituents. $\overline{a}$ is very close to a_2 and only the layers 1 are strained.

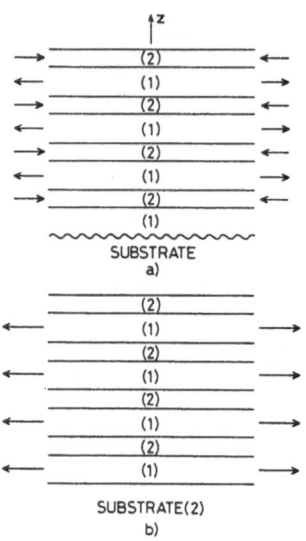

Figure 2. Two possible strain states for a superlattice
(see text)

II-2 PLASTIC DEFORMATION

When a lattice-mismatched layer is grown on a substrate, the mismatch is elastically accommodated, as described in the previous paragraph, up to a certain layer thickness d_c over which plastic strain relaxation occurs. This limit was estimated by many ways[7-9]. The calculations derive this thickness from detailed energy balance between the situations with and without mismatch dislocations. A complete analysis of this problem for S.L.S can be found in the series of papers by Matthews and Blakeslee [10]. The critical thickness can be viewed as the limit where the strain field will bend dislocation lines merging from the substrate into the interface planes, or where the existence of an array of misfit dislocations at the interface will decrease the total elastic energy. Figure 3 shows the value of the critical thickness for a single layer deposited on a thick substrate. It is roughly inversally proportional to the lattice mismatch f, independent of the elastic constants. Following this equilibrium approach, over the critical thickness the strain in the epilayer of thickness d should vary as d_c/d. In order to reach this equilibrium situation, misfit dilocation have to be created, for example from the expansion of a half dislocation loop from the free surface (Figure 4). In this process (examined in Ref. 9) there is an energetic barrier to overcome, and the elastic energy is

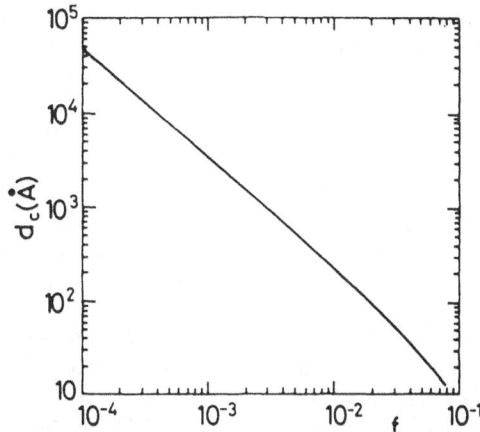

Figure 3. Critical thickness for a thin film grown on a thick
substrate as a function of their relative lattice
mismatch f

maximum for a radius of the half loop equal to R_c ($d > R_c$), which is a
function of the mismatch f and of the elastic constants. If the loop
radius R reaches about $2 R_c$, then there is a gain in energy, so
that $d_c \simeq 2 R_c$. This value is close to that given in Fig. 3, but
clearly, the creation of this defect is an activated process. The higher
the strain is, the lower the barrier $E(R_c)$. For moderate strains, one
structure for which the equilibrium should be obtained with some stress
relaxation, can still be elastically strained (metastable state). This
should explain why, in practice, the value given in Figure 3 can be
considered as underestimated.

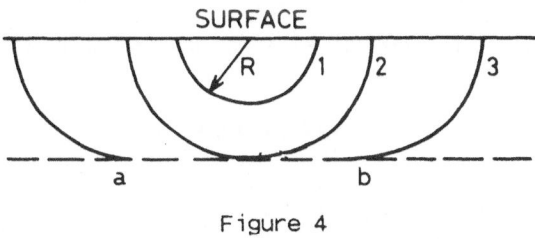

Figure 4

In the case of S.L.S, two critical thicknesses are relevant:
the first one concerns the relaxation of the superlattice, of mean
parameter $\bar{a}$, with respect to the substrate (a_s) and is given by Figure
3, where d is the total superlattice thickness and $f = (\dfrac{\bar{a} - a_s}{a_s})$. The
second concerns the intra-period relaxation by generation of defects at
the interfaces inside the superlattice. This critical thickness applies
to the thicknesses of the sublayers in one period and its value
(with $f = |\epsilon_{xx}|$)is identical to that of Figure 3, if one type of
sublayers only is strained (Fig.2 b))and to twice this value in the
situation of self-equilibrium.

The S.L.S. characteristics are selected so that no intra-period relaxation, which would result in a large number of interfacial defects, takes place. The strain state of the S.L.S. as a whole can be chosen, by a proper choice of the materials and thicknesses for the superlattice, and of the underlying buffer layer. This buffer layer can be mismatched with respect to the substrate, and thick enough to be nearly unstrained. The superlattice can then be either strained on the buffer-layer or lattice-matched to it, on the average. The last point to be emphasized is that the epitaxial layer are supposed to remain planar in this approach of plastic deformation. In fact, when the critical thickness is of the order of magnitude of the lattice parameter, the growth process can switch from bidimensional to three-dimensional: this process results in the formation of 3D islands in which the stresses can be relaxed elastically, by deformations not only in the growth direction but also in the plane of the layers. In this situation the strain fields are no longer homogeneous in the plane of the layers, and it is difficult to account quantitatively for their effects.

III STRAIN EFFECTS ON THE BAND STRUCTURE

III-1 Bulk III-V material

The quadratic deformation effects on the bulk materials have been thoroughly studied [11,12], in pressure experiments. The changes in the conduction and valence band extrema at Γ point are described by the deformation potentials a and b. The first gives the variation in the band gap due to a hydrostatic strain by:

$$E_g(\epsilon) - E_g(0) = a\, Tr(\epsilon) = a\, (\epsilon_{xx} + \epsilon_{yy} + \epsilon_{zz}) \quad . \tag{10}$$

The second one b describes the degeneracy lifting of the Γ_8 valence band imposed by the lowering of the symmetry, and the strain induced coupling of the light hole and split-off bands. With the origin of energies taken at the conduction band edge of the strained material, the valence band strain dependent Hamiltonian at k = 0, expressed in the basis $|J,m_J\rangle_z$ is

$$|3/2,+3/2\rangle_z \qquad |3/2,+1/2\rangle_z \qquad |1/2,+1/2\rangle_z$$

$$\begin{array}{ccc}
-E_g - a\, Tr(\epsilon) - b\,\delta\epsilon & 0 & 0 \\
0 & -E_g - a\, Tr(\epsilon) + b\,\delta\epsilon & -\sqrt{2}\, b'\delta\epsilon \\
0 & -\sqrt{2}\, b'\delta\epsilon & -E_g - \Delta - a'\, Tr(\epsilon)
\end{array}$$

where $\delta\epsilon = (\epsilon_{zz} - \epsilon_{xx})$ and the same expression for the negative m_J. In most cases $a' \simeq a$ (of the order of -10 eV) and $b' \simeq b$ (of the order of -2 eV). For a biaxial compression (tension), it results in an increase (decrease) of the band gap, the higher energy valence band being the heavy (light) hole band. With the quantization axis along the Z direction, the heavy hole states remain uncoupled from the others, whereas the split-off and light hole bands are coupled by the strain. In the following, we will suppose that the materials have large spin-orbits

Δ, so that the strain effects on the Γ point states are limited to the diagonal contributions. To the lowest order, and neglecting the linear k terms due to the lack of inversion symmetry of the III-V, the Hamiltonian which describes the dispersion relations in the vicinity of Γ for the valence bands is simply

$$H = H(\epsilon, k=0) + H_{Lutt}(\epsilon=0, k)$$

where H_{Lutt} is the Luttinger Hamiltonian [13] for the valence band. In our basis, for $k_x = k_y = 0$, $H_{Lutt}(\epsilon=0, k_z)$ is diagonal, and the effective masses in the Z direction are not affected by the strain. On the contrary, for a non zero in plane wave vector, H_{Lutt} contains non diagonal terms which couple the heavy and light hole states. When $\epsilon=0$, and (for example) $k_y = k_z = 0$, and $k_x \neq 0$, the eigen-states of H are of course $|J, m_J\rangle_x$. For $\epsilon \neq 0$, the effects of these non diagonal terms in H_{Lutt} are reduced due to the splitting of the Γ_8 valence band: as a result, for large strains, the in-plane effective masses are modified and given by the diagonal terms of H_{Lutt} in the $|J, m_J\rangle_z$ basis.

Taking strongly strained GaAs as an example, the effective masses are:
for the $|3/2, \bar{+}3/2\rangle_z$ states:
$m_z = .34\ m_0$ (unstrained material eff. mass) , $m_x = m_y = .11\ m_0$
and for the $|3/2, \bar{+}1/2\rangle_z$ states:
$m_z = .087\ m_0$ (unstrained material eff. mass) , $m_x = m_y = .2\ m_0$
The "heavy" hole band corresponding to $|3/2, \bar{+}3/2\rangle_z$ states at Γ point have a smaller in-plane effective mass than the "light" hole related band. This phenomenon, often qualified of mass-reversal is an interesting feature of the strained materials: for a layer under biaxial compression, the higher energy valence band holes have a low in-plane effective mass, and as a consequence, a higher mobility than in the unstrained situation. Figure 5 shows schematically the dispersion

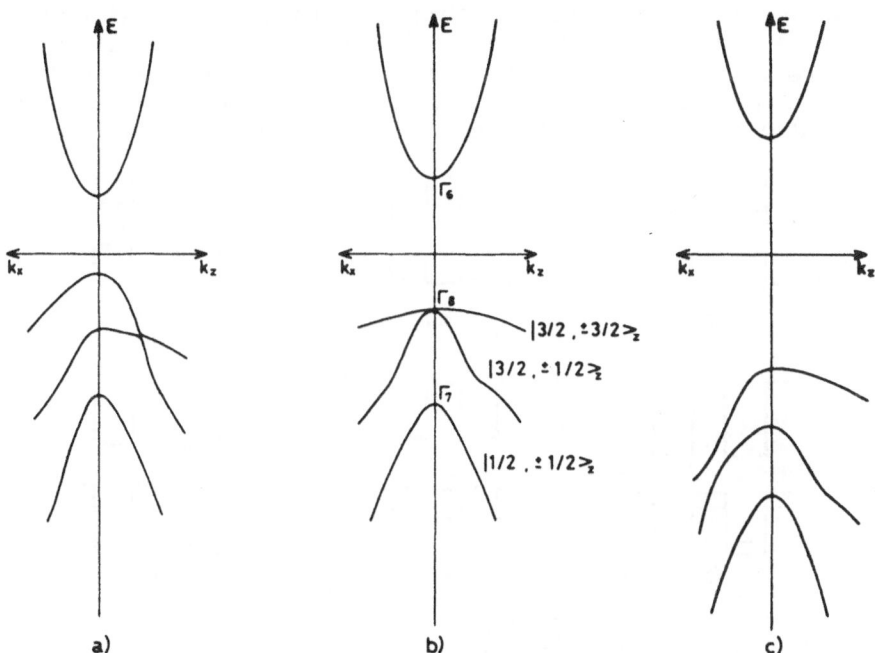

Figure 5. Schematic band structures in the vicinity of the Brillouin zone center for a bulk material under tension in (a), unstrained (b), and under compression (c)

relations for a strained III-V compound around the Γ point. The picture we have given is oversimplified: the masses along the Z direction are also modified due mainly to the change in energy of the other bands.

III-2 SUPERLATTICES

Detailed strained-layer superlattice band structure calculations are given in Ref 14 and 15, and, in this paragraph, we will discuss only schematically the specific features of strained-layer superlattices. One can note that the symmetry lowering induced by the strains in the constitutive materials is the same as that resulting from the formation of the superlattice which is quadratic of axis Z: due to this feature, the calculation of the S.L.S. band structure is very similar to that of the unstrained superlattices.

III-2-1 $k_x=k_y=0$

In the simplest approach [16], the quantized states with $k_x = k_y = 0$ for electrons, heavy and light holes are given by the solutions of three effective mass Hamiltonians:

$$H = P_z[\frac{1}{2 m_z(z)}] P_z + V(z)$$

where $m_z(z)$ is the effective mass for the considered particle in the z direction, which can be different in the two materials. $V(z)$ describes the profile, along the z direction, of the extremum of the corresponding band. The boundary conditions are the continuity of the envelope wave function F, which is the solution of the effecive mass Hamiltonian H, and of $1/m_z F$, at the interfaces.

In unstrained structures, the potential $V(z)$ is the same for the light and heavy holes, and due to this, the higher energy state for the valence band is the heavy hole first quantum level HH_1 . The strain effects modify this picture and the superpotentials for heavy and light holes are now different: this leads to a large number of possible configurations, and some of them are schematized in Figure 6.

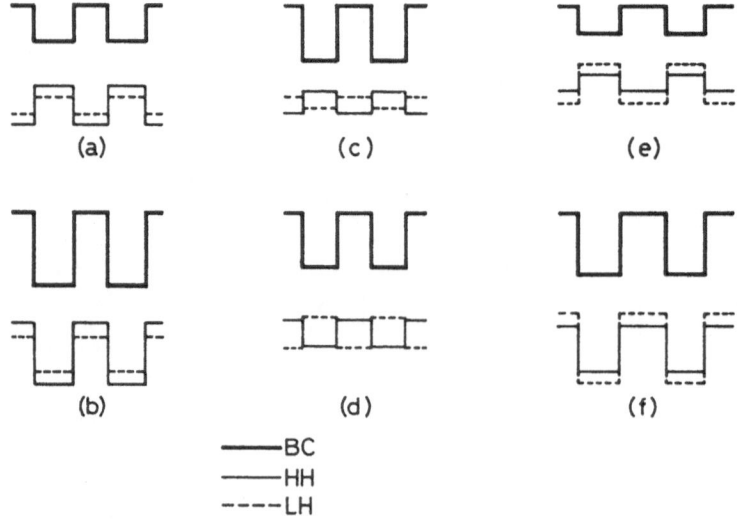

Figure 6. Some of the possible band configurations in the S.L.S.

The states for these particles may even be confined in different materials (case c or d). For some configurations, the first valence quantum level may now be LH_1 . This can be observed for example in GaSb/AlSb [17] system, where the smaller band gap material GaSb is under biaxial tension and the valence band configuration corresponds to case e. In this situation, the order in energy of the levels is determined by the competition of the confinement energies and of the strain induced splittings: for large GaSb wells thicknesses, the confinement energies are small and the first valence band level is LH_1. When the GaSb layer thickness is decreased, the confinement energy of LH_1 increases faster than the confinement energy of HH_1 (we have seen in the last paragraph that the masses along z are merely affected by the strain), and the first level is HH_1 for thin enough GaSb layers. One can play on these thickness effects (within the limits given by the critical thicknesses) to obtain a desired ordering of the levels.

III-2-2 k_x or $k_y \neq 0$

As the symmetry of the superpotential is the same as that of the strain, the heavy and light hole states HH_n and LH_m are coupled, in an unstrained superlattice. This leads to rather complicated valence band dispersion curves [18], which strongly depend on the ordering of the quantum levels, and on the energy between them, at $k_x = k_y = 0$. This situation may be simplified in some cases by the effects of strain, when they lead to a larger separation between the two first LH_1 and HH_1 levels. When it occurs, this tends to decouple the LH_1 and HH_1 bands: for the higher energy band (either HH_1 or LH_1 according to the valence band configuration), for example HH_1 , the in-plane dispersion is parabolic with a reversed mass (light in this case), as dicussed in the last paragraph; whereas LH_1, in this case, will be close in energy and strongly coupled to higher index HH_n levels. This shows, anyhow, that both strain and bidimensionality can be combined in principle, to obtain high mobility 2D hole gases, as long as the transport phenomena involve only the first hole subband.

IV OPTICAL PROPERTIES: moderately strained superlattices

IV-1 In$_x$Ga$_{1-x}$As / GaAs

This system has been thoroughly studied mainly because it allows to grow on a GaAs substrate a lower band gap material than GaAs and therefore can get interesting optoelectronic applications. Because the lower band gap material is under biaxial compression, an efficient mass reversal for the first heavy hole state occurs, and makes this system also promising for microelectronics.

Figure 7 shows the 77 K transmission spectra obtained on a series of S.L.S. On the GaAs substrate and buffer layer, 10 periods superlattices have been grown by M.B.E.. Each period consists of a 200 Å GaAs layer and a In$_{.15}$Ga$_{.85}$As layer which thickness has been varied from sample to sample from 50 to 300 Å. In this case, the superlattice is strained as a whole on the GaAs substrate: In$_x$Ga$_{1-x}$As layers only are strained and experience a biaxial compression whereas, if no plastic relaxation occurs, the GaAs layers remain unstrained. The sample design parameters have been corrected from the study of their X-ray double diffraction profiles. This technique can be shown to be specially helpful in the case of S.L.S.[19]

The transmission spectra consist in clear excitonic features merging from a staircase-like continuum which is characteristic of two-dimensional systems. The first exciton shifts to lower energy as the ternary well thickness increases as expected from the reduction of the confinement energies. Its strength indicates that the corresponding electron and hole levels are confined in the same material $In_xGa_{1-x}As$: the two possible valence band configurations are schematized in Figure 8. In both situations, the first exciton corresponds to HH_1-E_1. The position of the maximum of the photoluminescence line is also indicated in Figure 7, and corresponds to the first transition observed in transmission. The band gap of the unstrained alloy is around 1.3 eV at low temperature and the strain effects appear clearly on these spectra: for the sample with $L_z=100$ Å, for example, the sum of the electron and hole confinement energies is much smaller than the 70 meV blue shift of the absorption edge of the superlattice, with respect to the bulk material. Part of this shift (50 meV) is due to the strained induced increase of the heavy hole band gap of $In_xGa_{1-x}As$ (1.35 eV). Contrarily to what would be observed in GaAs-$Ga_xAl_{1-x}As$ structures with the same sublayers thicknesses (100 Å well), the first exciton HH_1-E_1 is not followed (some 10 to 20 meV higher in energy) immediately by the LH_1-E_1

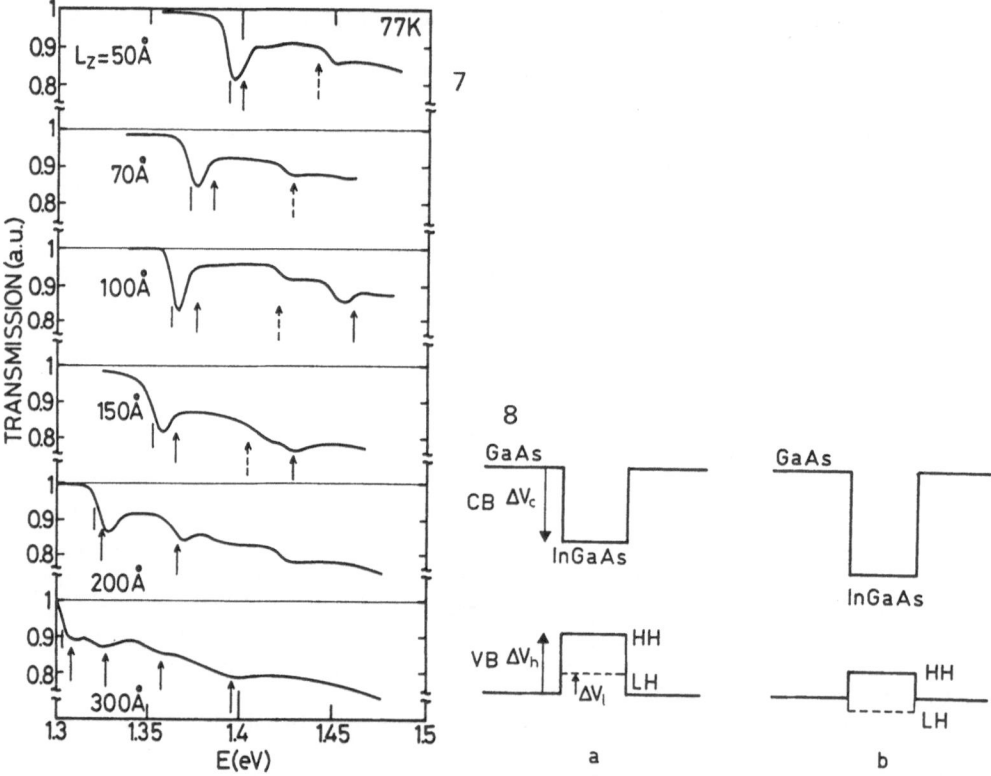

Figure 7. Transmission spectra for a series of samples. The vertical bars are associated with the P.L. peaks The arrows are the calculated transitions energies for the heavy (full lines) and light holes (dashed)

Figure 8. Possible valence bands configurations

exciton: this is due to the splitting of $In_xGa_{1-x}As$ valence band which tends to increase the energy separation between these two levels. On-edge photoluminescence excitation allows to assign the observed transitions from their selection rules [20], and these assignements are indicated in Figure 7. The conduction band discontinuity between the two constitutive materials can then be determined by fitting the energies of these transitions [20]. This has been done on samples with L_z smaller than 125 Å and leads to the conclusion that the valence band configuration corresponds to Figure 8 b), which is specific of strained systems. The light hole transitions are nevertheless observed because of the small height (15 meV) of the potential barriers which tend to confine the light holes levels in the GaAs layers. The arrows in Figure 8 show the transition calculated energies.

The spectra of samples with L_z greater than 150 Å illustrate the effects of a stress relaxation between the superlattice and the substrate. For smaller thicknesses, the superlattice is "well" strained on the substrate, as confirmed by X-ray diffraction analysis whereas, for $L_z > 150$ Å, noticeable plastic deformation takes place. Arrays of misfit dislocations, which can be shown by X-ray topography on bevelled structures [21] to be located at the first interfaces of the superlattice, relax part of the lattice mismatch between the substrate or buffer layer and the superlattice as a whole. The S.L.S then approaches its self equilibrium strain state. The experimental evidence for such a phenomenon is observed in the samples with $L_z = 200$ or 300 Å, where some transitions appear at lower energy than the $In_{.15}Ga_{.85}As$ strained band gap. In Figure 7, we have indicated the calculated heavy hole transitions for these two samples, assuming a self-equilibrium strain state, whereas in the sample with $L_z = 125$ Å, one third of the stresses between the substrate and the superlattice have been supposed to be plastically relaxed.

IV-2 $In_xGa_{1-x}As-In_yAl_{1-y}As$ on InP

Figure 9 illustrates the interest of this system: S.L.S. can be built unstrained as a whole on the InP substrate and the strain state of the smaller band gap $In_xGa_{1-x}As$ can be either a biaxial compression (for $x > 0.53$) or a biaxial tension (for $x < 0.53$). In the first case, these S.L.S lower the achievable band gap on InP and we will focus on the latter strain situation. Figure 10 shows the low temperature transmission and photoluminescence spectra of two samples with $x<0.53$. The samples are M.B.E. grown on InP substrates and lattice matched $In_xGa_yAl_{1-x-y}As$ quaternary alloy buffer layers. In these two samples the superlattice has 10 periods of $In_{.42}Ga_{.58}As$ (thickness L_x) and $In_{.59}Al_{.41}As$ (thicness L_y). The first excitonic transition corresponds now to LH_1-E_1, and this situation is again specific of S.L.S optical properties. Reducing the dimensions of the wells and barriers while keeping the same alloy compositions results in the reduction of the separation between LH_1-E_1 and HH_1-E_1, as dicussed in part III, the strain and confinement effects being in this case opposite. The calculated transition energies are indicated in Figure 10: they are not sensitive enough to the band offsets to determine precisely the discontinuities in this case.

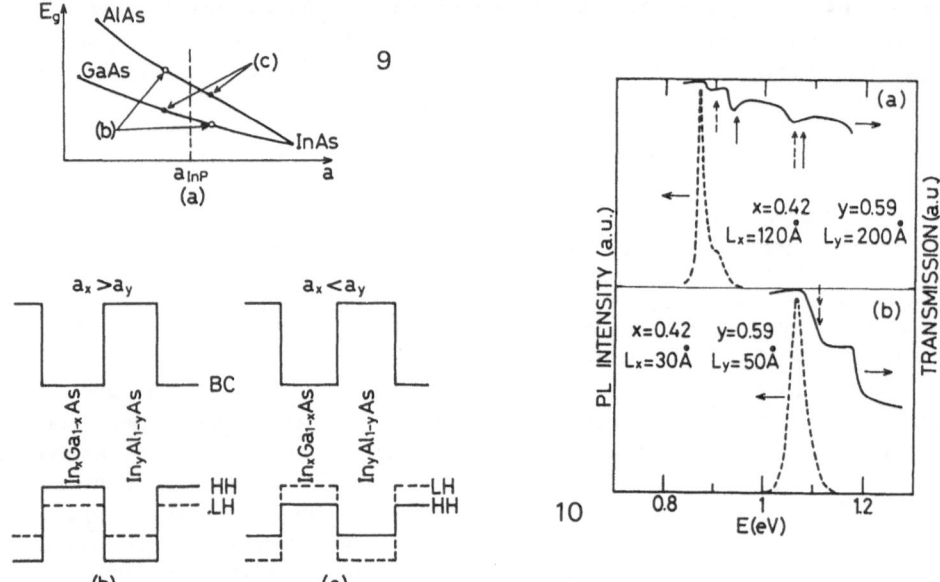

Figure 9. The possible choices for the alloys $In_xGa_{1-x}As$ and $In_yAl_{1-y}As$ used to build the superlattices (a) and the corresponding valence band configurations (b,c)

Figure 10. Transmission (full lines) and photoluminescence (dashed lines) at 10 K for two $In_xGa_{1-x}As-In_yAl_{1-y}As$ samples. The arrows indicate the calculated transitions associated to heavy (full line) and light (dashed line) hole levels

V HIGHLY STRAINED SYSTEMS: InAs-GaAs

The lattice mismatch between InAs and GaAs is 7%. Due to this large difference in the lattice parameters, the growth process itself can be modified by the strain field: though the critical thickness (calculated for planar layers) is 20 Å, it turns out to be difficult to deposit more than 3 bidimensional monolayers of InAs on GaAs [21]. For thicker layers, the growth becomes threedimensional and 3D InAs islands are observed. If a GaAs layer is deposited on an InAs layer, when trying to make a thin InAs quantum well with GaAs barriers, In-rich clusters, embedded in GaAs, are obtained instead when 3D growth occurs.

This growth transition can be seen in situ by the appearance of the bulk spotty contribution in the RHEED pattern, and the inhomogeneous strain fields due to the In-rich clusters allow to visualize them by TEM [22]. As far as optical properties are concerned, In rich clusters can trap the carriers and act as recombination centers, the corresponding emission being shifted to lower energy with respect with what is expected for a planar structure. This is illustrated in Figure 11, where the photoluminescence spectra of three structures containing InAs thin quantum wells in GaAs are shown. A few monolayers(1 to 3) of InAs were deposited to form the quantum well: in samples 1 and 2 the growth was 2D, whereas a spotty RHEED pattern was observed during the growth of InAs for sample 3: the photoluminescence line corresponding to this sample is shifted 100 meV towards lower energy and much broader

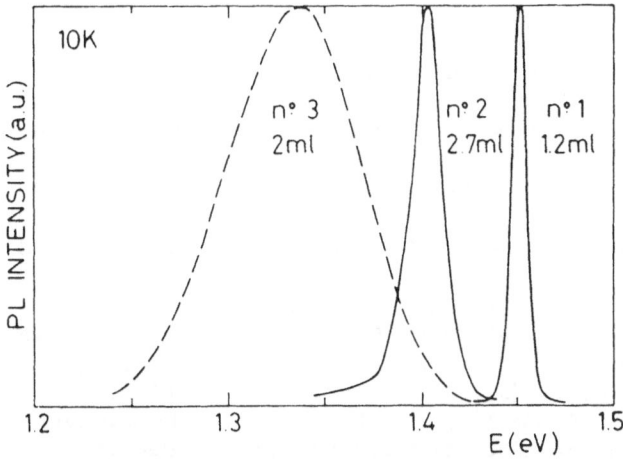

Figure 11. 10 K photoluminescence spectra of ultra-thin (3 monolayers) InAs quantum wells in GaAs. For sample 1 and 2, the growth was 2D. For sample 3, it switched to a 3D regime

than that of sample 2, though the wells in this sample are thicker. The integrated intensity for both samples remains however similar, showing that sample 2 contains very few non radiative defects: the composition inhomogeneities result mainly in an inhomogeneous strain, and the deformations are still elastic.

One important additional phenomenon occurs that confuses the situation, even when the InAs growth remains 2D: when GaAs is grown on top of InAs, In atoms tend to segregate at the surface of the epilayer and are slowly incorporated in GaAs, resulting in an exponentially decreasing In composition along the growth axis. Due to this In segregation, this interface is not abrupt whereas the reversed one formed when depositing InAs on GaAs is. The characteristic length of the In composition exponential decay have been shown [23] to be nearly independent of the growth conditions (in standard MBE growth) and of the strain state of the layers (InAs on GaAs substrate or InAs-GaAs structures on InP substrate) and is of the order of 12 Å. Although it is clear that such a mechanism cannot be neglected when very thin layers are investigated, quantitative estimates of its effects on the electronic or optical properties are still to be investigated.

VI CONCLUSION

In this paper, we have shown some of the effects of the lattice mismatches between the constituent materials of a superlattice on their optical properties. These effects can be accounted for by simple models including strain, and lead to interesting specific properties which are already used in some opto- or micro-electronic devices. Recent advances in the calculation of the band offsets [24] between such materials should allow to taylor the band structure of the superlattices, using built-in strains as an additional degree of freedom. The situation is not as clear in highly-strained systems, which growth can be perturbated by the large strains, and where the detail of the interface formation is an important feature because the layers are very thin in these structures.

ACKNOWLEDGMENTS

The author wishes to thank J.M. Gerard for the helpful comments and discussions during the preparation of this paper.
* Bagneux Laboratory is associated to the C.N.R.S. (LA 250)

REFERENCES

1- F.C. Frank and J.H. Van der Merwe, Proc. Roy. Soc. (London) A $\underline{198}$, 216 (1949)
2- J.H. Van der Merwe, J. Appl. Phys. $\underline{34}$, 117 (1963)
3- J.H. Van der Merwe in Single Crystal Films (Pergamon, New York), p. 139 (1964)
4- G. C. Osbourn, J. Vac. Sci. Tecnol. $\underline{21}$, 459 (1982)
5- M. Quillec, J.Y. Marzin, J.L. Benchimol, J. Primot and G. Le Roux, S.P.I.E. Vol 587 "Optical fiber sources and detectors", p. 62 (1985)
6- J.E. Schirber,I.J. Fritz and L.R. Dawson, Appl. Phys. Lett. $\underline{46}$, 187 (1985)
7- Van der Merwe, J. Appl. Phys. $\underline{41}$, 4725 (1970)
8- Van der Merwe, Thin Solid Films $\underline{74}$, 129 (1980)
9- J.W. Matthews, J. Vac. Sci. Technol. $\underline{12}$, 126 (1974)
10- J.W. Matthews and A.E. Blakeslee J. Cryst. Growth $\underline{27}$ 118, $\underline{29}$ 273, $\underline{32}$ 265 (1974-1976)
11- F.H. Pollack, Surf. Sci. $\underline{37}$, 863 (1973) and references therein
12- G.L. Bir and G.E. Pikus, "Symetry and strained induced effects in semiconductors", Wiley ed., New York (1974)
13- J.M. Luttinger, Phys. Rev. $\underline{102}$, 1030 (1956)
14- J.Y. Marzin, in "Semiconductor heterojunctions and superlattices", Proc. 1985 Les Houches winter school, Springer, Berlin, p. 161 (1985)
15- J.A. Brum, These, Paris (1986)
16- G. Bastard, Phys. Rev. B $\underline{24}$, 5693 (1981)
17- P. Voisin, C. Delalande, M. Voos, L.L. Chang, A. Segmuller, C.A. Chang and L. Esaki, Phys. Rev. B $\underline{30}$, 2276 (1984)
18- M. Altarelli, U. Ekenberg and A. Fasolino, Phys. Rev. B $\underline{32}$, 5138 (1985)
19- M. Quillec, L. Goldstein, G. Le Roux, J. Burgeat and J. Primot, J. Appl. Phys. $\underline{55}$, 2904 (1984)
20- J.Y. Marzin, M.N. Charasse and B. Sermage, Phys. Rev. B $\underline{31}$, 8298 (1985)
21- M.C. Joncour, R. Mellet, M.N. Charasse and J. Burgeat, J. Cryst. Growth $\underline{75}$, 295 (1986)
22- L. Goldstein, F. Glas, J.Y. Marzin, M.N. Charasse and G. Le Roux, Appl. Phys. Lett. $\underline{47}$, 1099 (1985)
23- C. Guille, F. Houzay, J.M. Moison and F. Barthe, Surf. Sci. $\underline{189}$, 1041 (1987)
24- C.G. Van de Walle and R.M. Martin, J. Vac. Sci. Technol. B $\underline{4}$, 1055 (1986)

Si/Ge MULTILAYERED STRUCTURES

E. Kasper, H.J. Herzog, and F. Schäffler

AEG Research Center, Ulm

Sedanstreet 10, 7900 Ulm, FRG

1. INTRODUCTION

Molecular Beam Epitaxy (MBE) /1, 2/ is a straightforward and powerful tool for creating semiconductor structures with novel physical properties. Besides homoepitaxy, which provides doping profiles of almost arbitrary shape, MBE heteroepitaxy appears as an extremely promising technique that allows the single crystalline growth of semiconductor structures consisting of more than one material. Early attempts were mainly limited to lattice matched semiconductors (especially GaAs/$Ga_yAl_{1-y}As$) and led to a large variety of new physical effects (e.g. the fractional quantum Hall effect /3/) and device applications (e.g. the high electron mobility transistor (HEMT) /2/). Meanwhile, increasing effort is dedicated to the growth of heterostructures consisting of materials with a more or less pronounced mismatch between their lattice constants. This general kind of heteroepitaxy is not restricted to semiconductor/semiconductor junctions /4-6/ but has also been applied to epitaxial semiconductor/insulator /7/ and semiconductor/metal /8/ junctions. In the following, we will discuss physical properties and some device applications for the case of SiGe heterostructures. This system is of particular interest, since it will allow monolithic integration of novel SiGe-based devices together with proven Si large scale integrated circuits, which might replace many of the present hybrid devices.

2. GROWTH OF SiGe STRUCTURES BY MBE

Essential for the epitaxial growth of SiGe heterostructures are (i) low growth temperature to avoid interdiffusion, (ii) precise control of layer composition and thickness, (iii) ability of selective doping with precise doping profiles. MBE is ideally suited for all these demands, since it is a low temperature process (≤ 600 °C) with excellent control over the relatively low material fluxes (typical: Ångströms per second), which become possible by ultra-high vacuum conditions in the growth chamber. An additional advantage of MBE is the non-equilibrium growth mode, which becomes important for the strain adjustment in lattice mismatched layers (see section 3).

The technical realization of SiGe-MBE requires some modifications compared to standard III-V-MBE, such as the use of an e-beam evaporator for the Si-source, because of the low vapour pressure and the high reactivity of Si. Special doping techniques, like DSI (doping by secondary implantation) /9, 10/ and SPE (solid phase epitaxy), have to be used to overcome the problem of dopant segregation at typical growth temperatures above 500 °C. The former consists of implanting a dopant adlayer by partly ionizing the Si-flux and accelerating the Si-ions towards the substrate. Precise doping control has been achieved with DSI, especially for the donor Sb. SPE is used for p-type layers with Ga doping $> 1 \times 10^{18}$ cm^{-3}: Reducing the growth temperature to below 150 °C (for Si layers) results in an amorphous layer which can be doped to very high levels without segregation. This layer is recrystallized at $\approx$ 600 °C to allow the continuation of epitaxial growth. A detailed review of Si-MBE concepts and technical solutions is given in two recent review articles /11, 12/.

3. STRAIN ADJUSTMENT

A prominent feature of SiGe-MBE is the relatively large lattice mismatch of 4.2 % between Si and Ge. $Si_y Ge_{1-y}$ alloys allow a continuous adjustment of the lattice constant between the values of the pure constituents. Epitaxial growth of overlayers with differing lattice constant starts initially with the lattice parameter of the substrate, provided the growth temperature is low enough to suppress 3-dimensional cluster growth, and high enough for epitaxial growth. This results in a lateral compressive or tensile strain within the overlayer, depending on the sign of the mismatch. This growth mode, which accommodates a lattice mismatch elastically, is called pseudomorphic. It is limited to a critical, mismatch dependent thickness t_c. t_c varies from a few monolayers (ML) for pure Ge on Si to several thousand ML for Si-rich alloys ($y \geq 0.8$) on Si. For overlayer thicknesses exceeding t_c, the built-in strain is more and more relaxed by the creation of a misfit dislocation network, which is - in the ideal case - located at the interface.

Van der Merwe has studied the relation between t_c and lattice mismatch in an energy minimization calculation based on thermodynamic equilibrium theory /13/. Although his results reproduce the trends correctly, MBE experiments result in considerably higher values of t_c for a given mismatch. This is shown in Fig. 1, which also reveals the substantial increase of t_c with decreasing substrate temperature. The latter is a clear indication for the limited applicability of equilibrium theory for MBE growth, but it also provides an additional degree of freedom in designing strained layers.

The physical properties of mismatch accommodation by elastic lattice distortion below t_c, and by dislocation creation above t_c, can be exploited for strain adjustment in a subsequent active layer or layered structure. This is usually achieved by means of a buffer layer on the substrate, which is optimized in composition, thickness and growth temperature to reach the desired lattice constant at the interface to the active heterostructure. Two distinct concepts for the buffer layer design are widely used: (i) A thick buffer with graded composition profile, and hence gradual adjustment of the lattice constant. (ii) A relatively thin buffer of homogeneous composition which retains a certain fraction of the built-in strain (up to 30 %) /14/. Essential for both methods is the proper joice of growth conditions to keep the strain-relaxing dislocation network within the buffer layer (concept (i)), or at the buffer/substrate interface (concept (ii)). In either case, excessive densities of threading dislocations, which can penetrate into the active layers, have to be

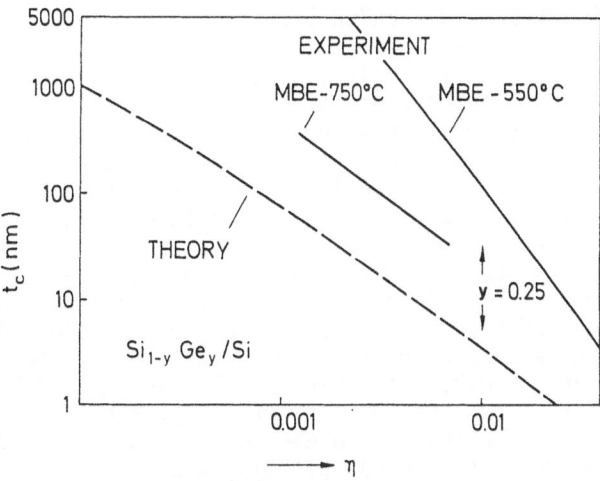

Fig. 1. Critical thickness t_c vs lattice mismatch $\eta = (a_{SiGe} - a_{Si})/a_{SiGe}$
for different growth temperatures. The theoretical curve
refers to van der Merwe's calculation /13/.

avoided. It is believed that a thin, homogeneous buffer layer (concept (ii)) is better suited for this purpose /14/. An example for a strain adjusted Si/SiGe superlattice grown on a Si-substrate with a $Si_y Ge_{1-y}$ buffer-interlayer is plotted in Fig. 2. The symmetric strain pattern shown in Fig. 2 has turned out to be of special interest and importance (see next sections).

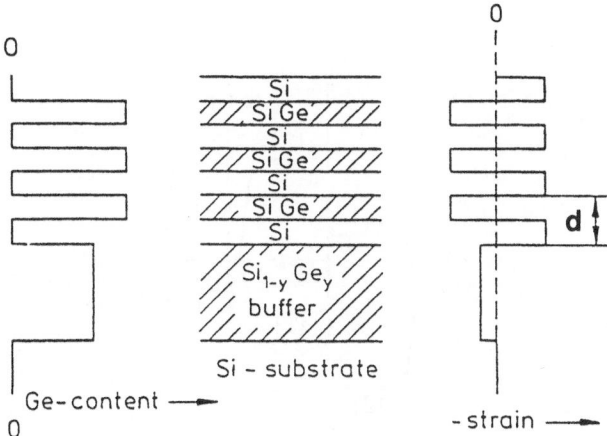

Fig. 2. Schematic, cross sectional view of a typical Si/SiGe superlattice.
The widely used situation of symmetric strain distribution is shown.

4. BAND OFFSETS IN THE STRAINED LAYER Si/SiGe SUPERLATTICE

The bandgap of Si_yGe_{1-y} is for all values of y smaller than that of Si. This leads to distinct band offsets at Si/SiGe interfaces, the values of which depend on the composition y and on the strain distribution in the layers. The composition dependence results mainly from the decreasing bandgap with decreasing y /15/, while the effect of strain is twofold: A compressive or tensile, biaxial strain in the layer is equivalent to the sum of a hydrostatic component and a uniaxial strain perpendicular to the layer, which has the opposite sign of the original biaxial value. The hydrostatic component results in a bandgap lowering /16/, while the uniaxial term leads to a reduction in the degeneracy of valence and conduction band. The latter effect is especially relevant for the 6-fold degenerate conduction band (for $y \geq 0{,}2$) which is split in 4-fold and 2-fold degenerate valleys /17/(Fig. 3). For the case of a symmetrically strained Si/$Si_{0.5}Ge_{0.5}$ superlattice, which is achieved by a $Si_{0.7}Ge_{0.3}$ buffer layer, the strain-induced conduction band splitting has been shown to result in a staggered (type II) band line-up, i.e. both valence and conduction band of the smaller-bandgap alloy lie energetically above the respective Si bands. This has been demonstrated in an elegant experiment by Jorke et al. /17, 18/ who measured the temperature dependent Hall mobility of selectively Sb doped, symmetrically strained Si/$Si_{0.5}Ge_{0.5}$ superlattices as a function of doping location: Maximal Hall mobility was found when the narrow (≈ 3 nm wide) n-type (Sb) doping spike was located in the center of the $Si_{0.5}Ge_{0.5}$ layers. This result is interpreted as evidence for electron transfer from the doped SiGe layers into adjacent Si layers, which is only possible in the case of type II band line-up. It is not clear yet, whether the strain-induced staggered line-up holds also for asymmetrically strained superlattices. Theoretical results of Van de Walle et al. /19/ predict almost flat conduction band alignment for the case of unstrained Si layers enclosed between fully strained SiGe layers, while Zeller et al. /20/ postulate type II line-up even for this case. Only few

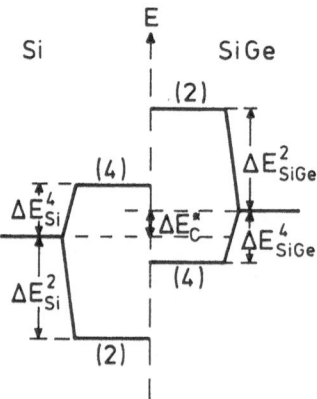

Fig. 3. Conduction band offset at a Si/SiGe heterojunction with tensily strained Si- and compressively strained SiGe layer. The strain induced conduction band splitting in 4-fold (ΔE^4) and 2-fold (ΔE^2) degenerate valleys and the virtual offset ΔEc^* for the unsplit case are shown. (After ref. 20).

experimental data are available for asymmetrically strained, n-doped samples /21/ which are not conclusive, because the doping was always located in the Si layers. This would not result in enhanced electron mobility, even in the case of a type II superlattice.

5. CARRIER MOBILITY IN MODULATION DOPED MULTILAYERS

As mentioned above, enhanced electron mobility was observed in selectively doped, symmetrically strained Si/SiGe superlattices. Fig. 4 shows temperature dependent Hall mobility curves for such samples consisting of 10 periods of $Si/Si_{0.5}Ge_{0.5}$ double-layers with a thickness of 6 nm for each layer. Sample VS 82 had the doping spike centered in the SiGe layers, which led to the highest mobilities. Lower enhancement is observed for doping close to the interfaces, but still within the SiGe layers (VS 83), while the mobility decreases with temperature when the Si layers were doped (VS 75). The latter behavior is typical for highly doped n-Si layers. The enhancement in the other two samples is attributed to the strong reduction in ionized impurity scattering, which is a consequence of the carrier transfer from the doped SiGe-layers into the undoped Si layers, i.e. of the spatial separation between mobile carriers and ionized impurity atoms. The electrons are confined to the narrow Si-layers and behave as a two-dimensional electron gas (2 DEG). This has clearly been demonstrated by Shubnikov-de Haas measurements, which showed oscillations in the magneto-resistance only if the magnetic field was applied perpendicular to the layers /17/.

A systematic investigation of p-doped Si/SiGe heterostructure has been performed by People et al. /21, 22/. In this case the carrier transfer is from the doped Si layers into the undoped SiGe layers. 2-D character and hole mobilities up to 3.3×10^3 cm^2/(Vs) at 4.2 K were reported.

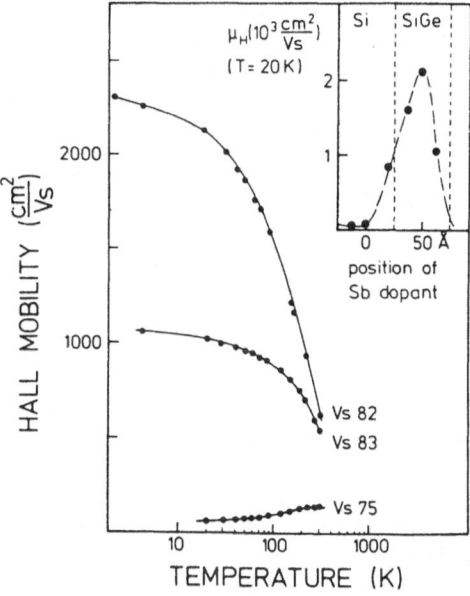

Fig. 4. Temperature dependent Hall mobility curves for various positions of the n-type doping within the superlattice (see text). The inset shows 20 K peak mobilities vs. doping location. From ref. 17.

6. ZONE FOLDING EFFECTS

Single crystalline superlattices are characterized by a periodic over-structure superimposed on the regular crystal structure. The period length d of the superlattice may vary from a few ML to several hundred ML. Solid state properties that depend on the periodicity of a crystal, like e.g. phonon dispersion curves or the electron bandstructure, can be affected by the presence of such a superstructure. A prominent example is the existence of zone-folded longitudinal acoustic (LA) phonons which can be described to first approximation by elastic continuum theory, assuming

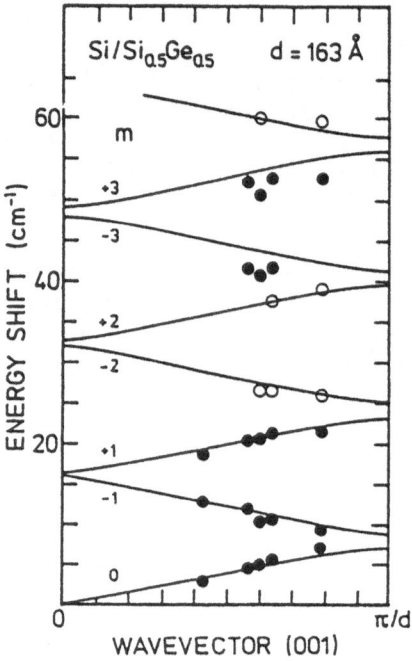

Fig. 5. Dispersion of acoustic LA phonons for an asymmetrically strained superlattice. From ref. 23.

alternating layers of different densities. Brugger et al. /23/ determined the dispersion of zone-folded LA phonons in Si/SiGe superlattices by Raman scattering. Fig. 5 shows results for an asymmetrically strained $Si/Si_{0.5}Ge_{0.5}$ superlattice with a period length d = 16.3 nm. The zone-folding results in a reduction of the Brillouin zone length in the [001] growth-direction by a factor 1/d. The agreement between the data and the simple continuum approximation is remarkably good and allows an accurate determination of the actual period length d. This can be used as an independent check for the flux calibration in the MBE facility.

More interesting for device applications are the zone-folding effects concerning the band structure and thus the electronic and optical properties of a superlattice. The most striking prediction is the conversion of the indirect gap semiconductors $Si_y Ge_{1-y}$ $(0 \leq y \leq 1)$ into quasi-direct gap semiconductors for period lengths of the order of a few ML /24, 25/. The required period of the superlattice is directly related to the position of the conduction band minimum in k-space. For the case of a Si-like conduction band structure with energy minima at about 80 % of the distance between Γ and X point, a superlattice period of d = $5a_0$ (a_0 = 5.54 Å, i.e. the lattice constant in (100) direction) would fold the Δ_1-point conduction band minimum into the Γ-point. First experimental results were reported by Pearsall et al. /26/ : In their electroreflectance measurements, which are only sensitive to direct optical transitions, they find new transitions near 0.76, 1.25 and 2.31 eV for a (pure) Si/(pure)Ge superlattice with individual layer thicknesses of 4 ML. However, no direct transitions in the 1.0 - 2.0 μm spectral range were found for similar superlattices with layer thicknesses of 1,2 and 6 ML, respectively. Hence, the new low energy transitions appear to be a unique feature of the 4 x 4 ML superlattice, that probably cannot be treated in terms of the simple description mentioned above, which assumes a Si-like bandstructure perturbed by the presence of a superstructure. Instead, an ab initio band structure calculation for the superlattice itself appears to be necessary. The technological relevance of an indirect/direct gap transition will surely trigger more detailed experiments and theoretical studies in the near future.

7. DEVICE APPLICATION

The high quality of Si/SiGe interfaces that has been reached by MBE growth, has led to a number of device applications. n- as well as p-channel modulation doped field effect transistors (MODFET) were reported /27, 28/. A schematic view of an n-channel MODFET is depicted in Fig. 6: The cross section reveals the growth sequence consisting of the high-resistive Si substrate, a 0.2 μm thick $Si_{0.7} Ge_{0.3}$ buffer layer for strain symmetrization (see section 2), the 20 nm thick, undoped, active Si layer, the selectively Sb doped $Si_{0.5} Ge_{0.5}$ layer, a graded SiGe layer to suppress a second electron channel, and an undoped Si cap that improves the properties of the Schottky gate. The ohmic source and drain contacts as well as the Schottky gate were added after mesa-isolation of the transistors, using standard techniques. Fig. 7 shows the variation of the conduction band minimum and the valence band maximum along the growth direction of the heterostructure. The 2 DEG results from charge transfer from the doped $Si_{0.5} Ge_{0.5}$ layer into the adjacent intrinsic Si-layer. Electron mobilities of more than 1500 cm^2/(Vs) at room temperature were derived from magnetotransconductance measurements. This value is somewhat higher than typical mobilities found in n-channel MOSFETs. Further improvement is expected upon optimization of the transistor design.

Another important device application is the realization of photodetectors with high sensitivity in the 1.3-1.55 μm wavelength range. Such devices would allow large-scale integration of photodetectors together with Si-VLSI readout and processing circuits on a single chip, which is expected to improve speed and cost-effectiveness as compared to conventional hybrid circuits. First successful attempts towards long-wavelength integrated optics have been reported by Luryi et al. /29/, who grew Ge p-i-n photodiodes on Si-substrates. Since the intrinsic layer has to be relatively thick (>1 μm) to achieve a reasonable efficiency in the near infrared range, pseudomorphic growth ($t_c \approx$ 1 nm) is impossible. Instead, a graded, almost completely relaxed buffer layer was grown to adjust the

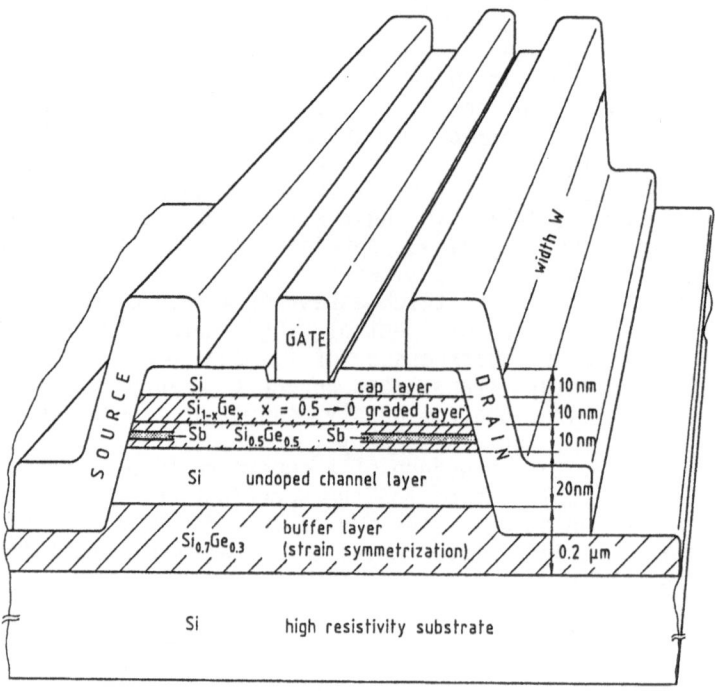

Fig. 6. Cross sectional view of a n-channel MODFET.

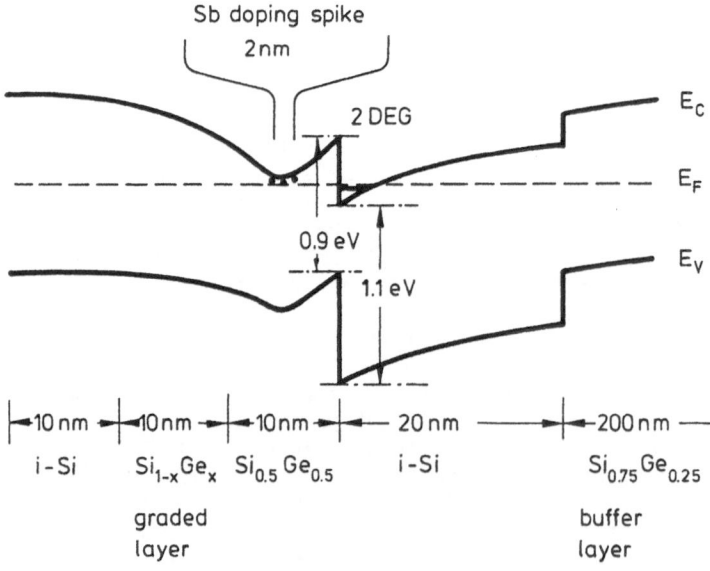

Fig. 7. Band diagram along the growth direction of the MODFET in Fig. 6.

lattice mismatch between the substrate and the Ge p-i-n-structure. Although the first test structures reached quantum efficiencies comparable to those of pure Ge photodiodes, further improvements in the buffer layer growth are necessary to reduce the density of threading dislocations, which cause leakage currents between the active layers. To overcome those problems, p-i-n-diodes have been tested which had an active layer made from a pseudomorphic, asymmetrically strained Si/SiGe superlattice /30/. It has been demonstrated that the strain-induced band gap lowering in the Si_yGe_{1-y} layers results in good photo-response at 1.3 µm for Ge-contents of > 40 %. Further improvement is expected from a combination of a strained-layer superlattice infrared absorber and a Si-avalanche diode, the latter leading to a higher gain and an improved noise figure. First teststructures showed the feasibility of that concept /31/.

The demonstrated ability to grow high-quality, strain-adjusted SiGe heterostructures allows to create novel devices with new or, compared to present solutions, superior properties. Work on several such concepts is in progress, e.g. Si-based light intensity modulators /32/, or hetero-bipolar transistors. The first steps towards a light emitting SiGe device, as mentioned in section 6, are another example.

8. CONCLUDING REMARKS

We gave a brief outline of the actual physical understanding and of present device applications and concepts of SiGe heterostructures. The basic understanding of strain effects on band line-up has provided the ability to affect electronic and optical properties by adequate strain adjustment. This basic concept has reached a high degree of realization by optimizing MBE growth procedures in order to get pseudomorphic growth on various buffer layers. The crystalline quality and the perfection of the interfaces in SiGe superlattices is already satisfactory, but additional effort is necessary to reduce the density of threading dislocations within the active layer. This is especially relevant for reverse biased optoelectronic devices, which require low leakage currents, and for devices, which are operated at extremely high power densities.

ACKNOWLEDGEMENTS

Cooperation with H. Kibbel, H. Jorke, H. Dämbkes and A. Casel are gratefully acknowledged. The project was supported in part by the Bundesministerium für Forschung und Technologie (Bonn).

REFERENCES

1) e.g. "The Technology and Physics of Molecular Beam Epitaxy" ed. E.H.C. Parker, Plenum Press, New York 1985
2) K. Ploog, "Molecular Beam Epitaxy of III-V Compounds", in "Crystals, Growth, Properties and Applications", Vol. 3, ed. H.C. Freyhardt, Springer 1980
3) H.L. Störmer, A. Chang, D.C. Tsui, J.C.M. Hwang, A.C. Gossard, and W. Wiegman, Phys. Rev. Lett. 50, 1953 (1983)
4) E. Kasper, H.J. Herzog, and H. Kibbel, Appl. Phys. 8, 199 (1975)
5) G.C. Osbourn, R.M. Biefeld, and P.L. Gourley, Appl. Phys. Lett. 41, 172 (1982)
6) M. Ludowise, W. Dietz, C. Lewis, M. Camras, N. Holonyak, B. Fuller, and M. Nixon, Appl. Phys. Lett. 42, 487 (1983)
7) H. Zogg, P. Maier, H. Melchior, J. Cryst. Growth, 80, 408 (1987) and references herein
8) S. Saitoh, M. Ishiwara, S. Furukawa, Appl. Phys. Lett, 37, 203 (1980); J.C. Bean, J.M. Poate, Appl. Phys. Lett. 37, 643 (1980)
9) H. Jorke, H.J. Herzog, and H. Kibbel, Appl. Phys. Lett. 47, 511 (1985)

10) R.A.A. Kubiak, W.Y. Leong, and E.H.C. Parker, Appl. Phys. Lett. 46, 565 (1985)

11) S. Shiraki, "Silicon Molecular Beam Deposition in the Technology and Physics of MBE", ed. E.H.C. Parker, Plenum Press, New York 1986

12) Silicon-Molecular Beam Epitaxy, eds. E. Kasper, and J.C. Bean, CRC Press, Boca Raton (USA), in press

13) J.H. Van der Merwe, Surf. Sci. 31, 198 (1978)

14) E. Kasper, H.J. Herzog, H. Jorke, and G. Abstreiter, Superlattices and Microstructures 3, 141 (1987)

15) R. Braustein, A.R. Moore, F. Herman, Phys. Rev. 109, 695 (1958)

16) R. People, Phys. Rev. B 32, 1405 (1985)

17) G. Abstreiter, H. Brugger, T. Wolf, H. Jorke and H.J. Herzog, Phys. Rev. Lett. 54, 2441 (1985)

18) H. Jorke, and H.J. Herzog, Proc. 1st Symp. on Si Molecular Beam Epitaxy ed. J.C. Bean, The Electrochemical Society, Electronics Division, Proc. Vol. 85-7, Pennington 1985, p. 352

19) C.G. Van de Walle, and R.M. Martin, Phys. Rev. B 34, 5621 (1986)

20) Ch. Zeller, and G. Abstreiter, Z. Phys. B 64, 137 (1986)

21) R. People, J.C. Bean, O.V. Lang, A.M. Sergent, H.L. Störmer, K.W. Wecht, R.T. Lynch, and K. Baldwin, Appl. Phys. Lett. 45, 1231 (1984)

22) R. People, J.C. Bean, and D.V. Lang, J. Vac. Sci. Technol. A3, 846 (1985)

23) H. Brugger, G. Abstreiter, H. Jorke, H.J. Herzog, and E. Kasper, Phys. Rev. B 33, 5928 (1986); H. Brugger, H. Reiner, G. Abstreiter, H. Jorke, H.J. Herzog, and E. Kasper, Superlattices and Micro-structures 2, 451 (1986)

24) U. Gnutzmann, and K. Clausecker, Appl. Phys. 3, 9 (1974)

25) J.A. Moriarty, and S. Krishnamurthy, J. Appl. Phys. 54, 1892 (1983)

26) T.P. Pearsall, J. Bevk, L.C. Feldman, J.M. Bonar, J.P. Mannearts, and A. Ourmazd, Phys. Rev. Lett. 58, 729 (1987)

27) H. Dämbkes, H.J. Herzog, H. Jorke, H. Kibbel, and E. Kasper, IEEE Trans. Electron Devices, ED-33, 633 (1986)

28) T.P. Pearsall, and J.C. Bean, IEEE Electr. Device Lett., EDL-7, 308 (1985)

29) S. Luryi, A. Kastalsky, and J.C. Bean, IEEE Trans. Electron Devices, ED-31, 1135 (1984)

30) H. Temkin, T.P. Pearsall, J.C. Bean, R.A. Logan, and S. Luryi, Appl. Phys. Lett. 48, 963 (1986)

31) H. Temkin, A. Antreasyan, N.A. Olsson, T.P. Pearsall, and J.C. Bean, Appl. Phys. Lett. 49, 809 (1986)

32) D.A.B. Miller, D.S. Chemla, T.C. Damen, A.C. Gossard, W. Wiegman, T.T. Wood, and C.A. Burrns, Phys. Rev. B 53, 2173 (1984)

PART III

APPLICATIONS OF MULTILAYER STRUCTURE

ELECTRONIC DEVICES USING MULTILAYER STRUCTURES

Serge Luryi

AT&T Bell Laboratories
Murray Hill
NJ 07974, USA

1. *RESONANT TUNNELING HETEROJUNCTION DEVICES*

The field of resonant tunneling in quantum-well structures is in the state of renaissance. The basic physical phenomena anticipated in such structures were qualitatively understood in the earlier period (1970's), but their experimental realization had to wait until the maturity of modern epitaxial techniques. Since the early reports, substantial progress has been achieved in the material quality of heterojunction-barrier structures grown by MBE and OMCVD techniques. The interest in such structures has risen further after the remarkable recent experiments of Sollner and coworkers who studied the microwave activity in double-barrier (DB) quantum-well (QW) diodes. These workers have demonstrated a negative differential resistance (NDR) in these diodes directly in the current-voltage characteristics at 77 K (rather than in the derivative of the current as was the case with the first reports) and obtained active oscillations from a DBQW diode mounted in a resonant cavity (Sollner et al., 1984). The material quality of DBQW diode structures has steadily improved to the point that a pronounced NDR can now be observed at room temperature. A review of resonant-tunneling and other perpendicular quantum transport phenomena in double barriers and superlattices, as well as some of their device applications, was recently given by Capasso et al. (1986) and Luryi (1987). Active research going on in many laboratories can be expected to culminate in the implementation of new and exciting devices to be used in the future high-speed electronics.

1.1 *Double-Barrier Quantum-Well Structures*

The NDR in DBQW diodes is a consequence of the dimensional confinement of states in a QW, and the conservation of energy and lateral momentum in tunneling. In addition to that, the operation of these structures has often been discussed in connection with a resonant tunneling effect analogous to that in a Fabry-Pérot resonator. That effect occurs when the energies of incident electrons in the emitter match those of unoccupied states in the QW and the wave function of resonant electrons is coherent across the entire double-barrier structure. Under such conditions, the amplitude of the de Broglie waves in the QW builds up to the extent that these waves leaking in both directions cancel the reflected waves and enhance the transmitted ones. In the absence of scattering, a system of two identical barriers is completely transparent for electrons entering at the resonant energies and the total transmission coefficient plotted against the incident energy has a number of sharp peaks, cf. Fig. 1.

Let us review the mechanism of NDR in double-barrier QW structures — without invoking a resonant Fabry-Pérot effect. This mechanism is illustrated in Fig. 2. Consider the Fermi sea of electrons in a degenerately doped emitter (the bottom figure). Assuming that the AlGaAs barrier is free of impurities and inhomogeneities, the lateral electron momentum (k_x, k_y) is conserved in tunneling. This means that for $E_C < E_0 < E_F$ (where E_C is the bottom of the conduction band in the emitter and E_0 is the bottom of the subband in the QW) tunneling is possible only for electrons whose momenta lie in a disk corresponding to $k_z = k_0$ (shaded disk in the figure), where $\hbar^2 k_0^2 / 2m = E_0 - E_C$. Only those electrons have isoenergetic states in the QW with the same k_x and k_y. This is a general feature of tunneling into a two-dimensional system of states. As the emitter-base potential rises, so does the number of electrons which can tunnel: the shaded disk moves downward to the equatorial plane of the Fermi sphere. For $k_0 = 0$ the number of tunneling electrons per unit area equals $mE_F / \pi \hbar^2$. When E_C rises above E_0, then at $T = 0$ there are no electrons in the emitter which can tunnel into the QW while conserving their lateral momentum. Therefore, one can expect an abrupt drop in the tunneling current. The effect is conceptually similar to that in the Esaki tunnel diode. Extension of this picture to the case of several subbands in the QW is straightforward. Similar arguments, of course, apply to systems of lower dimensionality, e.g., to tunneling through a "quantum wire" (Luryi and Capasso, 1985).

It is clear from the above that a similar effect should also be observable in various *single-barrier* structures in which tunneling occurs into a two-dimensional system of states. Indeed, according to the described model, in DBQW structures the removal of electrons from the QW occurs via sequential tunneling through the second barrier but other means of electron removal can also be contemplated, for example, *recombination*. Rezek et al. (1977) studied electron tunneling through a single barrier into a QW located in a *p*-type quaternary material. In these experiments the diode current resulted from the subsequent recombination of tunneling electrons with holes in the direct-gap QW. The observed structure in the dependences of the current and the intensity of the recombination radiation on the applied bias can be explained in terms of the above picture based on the momentum conservation.

The NDR effect of a similar nature can also be observed in a unipolar single-barrier structure, as was proposed by Luryi (1985b) and demonstrated by Morkoç et al. (1986). Let the emitter be separated by a thin tunneling barrier from a QW which is confined on the other side by a thin but impenetrable (for tunneling) barrier. The drain contact to the QW, located outside the emitter area, is electrically connected to a conducting layer underneath. Application of a negative bias to the emitter results in the tunneling of electrons into the QW and their subsequent drift laterally toward the drain contact. There will be no steady-state accumulation of electrons in the QW under the emitter if the drift resistance is made sufficiently small. Since the drain contact is shorted to the conducting layer underneath the collector barrier and its lateral distance from the edge of the emitter much exceeds the combined thicknesses

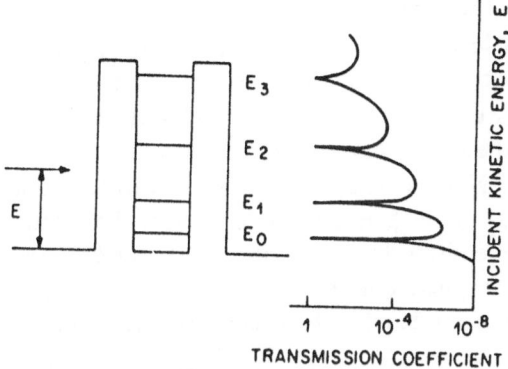

INCIDENT KINETIC ENERGY, E

TRANSMISSION COEFFICIENT

FIGURE 1. Schematic illustration of a double-barrier electron resonator. The intensity transmission coefficient plotted against the incident kinetic energy in the direction normal to the resonator layers has a number of sharp peaks. In the absence of scattering, a symmetric resonator is completely transparent for electrons entering at the resonant energies (the Fabry-Perot effect).

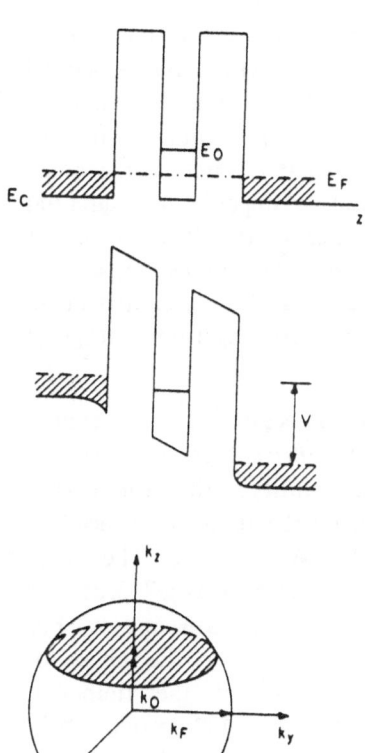

FIGURE 2. Illustration of the operation of a double-barrier resonant-tunneling diode, after Luryi (1985a). The bottom figure illustrates the Fermi surface for a degenerately doped emitter. In an ideal DBQW diode at zero temperature the resonant tunneling occurs in a voltage range during which the shaded disk moves down from the pole to the equatorial plane of the emitter Fermi sphere. At higher V, resonant electrons no longer exist, which results in a sharp drop in the current.

of the two barriers and the QW, application of a drain-emitter voltage results in nearly vertical electric field lines under the emitter, which allows one to control by the applied voltage the potential difference between the emitter and the QW. Of course, this control is much less effective (by the lever rule) than it would be if the second barrier were as thin as the tunnel barrier. Experimentally, Morkoç et al. (1986) were able to see a pronounced NDR already at room temperature and at 77 K the observed PTV ratio in current was more than 2:1. As expected, the NDR was seen only for a negative polarity of the emitter bias — with a peak current occurring at a voltage which is higher than that observed in a control symmetric DBQW structure by a factor given by the ratio of the barrier thicknesses (the lever rule).

In the presence of scattering, the sequential-tunneling mechanism of NDR should be experimentally distinguishable from the Fabry-Pérot model. However, it is not clear whether or not such a distinction can be made on the basis of *dc* current-voltage characteristics. Recently, Weil and Vinter (1987) argued against this possibility. They have pointed out that in all practical DBQW diodes, for a given lateral momentum, the energy distribution of electrons in the emitter is much wider than the Fabry-Pérot resonance peak. In this case they showed that both the sequential and the resonant-tunneling pictures lead to similar predictions for the peak current — which is limited by the transmission coefficient of the *single* least transparent barrier. Most of the experiments on the DBQW structure published to-date had concentrated on the demonstration of NDR and the microwave activity. As described above, all of these data — including dependences of the static current-voltage characteristics on the thickness of the barriers and the QW width (Tsuchiya and Sakaki, 1986), shift of these characteristics by a fixed charge stored in the barriers in the persistent photoconductivity effect (studied in the DBQW structure by Sollner et al., 1985), as well as the position and shape of the current peaks associated with tunneling into the excited states of the QW — can be adequately explained without invoking resonant transmission, so that caution is required in the interpretation of such data. It remains an open question what fraction of the diode current in a particular experiment is due to the resonant as opposed to sequential tunneling. It appears, however, that claims of an observation of the "true" resonant tunneling should be accompanied by an experimental proof. An interesting, though still inconclusive, possibility in this regard was recently discussed by Goldman et al. (1987) who studied an intrinsic bistability of DBQW diodes resulting from storage of the mobile charge in QW.

It should be clear that without the inclusion of scattering one cannot "theoretically distinguish" between the sequential and the coherent processes simply because only coherent processes are possible. The critical quantity is the ratio τ_0/τ_Θ, where τ_0 is the lifetime of the resonant state, as limited by the tunneling processes, and τ_Θ is the total phase relaxation time inside the DBQW structure. The latter includes contributions of all inelastic processes, as well as those "elastic" processes which redistribute the electron energy between the lateral directions x and y. For example, a collision of two electrons inside the QW will destroy the phase of individual particles. The Fabry-Pérot resonant enhancement of the tunneling probability is observable when $\tau_0/\tau_\Theta \lesssim 1$. In the opposite limit, $\tau_0/\tau_\Theta \gg 1$, electrons will tunnel incoherently through the intermediate states in the QW, and only the sequential process is observable.

A large body of theoretical work has been devoted to the difficult question of the time development in tunneling. As emphasized by Thornber et al. (1967), the main difficulty often is in the proper posing of the problem, resulting in physically different delay times associated with the tunneling process. For recent discussions of this problem the reader is referred to the papers by Büttiker and Landauer (1982), Büttiker (1983), Stevens (1983), and references therein.

In the instance of the resonant transmission through double barriers the problem was extensively discussed by Ricco and Azbel (1984). They pointed out that the ideal conductance with a near-unity transmission coefficient occurs only in the steady state − when the wave function inside the QW has already attained its appropriate amplitude. The establishment of such a stationary situation (say, in response to a suddenly imposed external field) must be preceded by a transient process during which the amplitude of the resonant mode is built up inside the well. The transient time is of the order of the resonant-state lifetime τ_0 and hence can be expected to increase *exponentially with the barrier phase area*. It is this transient time which gives the fundamental speed limit for the active oscillations of most practical DBQW diodes.

A simple estimate for τ_0 can be derived (Luryi, 1985a) by regarding the transient process as a modulation of charge in the "quantum capacitor" formed between the QW and the controlling electrodes. During the operation of a DBQW diode, this capacitor is being charged or discharged by a transient difference between the emitter and the collector currents. Even for a single electronic mode it is permissible to speak of a "capacitor" and its RC time constant since any change in the wave-function amplitude $|\Psi|^2$ is accompanied by a displacement current. For a single barrier of average height Φ and thickness d we find

$$\tau_0 \equiv RC = \epsilon\, \alpha^{-1}\, (\lambda / c)\, e^{4\pi d / \lambda}\,, \qquad (1)$$

where $\lambda \equiv h/\sqrt{2m\Phi}$ is the de Broglie wavelength of an electron tunneling under the barrier, $\alpha^{-1} \equiv \hbar c / e^2 \approx 137$ and c is the speed of light. The quantity $\Gamma \equiv \hbar/\tau = \Gamma_0 \exp(-4\pi d /\lambda) \equiv \Gamma_0 T_{max}$ corresponds to a homogeneous broadening of the energy levels in the QW due to electron tunneling in and out; we can interpret τ_0 as a lifetime of the resonant state limited by the tunneling processes.[†] Recently, Brown et al. (1987) reported millimeter-band oscillations at room temperature and frequencies up to 43.7 GHz from a DBQW diode with 30 Å-thick $Al_{0.3}Ga_{0.7}As$ barriers and a 45 Å-thick QW. In that structure, the effective tunneling barrier height is $\Phi \approx 0.15\,eV$, resulting in $\lambda \approx 110\,\text{Å}$. Equation (1) then gives $\tau_0 \approx 2\,ps$ and $f_{max} \approx 86\,GHz$. With sufficiently narrow barriers, DBQW oscillators are capable of operating in a subpicosecond regime.

[†] It should be noted that all WKB estimates give similar predictions. The linewidth $\Gamma = \Gamma_0 T_{max}$ is not different from $\Gamma = E_0 T_{max}$ used by Coon and Liu (1986) and Weil and Vinter (1987). The quantity E_0, which in those estimates must be understood as the zero-point-motional energy in the QW (so that E_0/h is a quasi-classical "attempt frequency") is practically close to the Γ_0 following from (1). Minor differences in the pre-exponential factor are irrelevant when T_{max} is evaluated quasi-classically. Quasi-classical predictions seem to fail when applied to DBQW structures with AlAs barriers (Sollner et al., 1987) which led these authors to question the applicability of the WKB method to estimating the escape time of an electron from a metastable state in a narrow QW.

Many workers have appreciated the various attractive possibilities which arise from the integration of a resonant-tunneling (RT) structure in a three-terminal transistor device. Bonnefoi et al. (1985a) discussed the possible integration of a DBQW structure with various analog and field-effect transistors and theoretically analyzed some of the expected characteristics. Yokoyama et al. (1985) proposed to employ a DBQW structure for injecting hot electrons in a ballistic transistor. The existence of a peak in the output current with respect to the base-emitter voltage enabled them to implement functional logic gates, such as a frequency multiplier and an exclusive-NOR gate with one transistor. Capasso and Kiehl (1985) considered the operation of a bipolar transistor with a resonant-tunneling structure (double-barrier or superlattice) built into its base. That device can be viewed as a switch, in which the input base-emitter voltage controls the transistor α, so that when the RT structure is off resonance most of the emitter current flows into the base.

Jogai and Wang (1985) discussed a DBQW device in which an independent contact is made to a heavily doped QW base. Useful operation of such a transistor requires that the base sheet resistance should be less than $100 \ \Omega /\square$, which may be nearly impossible to achieve in practice — without introducing an intolerably large loss of α due to scattering in the base (note that such a device must rely on the "true" resonant transmission, since in a sequential two-step process most of the current will flow into the base contact). It may be worthwhile to consider a variation of this device, in which the base is undoped with the base conductivity provided by a two-dimensional electron gas induced in the QW by the collector field (similar to the IBT device, discussed in Sect. 2.1.2).

Bonnefoi et al. (1985b) proposed a tunnel transistor in which tunneling from a doped emitter into a QW collector is controlled by a third "base" electrode, separated from the well by a non-tunneling barrier. The control is effected by the electric field which shifts the QW energy levels via the Stark effect. Another way of controlling the tunneling current by a third electrode was proposed by Luryi and Capasso (1985), who described a transistor in which both the emitter and the collector are 2-dimensional and resonant tunneling occurs through a 1-dimensional "quantum wire" whose potential relative to the emitter is controlled by a fringing electric field emanating from a planar gate. These tunnel transistors can be expected to exhibit ranges of both *positive and negative transconductance* — an unusual feature for a unipolar device. A negative-transconductance transistor can perform the functions of a complementary device analogous to a *p*-channel transistor in the silicon CMOS logic. A circuit formed by such a device and a conventional *n*-channel field-effect transistor can act like a low-power inverter in which a significant current flows only during switching. This feature can find applications in logic circuits.

2. HOT-ELECTRON HETEROSTRUCTURE DEVICES

As the dimensions of semiconductor devices shrink and the internal fields rise, a large fraction of carriers in the active regions of the device during its operation are in states of high kinetic energy. At a given point in space and time the velocity distribution of carriers may be narrowly peaked, in which case one speaks about "ballistic" electron packets. At other times and locations, the non-equilibrium electron ensemble can have a broad velocity distribution — usually taken to be

Maxwellian and parameterized by an effective electron temperature $T_e > T$, where T is the lattice temperature. Hot-electron phenomena have become important for the understanding of all modern semiconductor devices. Moreover, a number of devices have been proposed whose very principle is based on such effects. This group of devices will be reviewed in the present chapter.

We shall be concerned only with the hot-electron *injection* devices, i.e. such devices in which hot carries are physically transferred between adjacent semiconductor layers. Depending on the employed mode of hot-electron transport, these device structures can be classified as either *ballistic-injection* or *electron-temperature* devices. Two distinct classes of hot-electron-injection devices can be identified — depending on which of the two hot-electron regimes is essentially employed (the ballistic or the T_e regime). In the *ballistic* devices, electrons are injected into a narrow base layer at a high initial energy in the direction normal to the plane of the layer. Performance of these devices is limited by various energy-loss mechanisms in the base and by the finite probability of a reflection at the base-collector barrier. In the *electron-temperature* (RST) devices the heating electric field is applied parallel to the semiconductor layers with hot electrons then spilling over to the adjacent layers over an energy barrier. This process is quite similar to the usual thermionic emission — but at an elevated effective temperature T_e — and the carrier flux over a barrier of height Φ can be assumed proportional to $\exp(-\Phi/kT_e)$. Even though a small fraction of electrons — those in the high-energy tail of the hot-carrier distribution function — can participate in this flux, their number is replenished at a fast rate determined by the energy relaxation time, so that the injection can be very efficient.

2.1 Ballistic Injection Transistors

For a review of the early developments in the history of the ballistic hot-electron transistors the reader is referred to Heiblum (1981) or Luryi (1987). Most of the proposed devices in this group are semiconductor analogs of the well-known metal-base transistor (MBT). Experimental studies of the MBT device itself are actively pursued to this day: recent advances have been associated with the development of epitaxial techniques for the growth of monolithic single-crystal silicon-metal silicide-silicon (SMS) structures. This continued interest is explained not only by the scientific importance of the SMS structure, but also by lingering hopes to produce a transistor which is faster than the bipolar or FET devices.

The potential merits of the MBT transistor had been appraised long ago by Sze and Gummel (1966). They predicted this device would hardly ever replace the bipolar junction transistor. The problem which has plagued the metal-base transistors is their poor transfer ratio α (the common-base current gain). Even assuming an ideal monocrystalline SMS structure and extrapolating the base thickness to zero, the typical calculated values of α are unacceptably low — mainly due to the quantum-mechanical (QM) reflection of electrons at the base-collector interface. Nevertheless, there have been recent reports of a transistor action in monocrystalline $Si/CoSi_2/Si$ structures with α as high as 0.6. One cannot rule out some "accidental" resonance which aids the QM transmission of hot electrons in these devices; such an interpretation, however, appears unlikely. A more probable explanation is related to the existence of pinholes in the base metal film, i.e. continuous silicon "pipes" between the emitter and the collector. In our opinion, the thermionic emission through a permeable base (Lindmayer, 1964) has a greater device potential than the hot-electron transport through a metal base, and this silicide films may offer an attractive way of fabricating a permeable-base transistor — if one learns how to control the statistics of pinhole sizes, making it sharply peaked at a desired area scale.

The problem of QM reflections is largely avoided in the all-semiconductor ballistic hot-electron transistors. A number of such devices have been manufactured recently. Figure 3 shows the schematic energy-band diagrams of several ballistic hot-electron transistors.

2.1.1 Doped-Base Transistors

Examples of these are shown in Figs. 3a and 3b. One should understand the trade-off involved in the design of all hot-electron transistors with a doped base: cooling of hot-electrons by phonon emission and other inelastic processes (minimized by thin base layers) against the increasing base resistance for thinner layers. It is easy to estimate the RC delay associated with charging the working base-emitter capacitance and the parasitic base-collector capacitance through the lateral base resistance:

$$RC = \tau_B = \frac{\epsilon L^2}{\ell \mu \sigma} , \tag{2}$$

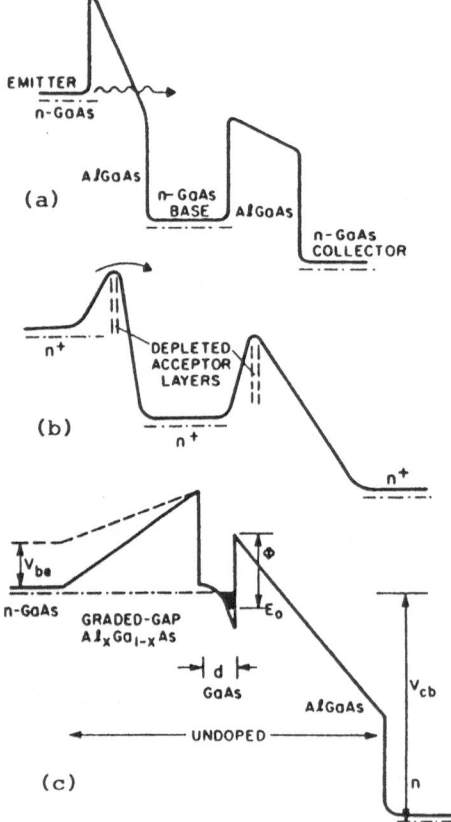

FIGURE 3. Ballistic hot-electron transistors with a monolithic all-semiconductor structure.
a) tunnel-emitter (THETA) transistor, after Heiblum (1981);
b) planar-doped barrier (PDB) transistor, after Malik et al. (1981);
c) induced-base transistor (IBT), after Luryi (1985c).

where ℓ is the thickness of the emitter or the collector barriers, $\ell \sim 10^{-5}$ cm, L the characteristic lateral base dimension (shortest distance to the base contact from the geometric center of the base), $L \sim 10^{-4}$ cm, μ the mobility in the base, σ the mobile charge density per unit base area, and ϵ the dielectric permittivity. For a hot-electron transistor to be competitive, one must have $\tau_B \approx 1$ psec, which means that the sheet resistance in the base must be $(\mu \sigma)^{-1} \lesssim 1\, k\Omega\, / \square$. On the other hand, one cannot make the base too thick, say thicker than $1\,000$ Å, since that would lead to a degradation in α due to various energy-loss mechanisms: for example, hot electrons in GaAs lose energy at the rate of about 0.16 eV/psec due to the emission of optic phonons. The limitation (2) is rather severe. The minimum value of L is governed by the lithographic resolution. One cannot really make the barrier thicknesses ℓ much larger than 1000 Å, since this would introduce the emitter and the collector delays of more than 1 psec. It may appear that increasing the base doping can resolve all the difficulties; however if it is increased much beyond 10^{18} cm^{-3}, then one can expect a strong degradation in the transfer ratio due to plasmon scattering. For $N_D \geq 10^{18}$ cm^{-3} the maximum mean free path is of order 300 Å. On the basis of such estimates, Levi et al. (1986) declared the GaAs/AlGaAs system unsuitable for the fabrication of a viable doped-base ballistic transistor. They have also briefly discussed alternative semiconductor systems, suggesting two routes for possible improvement. One can either look for material with a higher satellite-valley separation (to take advantage of the decrease in the elastic scattering rate with increasing injected-electron energy) or for a material with lower effective mass (the latter corresponds to a lower density of states and therefore also a lower scattering rate).

2.1.2 Induced Base Transistor

An attempt to circumvent (2) was made in a recent proposal by Luryi (1985c) of an induced-base transistor (IBT). In this device, illustrated in Fig. 3c, the base conductivity is provided by a 2-dimensional (2-D) electron gas induced by the collector field at an undoped heterointerface. The density of the induced charge is limited by a dielectric breakdown in the collector barrier. For a GaAs/AlGaAs system this means $\sigma/e \leq 2 \times 10^{12}$ cm^{-2}. The IBT operation requires little or no lateral electric field in the base, so the device can take a direct advantage of the high electron mobility in a two-dimensional metal at an undoped heterojunction interface. The low-field electron mobility parallel to the layers is greatly enhanced (especially at lower temperatures) because of the suppressed Coulomb scattering of electrons by ionized impurities — due to i) spatial separation from the scatterers and ii) higher than thermal electron Fermi velocity in a degenerate 2-D electron gas, which reduces the scattering cross-section in accordance with the Rutherford formula. At room temperature μ is limited by phonon scattering, $\mu \leq 8000$ cm^2/V·sec, giving $(\mu \sigma)^{-1} \approx 400\ \Omega\,/\square$ at the highest sheet concentrations in the base.† The base sheet resistance is still much lower at 77 K.

---

† The maximum breakdown-limited charge concentration in a 2-D electron gas confined at an AlGaAs barrier is not a very well established quantity, as it seems to depend on the way of material preparation. The above quoted value of $\sigma/e \approx 2 \times 10^{12}$ cm^{-2} was based on our experimental results with the charge-injection transistor fabricated by MBE (Luryi et al., 1984a). Recently, Kastalsky et al. (1986b) were able to apply voltages up to ~ 11 V across a $2\,000$ Å OMCVD-grown AlGaAs barrier — without a breakdown at room temperature — which implies, in principle, the possibility of obtaining an induced charge as high as $\sigma/e \approx 4 \times 10^{12}$ cm^{-2}.

The base conductivity in the IBT is virtually independent of its thickness down to $d \leq 100$ Å. At such short distances the loss of hot electrons in the base due to scattering is small. Indeed, injected hot electrons, traveling across the base with a ballistic velocity of order 10^8 cm/sec, lose their energy mainly through the emission of polar optic phonons. For $d = 100$ Å the attendant decrease in α is estimated to be about 1%. The IBT can be regarded as a metal-base transistor — with the notable difference that the base "metal" is two-dimensional. This permits a dramatic improvement in the transfer ratio α — mainly owing to a low quantum-mechanical reflection coefficient R at the collector barrier interface. Model calculations on the basis of Eq. (1) of the above-barrier reflection give $R < 0.02$, so that the total α can be as high as 97%. Recently, Chang et al. (1986) have manufactured a first induced-base transistor by MBE. Their preliminary results showed a differential $\alpha_{max} \approx 0.96$ (in a relatively narrow range of applied biases).

2.1.3 Satellite-Band IBT (a Speculation)

Some heterostructure combinations may offer a fascinating possibility of employing *different conduction bands* for transporting injected carriers across the base and the lateral conduction in the base. The band offsets in AlGaAs/GaAs heterojunctions at high aluminum concentrations $(x \rightarrow 1)$ are not as well established at present as those for $x \leq 0.45$. This gives me the liberty to illustrate the idea with an AlAs/GaAs structure by assuming that the band discontinuities $\Delta E_C^{(\Gamma)}$ and ΔE_V are split in the proportion 83 : 17.

Consider a device whose band diagram is illustrated in Fig. 4. Conduction-band bottom in the AlAs emitter is in the X valleys. If the above band offset rule is obeyed, then there is no discontinuity in the X band at the GaAs interface and, as far as the X electrons are concerned, the structure is not different from a homogeneous *nin* diode. In the absence of a base voltage the emitter to collector current is space-charge limited. Electrons sail through the base without noticing the quantum well

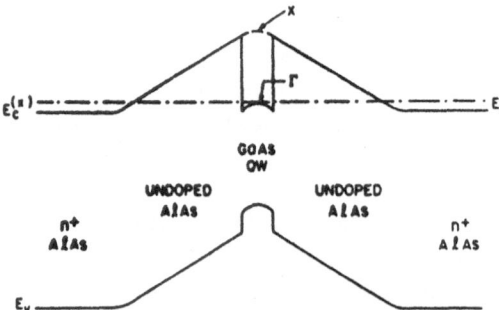

FIGURE 4. Schematic band-diagram of a satellite-band IBT. The diagram is drawn assuming a hypothetical AlAs/GaAs heterostructure in which the discontinuities in the Γ valleys of the conduction band and the valence band are assumed split in the proportion 83:17. This rule would imply a continuous X-band minimum.

(apart from a finite probability of making an intervalley transfer by spontaneous phonon emission). On the other hand, electrons in the GaAs quantum well (which may or may not be there at equilibrium and, when induced by the collector field, come from the base contact) are in the Γ valley. The collector current can be exponentially quenched by applying a base-emitter reverse bias. The device thus operates much like a bipolar junction transistor (of infinitesimal base thickness but good lateral conductivity) — with carriers in the subsidiary branch of the conduction band replacing the holes of an *npn* structure. It would be very neat indeed if a heterostructure could be found to implement this idea. Unfortunately, the $Al_xGa_{1-x}As/GaAs$ combinations would be "ruled-out", if the 60 : 40 proportion (or close to it) is found to persist for $x \rightarrow 1$, since in that case the X-band is strongly discontinuous at all values of x.

2.1.4 Speed of Ballistic Transistors

Before leaving the subject of ballistic transistors, let us briefly discuss their potential frequency performance. It is sometimes stated that hot-electron transistors are capable of subpicosecond operation because such is the time of flight of ballistic electrons across the base — a much too often repeated fallacy. The time of flight through the base has little to do with the intrinsic device speed. Like the bipolar, the FET, and most other transistors, hot-electron transistors have a regime in which their output current I rises exponentially with the input (base-emitter) voltage. In this regime, the maximum speed of operation is proportional to I. However, like every exponent in nature, this dependence eventually saturates and goes over into a linear law. One gains no further advantage in speed by increasing I, since the charge stored in all input capacitances will rise proportionally. *Ultimately, the speed of a transistor is determined by the current level at which one has a crossover between the exponential and the linear regimes.* In transistors with a thermionic emitter this crossover occurs because of the accumulation of the mobile charge diffusing up the emitter barrier and drifting down the collector barrier. A rigorous g_m/C analysis leads to the characteristic delays $\tau_E = \ell_E/v_T$ and $\tau_C = \ell_C/v_S$, where ℓ_E and ℓ_C are the thicknesses of the emitter and the collector barriers, respectively, $v_T = (kT/2\pi m)^{1/2}$ is the thermal velocity of carriers, and $v_S \sim 10^7$ cm/sec their saturated drift velocity. Of course, neither of the ℓ's can be shrunk much below, say, $1000\,\text{Å}$ — because of the complementary limitation (2). We conclude that an ideally optimized ballistic transistor will be a roughly 3 picosecond device.

2.2 Real-Space Transfer Devices

The term "real-space transfer" (RST) was coined by Hess et al. (1979) to describe a new mechanism for NDR they proposed and subsequently discovered in layered heterostructures. The original RST structure is shown in Fig. 5. In equilibrium the mobile electrons reside in undoped GaAs quantum wells and are spatially separated from their parent donors in AlGaAs layers. Guided by an analogy with the momentum-space intervalley transfer, Hess et al. (1979) suggested that carriers, heated by an electric field applied parallel to the layers, will move to the adjacent layers by thermionic emission, causing an enhancement of the mobile charge concentration in one set of layers and depletion in the other. Since the layers had different mobilities, the RST process was predicted to result in an NDR in the two-terminal circuit. If the device is used as an oscillator, electrons must cycle back and forth between the high and low mobility layers. The maximum oscillation frequency is limited by the delay due to "cold" electrons returning from the potential "pockets" in the wide-gap layers, cf. Fig. 5. For a modulation-doped AlGaAs/GaAs heterostructure at room temperature one can estimate the return time to be at least 10^{-11} sec and still longer at lower temperatures. On the other hand, the time constants involved in the initial transfer of hot electrons are considerably shorter (Hess, 1983).

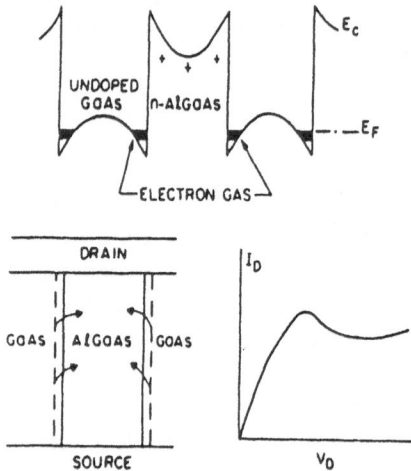

FIGURE 5. The real-space-transfer diode (after Hess et al., 1979). Electrons, heated by an applied electric field, transfer into the wide-gap layers, where their mobility is substantially lower, giving rise to an overall negative differential mobility. At low temperatures the effect may be hysteretic, since even when the electric field is removed, the transferred electrons remain trapped in the wide-gap layers.

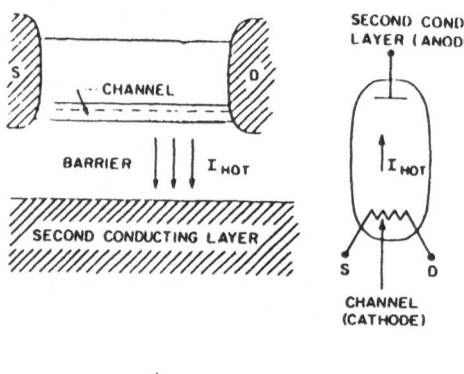

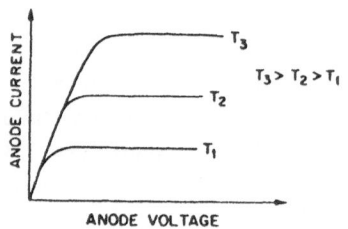

FIGURE 6. Illustration of the principle of three-terminal real-space-transfer devices (after Luryi et al., 1984a). The channel serves as a cathode whose effective electron temperature T_e is controlled by the source-to-drain field. The second conducting layer, separated by a potential barrier, serves an anode and is biased positively. To the extent that the barrier lowering by the anode field can be neglected, the anode current as a function of the anode voltage exhibits quasi-saturation at a value determined by T_e.

The idea of real-space transfer was further developed by Kastalsky and Luryi (1983) who proposed a *three-terminal* hot-electron device structure. In this structure the RST effect gives rise to charge injection between two conducting layers isolated by a potential barrier and contacted separately. The idea can be best illustrated by a glow-cathode analogy, displayed in Fig. 6. In a vacuum diode the anode current as a function of the anode voltage saturates at a value determined by the cathode work function and the temperature. One can think of a hypothetical amplifier in which an input circuit controls the cathode temperature and thus the output current, but that would be a slow device. In a three-terminal RST structure the input circuit controls the T_e which, unlike the temperature of a material, can be rapidly varied in one of the conducting layers ("the channel"), resulting in an efficient charge injection into the other layer. Based on this principle, several new device concepts were suggested, most of which by now have been demonstrated experimentally.

2.2.1 Three-Terminal RST Devices

The basic structures used for three-terminal RST devices are illustrated in Fig. 7. In the original structure the second conducting layer was implemented as a conducting GaAs substrate separated by a graded-gap AlGaAs barrier from the channel of a modulation-doped FET with source (S) and drain (D) contacts, Fig. 7a. This device had an auxiliary fourth electrode (gate) which concentrated the lateral electric field under a 1 μm wide notch. In the more recent work both the gate electrode and the modulation-doping were eliminated, Fig. 7b, and the channel was induced at the undoped heterointerface by a back-gate action of the second conducting layer. Also in the new structure the rectangular potential barrier provides a better insulation between the two conducting layers.

The charge injection transistor (CHINT) is a solid-state analog of the hypothetical vacuum diode with controlled cathode temperature, discussed above in connection with Fig. 6. Application of a voltage V_{SD} produces a lateral electric field which heats the channel electrons and leads to an exponential enhancement of charge injection over the barrier. By the physical principle involved, the operation of CHINT is different from all previous three-terminal devices — which were based either on the potential effect, i.e., the modulation of a potential barrier by an applied voltage (vacuum triode, bipolar transistor, various analog transistors), or on the field effect, which is the screening of an applied field by a variation of charge in a resistive channel. In CHINT the control of output current is effected by a modulation of the electron temperature resulting in charge injection over a barrier of fixed height. The *SUB* electrode serves as an anode and the channel as a hot-electron cathode, whose effective temperature is controlled by the source-to-drain field.

Figure 8 displays the collector characteristics of CHINT as a function of the heating voltage V_{SD} with the collector voltage V_{SUB} as a parameter. The existence of power gain in this device has been demonstrated experimentally — both in *dc* operation and at high frequencies (up to $\sim 10\,\text{GHz}$). The value of the mutual conductance g_m obtained in CHINT (over 1000 mS/mm) compares favourably to the best field-effect transistors.

As evident from Fig. 8 (dashed lines), the hot-electron injection in CHINT is accompanied by a strong NDR in the channel circuit. It occurs because part of the source current is diverted away from the drain; an important additional factor is the dynamical screening effect.[†] Device based on this NDR is called the NERFET

[†] This effect, discussed in detail by Luryi and Kastalsky (1985a), arises due to the space-charge of injected carriers drifting downhill in the collector barrier. The negative space charge dynamically stored (i.e., stored while in transit) in the AlGaAs barrier layer screens the backgate field and thus depletes the channel. The associated space-charge potential can be regarded as a threshold shift in

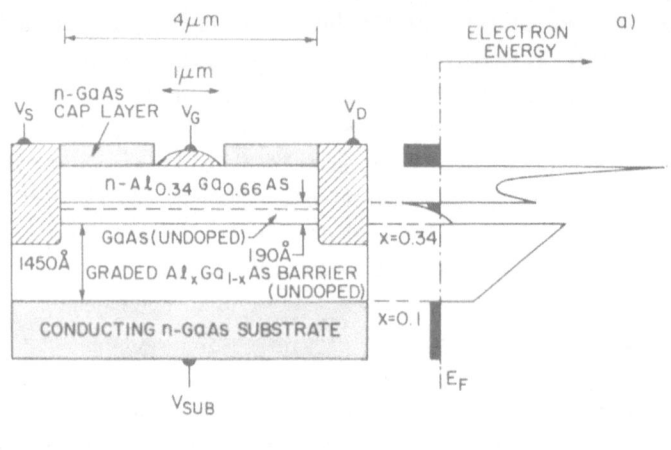

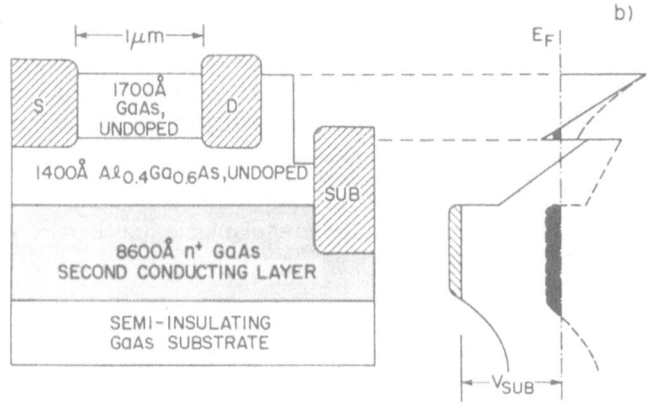

FIGURE 7. Cross-section and the energy band diagram of three-terminal RST devices.
a) Type-1 structure: a MODFET-like channel is separated from the second conducting layer (the SUB electrode) by a graded $Al_xGa_{1-x}As$ barrier. Two-dimensional electron gas is present in the channel even at $V_{SUB} = 0$ as well as with a floating SUB.
b) Type-2 structure: no electrons in the "channel" — until induced by a positive $V_{SUB} > V_T$. The barrier is ungraded, which helps a better insulation of the second conducting layer from the channel.

(negative-resistance FET). Typical NERFET characteristics are shown in Fig. 9. We note that the NDR is strongly affected by V_{SUB}. It is clear that higher V_{SUB} enhances the electron concentration in the channel (backgate action), but it also

a field-effect transistor in which V_{SUB} plays the role of a gate bias. This effect becomes important when the collector current exceeds the drain current, i. e., when the areal density of the injected mobile charge becomes comparable to that of channel electrons. The dynamical screening mechanism of the NDR is extremely fast — intrinsically limited by the time of flight of injected electrons toward the second conducting layer.

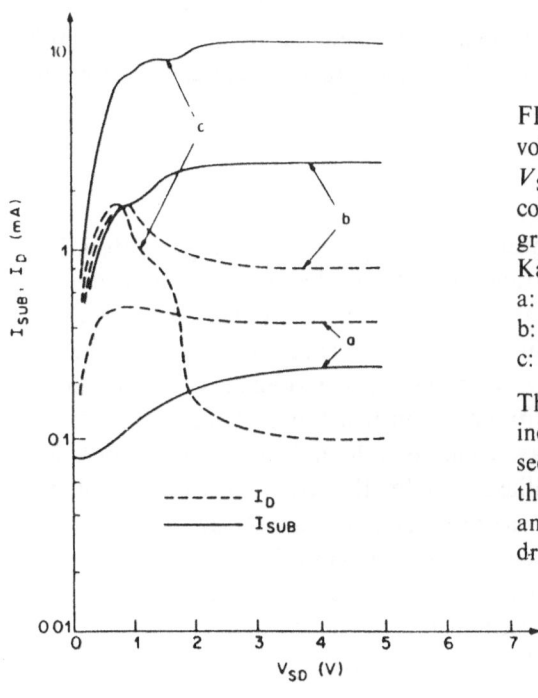

(a)

FIGURE 8. Typical experimental current-voltage characteristics, I_{SUB}-V_{SD} and I_D-V_{SD} at room temperature and different collector voltages $V_{SUB} - V_T$ in OMCVD-grown CHINT/NERFET devices (after Kastalsky et al., 1986).
a: $V_{SUB} - V_T = 1\,V$;
b: $V_{SUB} - V_T = 2\,V$;
c: $V_{SUB} - V_T = 2.7\,V$.

The threshold voltage $V_T \approx 5\,V$. With increasing heating voltage V_{SD}, one clearly sees a rapid rise and subsequent saturation of the collector current I_{SUB}, accompanied by an increase and subsequent sharp drop in the drain current I_D.

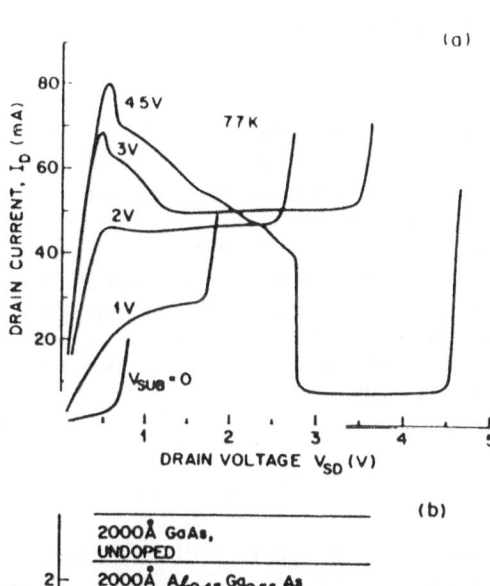

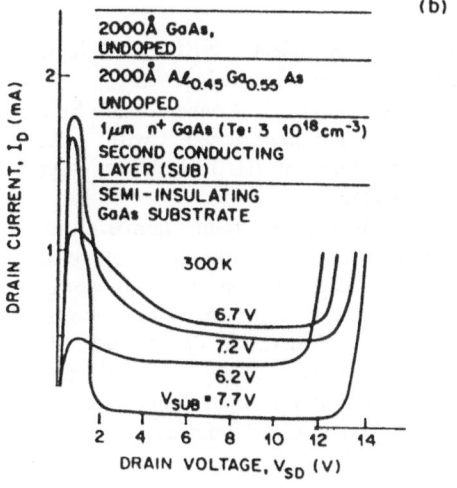

FIGURE 9. The NERFET characteristics.
a) Type-I structure at 77 K (after Kastalsky et al., 1984a). Gate dimensions: $1\,\mu m \times 250\,\mu m$;
b) Type-II structure at 300 K (Kastalsky et al., 1986b). Channel dimensions: $2\,\mu m \times 100\,\mu m$;
Note the existence of a region of a strongly pronounced *negative transconductance* $g_m \equiv (\partial I_D / \partial V_{SD})|_{V_{SUB}}$ in the quasi-saturation regions of the characteristics for both types of structures. This important effect is not yet fully understood, cf. the discussion in Sect. 2.2.4.

affects the magnitude of the hot-electron flux corresponding to a given T_e (by lowering the collector barrier). The highest peak-to-valley ratio in current obtained in these devices was 160 at room temperature.

2.2.2 Logic Applications

The main advantage of NERFET over two-terminal microwave generators lies in the possibility of controlling the oscillations by a third electrode. This advantage can also be used in logic applications, as discussed below. When two negative-resistance devices (like tunnel diodes or Gunn diodes) are connected in series and the total applied voltage V_{DD} exceeds roughly twice the critical voltage for the onset of NDR in the single device, then an instability occurs in which one of the devices takes most of the applied voltage, that is to say, contains a high-field domain, while the other is in the low-field mode. This is illustrated by the usual load-line graphical construct, Fig. 10. As is well known, the operating points A and C are stable, while B is unstable. Which of the devices contains the domain is determined by an accidental fluctuation or, if the system is prepared in one of the stable states, by the history. Various schemes have been proposed to utilize this bistability. Due to the existence of a controlling electrode, NERFET offers new possibilities for logic.

Room-temperature operation of a simplest NERFET logic circuit is illustrated in Fig. 11. Two type-2 NERFETs with nearly identical characteristics were connected

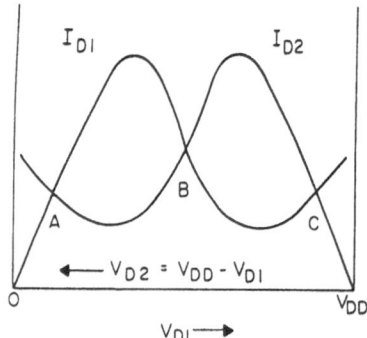

FIGURE 10. Graphical construct for determining the operating points of a circuit formed by two identical NDR elements in series. Points A and C are stable, point B unstable. For a pair of NDR diodes, the actual operating point is determined by a symmetry-breaking fluctuation. Using the third (SUB) electrode of the NERFET, it is possible to shift the operating point between A and C at will.

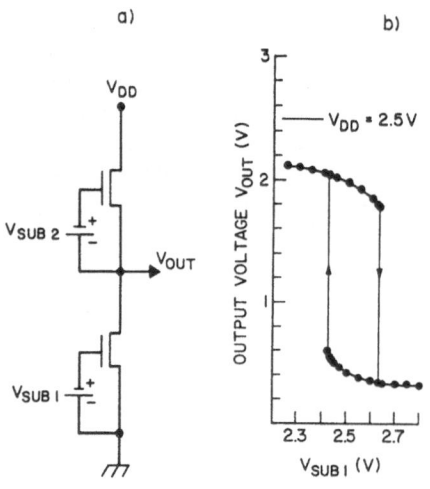

FIGURE 11. Simplest NERFET logic circuit (Kastalsky et al., 1985a).
a) Schematic diagram. Conventional FET circuit symbols are used with the understanding that the SUB electrode plays the role of a gate.
b) Logic transitions at room temperature. The output voltage was measured at fixed $V_{SUB2} = 2.5\,V$ as a function of V_{SUB1} slowly varied in the direction shown by the arrows. An interesting observation was that the switch points were sharply defined and repetitive to within less than 1 mV (i.e., to within $\ll kT/e$) in a given device pair.

256

in series, as shown in Fig. 11a. One of the controlling voltages was fixed, $V_{SUB2} = 2.5V$, and the output voltage V_{OUT} was measured as a function of V_{SUB1}, Fig. 11b. As the controlling voltage V_{SUB1} is varied, the system smoothly approaches the switch points (sharply defined and repetitive within 1 mV), at which V_{OUT} jumps between the low and the high values. Such a behaviour with a strong hysteresis is reminiscent of a phase transition.

Two types of logic operation can be thought of in this configuration. Firstly, we can dc pre-bias our input voltage to a value in the middle of the hysteretic loop, say $V_{SUB1} = 2.5V$. Applying controlling signals $\Delta V_{SUB1}(t)$ in the form of short low-amplitude ($|\Delta V| \geq 0.15V$) pulses of varying polarity, we have a *bistable element*: the system will "remember" the sign of the last pulse, viz. $V_{OUT} =$ high for $\Delta V < 0$ and $V_{OUT} =$ low for $\Delta V > 0$. A second type of logic operation — *inverter* action with amplification — can be obtained by dc pre-biasing V_{SUB1} to high enough voltages ($V_{SUB} \geq 2.7V$) to ensure a stable low state. The system will then switch to its high state only during a pulse of negative polarity $|\Delta V_{SUB1}| \geq 0.3V$. Both operations have been demonstrated using pulse-mode experiments.

2.2.3 Hot-electron Memory

Another RST logic device can be based on a memory effect which obtains in the same structure (Fig. 7a) — but with an *unbiased second conducting layer*. In this case, the hot electron injection leads to a charge accumulation in the floating layer and a drop in its electrostatic potential Ψ_{SUB}, which persists for a long time after the heating voltage V_{SD} is set to zero. The negative Ψ_{SUB} depletes the channel and leads to an NDR in the drain circuit. We should note, however, that this is a hysteretic NDR, not capable of generating oscillations. The "thermoelectric force" developed between the two conducting layers has a characteristic decay time determined by the ambient temperature and the barrier height. However, the transferred electrons remain mobile and can be rapidly discharged by grounding the second conducting layer. Based on this effect, Luryi and Kastalsky (1985a) proposed a memory device, which allows a fast operation of all functions: *write, read,* and *erase*. It received the name HE^2PRAM (hot-electron erasable programmable random access memory).

2.2.4 Theory of Charge Injection in Three-Terminal RST Devices

The problem of electron-gas heating and determination of the distribution function of hot electrons is quite involved in general. In the 2-D case it is further complicated by the subband structure which requires the consideration of not only intra- but also inter-subband transitions, as discussed in the extensive literature on the subject, cf. the review by Ando et al. (1982). In a phenomenological treatment of the device characteristics, it is unreasonable to include all the details of the electron heating problem — which would be appropriate in a study of the heating effect itself.

Recently, Grinberg et al. (1987) developed an analytical theory of charge injection in the CHINT/NERFET devices. They used the simplest and most common approximation in which the non-equilibrium distribution is assumed to differ from the equilibrium case only by the temperature T_e, and the latter is evaluated from the energy balance equation. The establishment of such a quasi-equilibrium distribution can be ensured by a sufficiently strong electron-electron interaction, which in turn is realized at high enough electron concentrations. Although, the electron concentration decreases substantially along the channel due to the injection effect, nevertheless the main portion of the collector current flows from the channel region where the concentration is still high enough ($\geq 10^{11}$ cm^{-2}). One can expect, therefore, that the electron temperature approximation is satisfactory for the purpose of calculating the injection current (see, however, the discussion below).

Some of the major features of the operation of three-terminal RST devices are adequately described by the theory. Calculations correctly predict the existence of a strong NDR in the drain circuit accompanied by a rapid increase and subsequent saturation of the collector current. Moreover, the theory describes the experimental fact that at high enough heating voltages V_{SD} the collector current substantially exceeds the drain current.

However two important experimental features were not captured by the model. Firstly, the theory does not describe the observed saturation of the drain current from below — after the NDR region. Secondly, although the theory correctly shows the increase in the peak drain current with higher collector voltage V_{SUB} (which is simply a back-gate field effect), it does not describe the salient (and quite puzzling) experimental feature of the *negative transconductance* in the saturation region. This effect (drop in the saturated I_D with increasing V_{SUB}) is clearly evident in Figs. 8 and 9. Grinberg et al. (1987) suggested that the major deficiency of the theory consisted in the assumption of a uniform field in the channel as well as the neglect of the diffusion component of the channel current and the dynamical screening. It is conceivable that the proper inclusion of these effects may account for the experimentally observed saturation of the drain current and the negative transconductance.

I believe, however, that in order to correctly describe these phenomena, one would have to go beyond the assumed electron-temperature approximation — especially in the high-energy tail of the hot-electron distribution, which is responsible for the injection current. In general, the electron temperature T_e different from the lattice temperature T is established when the electron-electron (e-e) scattering time is substantially shorter than the relaxation time associated with the lattice. The low-energy part of the distribution function is well described by a T_e already at low electron concentrations, as soon as the e-e scattering is faster than relaxation on acoustical phonons. The high-energy portion of the distribution, however, can be strongly distorted (depressed) compared to a Maxwellian curve — because of the emission of optical phonons. Recently, Esipov and Levinson (1986) considered the electron temperature in a two-dimensional electron gas taking into account the e-e scattering and the optical phonon emission. They showed that although the integrated distortion of the distribution function is weak, as it affects only the tails of the distribution (above the optical phonon threshold, where the number of electrons is small), nevertheless in those very tails the distortion can be quite strong.

In my opinion similar effects will be even more important in the operation of CHINT where one is interested in the tails of the distribution above the collector barrier height $\Phi \sim 0.3$ eV, which are constantly depleted by the injection current. The effect of channel depletion itself may have a significant influence on hot-electron distribution. Indeed, as the carrier density rapidly decreases in the direction from the source to the drain,† the hot-electron distribution function will deviate from the

Maxwellian form — with its high-energy tail further suppressed compared to value predicted by the electron-temperature approximation. This means that one can expect a *self limitation* of the RST process — with the channel concentration never dropping below a critical level — determined, for a given collector barrier height, by the concentration dependence of the electron-electron interaction. This effect should lead to the experimentally observed saturation from below of the NERFET I_D versus V_{SD} characteristic. Moreover, it may also describe the negative transconductance — since the lower the barrier the smaller is the expected critical concentration for equilibrating the electron distribution at energies above the barrier height. In my opinion, investigation of these effects is of considerable theoretical and practical importance.

2.2.5 The Ultimate Speed of CHINT/NERFET Devices

Fundamental limitations on the intrinsic speed of three-terminal RST devices arise due to the time-of-flight delays characteristic of a space-charge-limited current and because of a finite time required for the establishment of an electron temperature.

Consider the latter limitation first. Energy relaxation of hot carriers in bulk semiconductors has been a subject of a considerable number of studies. The dominant mechanisms for the maxwellization of the hot-electron energy distribution function are the polar optic phonon scattering and the electron-electron interaction. The phonon mechanism is expected to be not too different in the CHINT/NERFET structure compared to the bulk. Monte-Carlo studies indicate that the energy loss rate due to polar optic phonon emission by electrons in GaAs is nearly constant for electron energies above 0.1 eV and is of the order of 2×10^{11} eV/sec. This translates into about 1 psec equilibration time for $T_e \sim 1500$ K. The influence of e-e scattering (which, as discussed above, may be of primary importance for the establishment of the quasi-equilibrium in high-energy tails of the electron distribution) is much more difficult to take into consideration. It is probably a safe bet to assume that at the operating voltages of CHINT and NERFET the hot-electron ensemble equilibrates in less than 1 psec.

The second fundamental limitation of the speed arises due to the space-charge capacitance associated with the mobile charge drifting in the high-field regions of the device. It reduces to the time of flight of electrons over these regions — the high field portion of the channel and the downhill slope of the potential barrier. At high T_e both regions are of order 10^{-5} cm and the corresponding delay is about 1 picosecond. It should be emphasized that these time-of-flight limitations are different from the time-of-flight-under-the-gate limitation characteristic of an FET. The latter results from charging the channel by the gate field through the output resistance of a previous identical device which necessarily gives $\tau = L/v$ with L being the gate length. In the CHINT the controlling electrode is the drain and L is the total length of the space-charge-limited current regions — which can be substantially shorter than the source-to-drain distance. The same limitation governs the frequency cut-off in NERFET.

---

† In a simple model, Luryi and Kastalsky (1985a) had shown that the concentration drop along the channel is approximately exponential with a characteristic length which can be as short as 1000 Å at high T_e. (In this calculation the channel depletion is due to the "current stealing" effect of charge injection alone — neglecting the dynamical screening which would further magnify the effect.)

Although in principle the CHINT and NERFET are picosecond devices, their real speed limit at present arises from the RC delay due to large contact pads and the series channel resistance. The main parasitic capacitance — between the D and SUB electrodes — can be reduced by shrinking the contact pads and/or by ion implantation of oxygen underneath the pads. (The latter should also help reduce the leakage in CHINT.) At the same time one should try to minimize the channel resistance (at the peak prior to onset of the NDR), which includes the series contact resistance. Experimentally, Kastalsky et al. (1986) demonstrated the operation of CHINT with a current gain at frequencies up to 32 GHz at room temperature and microwave generation in NERFET up to 7.7 GHz.

3. NOVEL INFRARED PHOTODETECTOR CONCEPTS

3.1 Infrared Photodetectors on a Silicon Chip

As is well known, the celebrated silicon technology has not been able to produce an on-chip infrared photodetector for long-wavelength fiber-optics communications. The obvious difficulty lies in the fact that silicon bandgap E_G is wider than the photon energy in the range of silica-fiber transparency ($\lambda = 1.3 - 1.55 \, \mu m$). So far, the only practical way of employing silicon technology for fiber-optics communications has been to combine Si integrated circuits with Ge or InGaAs-on-InP detectors on a separate chip. The use of separate chips introduces additional parasitics of the interconnection and consequently leads to poorer noise performance.

One demonstrated approach to this problem has been based on growing single-crystal germanium pin junction on a silicon substrate. Molecular beam epitaxy (MBE) grown diodes have been reported (Luryi et al, 1984c), which had an absorption spectrum similar to that in bulk germanium with a quantum efficiency $\eta \approx 40\%$ at $\lambda = 1.3 \, \mu m$. However, the devices suffered from a relatively high reverse-bias parasitic leakage at room temperature. This leakage resulted from misfit dislocations originating at the Ge/Si interface and propagating through the germanium pin junction. In the subsequent work the dislocation density was reduced by inserting a Si/Ge superlattice in the epitaxially grown germanium film ("glitch grading"). It turned out that dislocations tend to be trapped in the strained superlattice region and do not propagate up into the active diode region. The measured dislocation density above the glitch region was reduced by two orders of magnitude. Similar improvement was seen in the room-temperature current-voltage characteristics. At comparable reverse-bias voltages the leakage current dropped by a factor of more than 100 and became within an order of magnitude from the theoretical diffusion-limited saturation current ($\sim 4 \times 10^4$ A/cm^2) of an ideal Ge pn junction. Ultimately, one should be able to produce on a silicon substrate Ge or even InGaAs layers comparable in quality to bulk samples. In this approach, no use is made of the electronic properties of the heterointerface, nor of the Si substrate itself. The latter merely serves as a carrier vehicle for an *incommensurate* growth of a useful single-crystal foreign semiconductor.

However, there is one property of silicon, which is very attractive for use in fiber-optics communications and whose utilization requires *commensurate* epitaxy. Silicon is an ideal material for avalanche multiplication of photogenerated signals. Neither Ge nor InGaAs are ideal avalanche photodetector (APD) materials from the point of view of the so-called *excess noise factor F*, which describes the stochastic nature of avalanche multiplication. The F factor generally depends on the avalanche gain M and the ratio of the impact ionization coefficients $K = \alpha_n / \alpha_p$ for electron and holes. If $K \approx 1$ then $F \approx M$ and the total noise power scales $\propto M^3$. Such is the situation for Ge with $\alpha_p / \alpha_n \lesssim 2$ and InGaAs, where $\alpha_n / \alpha_p \lesssim 2$. On the other hand, if $K \gg 1$ or $K \ll 1$, then $F \approx 2$ even for $M \gg 1$, provided avalanche is initiated by the type of carrier with higher α. It is well established that in Si at not too high electric fields ($\lesssim 3 \times 10^5$ V/cm) the electron ionization coefficient is substantially greater than the hole ionization coefficient. Thus, properly designed Si APD's can have the noise performance near the theoretical minimum. At present, there are commercially available silicon devices with $K \approx 20 - 100$ (of course, these APD's do not operate in the range of interest for fiber-optical communications). It would be very attractive to implement a heterostructure device with *separate* absorption and multiplication regions (SAM APD), in which electrons photogenerated in a Ge or InGaAs layer would subsequently avalanche in Si. Considerable research has been devoted to the use of III-IV compound-semiconductor SAM APD's for fiber-optics communications. Excellent performance has been demonstrated by InP/Ga$_{0.47}$In$_{0.53}$As APD's of this type; however, the implementation with Ge/Si heterojunctions requires far better material quality in the interfacial layers than that presently available with any crystal growth technique.

3.2 Waveguide Infrared Detectors Based on Ge/Si Superlattices

A novel infrared photodetector structure was recently proposed (Luryi et al. 1986), which utilizes the silicon advantage. It represents a waveguide in which the core is a strained-layer Ge$_x$Si$_{1-x}$/Si superlattice (SLS) sandwiched between Si layers of a lower refractive index. Absorption of infrared radiation occurs in the core region due to interband electron transitions, and photogenerated carriers are collected in the Si cladding layers. Due to the recently discovered effect of bandgap narrowing by the strain in Ge$_x$Si$_{1-x}$ alloy layers the fundamental absorption threshold of the SLS is shifted to longer wavelengths,[†] so that the detector can be operated in the range of silica-fiber transparency. If the alloy layers are sufficiently thin, the SLS can be grown by MBE without nucleating dislocations. Experimentally, such structures were recently manufactured and tested. The first SLS waveguide *pin* diodes showed an internal quantum efficiency of 40% at $\lambda = 1.3\,\mu$m and a frequency bandwidth of close to 1 GHz. The first APD structure showed an avalanche gain as high as $M = 50$ and a quantum efficiency of 100% at $M = 10$. The waveguide-detector approach is entirely compatible with the Si integrated circuit technology and offers the possibility of fabricating a complete receiver system for long-wavelength fiber-optics communications on a silicon chip. Introduction of optoelectronic elements into Si VLSI technology appears to be one of the most gainful possible applications of Si MBE (see the review by Luryi and Sze, 1987).

[†]   An important development was the theoretical calculation by People (1985) who considered the strained Ge$_x$Si$_{1-x}$ alloys grown on Si (100) substrates, and found that their bandgap is substantially reduced in comparison with the unstrained alloy. Subsequently, Lang et al. (1985) measured the fundamental absorption threshold in the Ge$_x$Si$_{1-x}$/Si SLS as a function of the Ge content in the alloy layers and found a good agreement with the theoretical predictions. At $x = 0.6$ the bandgap E_G is narrower than that of pure unstrained Ge, and for $x \geq 0.5$ one has $E_G \leq 0.8$ eV. The absorption edge is thus brought down by the strain to below the photon energy at wavelengths of silica-fiber transparency.

Below we shall briefly discuss the optimum composition of an SLS core, as determined by the trade-off between the confinement of radiation and the stability requirements for a Ge_xSi_{1-x}/Si SLS.

3.2.1 Material Properties of Strained-Layer Ge_xSi_{1-x}/Si Systems

Let us first discuss the questions of stability. The maximum thickness h_c of a single strained Ge_xSi_{1-x} alloy layer which can be grown pseudomorphically on Si depends on the germanium content, decreasing with x. People and Bean (1985) have calculated $h_c(x)$ on the assumption that the film grows initially without dislocations, which are then generated at the interface, as the strain energy density per unit area of the film exceeds the areal energy density associated with an isolated dislocation. Raman scattering studies have shown that most of the strain in such structures resides in the alloy layer, with Si cladding layers being nearly unstrained. A second Ge_xSi_{1-x} layer can then be grown on the Si cap layer, and the sequence can be repeated many times without a noticeable incommensurate growth (as many as 100 periods have been reported). The maximum total thickness of such strained layer superlattices (SLS) can be estimated from the semi-empirical rule that the stability of the SLS against the formation of dislocations is equivalent to that of a single alloy layer of same thickness but average Ge content. This rule can be represented by the following expression: $h_{SLS}^{max}(x, r, T) \approx h_c(xr)$, where $r \equiv h/T$ is the ratio of the thickness of the alloy layer to the superlattice period (i.e. the "duty cycle"), and h_{SLS} is the total thickness of the superlattice.

3.2.2 Design of Waveguide Detectors and APD's

Consider first a waveguide-detector structure in which the core represents a single alloy layer, Fig. 12a. We assume that the Ge content in this layer is $x > 0.5$, and that the absorption coefficient at wavelengths of interest is $\alpha \approx 10^2 \, cm^{-1}$. To be in the range of commensurate growth, the alloy thickness h must be less than the critical $h_c(0.5) = 100 \, Å$. To a good approximation, the fraction Γ of the integrated intensity of the light wave which falls within the absorbing core, is given by:

$$\Gamma = 2\pi^2 \left[\frac{h}{\lambda} \right]^2 (n_{core}^2 - n_{clad}^2) . \tag{3}$$

For $x = 0.5$, $h = 100 \, Å$, and $\lambda = 1.3 \, \mu m$ one finds $\Gamma = 2.3 \times 10^{-3}$. The effective absorption coefficient of such a waveguide, $\alpha_{eff} = \alpha \Gamma \sim 0.2 \, cm^{-1}$, $\alpha_{eff} = \alpha \Gamma \sim 0.2 \, cm^{-1}$, is too low for a practical use (a detector would have to be several centimeters long and even the speed of light is not fast enough over such distances).

The use of a superlattice is thus imperative. Consider the structure illustrated in Fig. 12b. Ignoring in first approximation the influence of strain on the dielectric constant, the refractive index of an SLS can be estimated as an average of $n^2(x)$ and n_{Si}^2 over one period and the effective absorption coefficient of an SLS core is given by $\alpha_{eff} = r \Gamma \alpha$. Analysis shows that α_{eff} is maximized by smaller r, which for $h_{SLS} = h_{SLS}^{max}$ implies maximizing the superlattice width, and that α_{eff} has its optimum value for those r which correspond to $\Gamma \approx 1/2$. Physically, as the superlattice is made thicker to absorb the wings of the light intensity distribution, the SLS requirement of

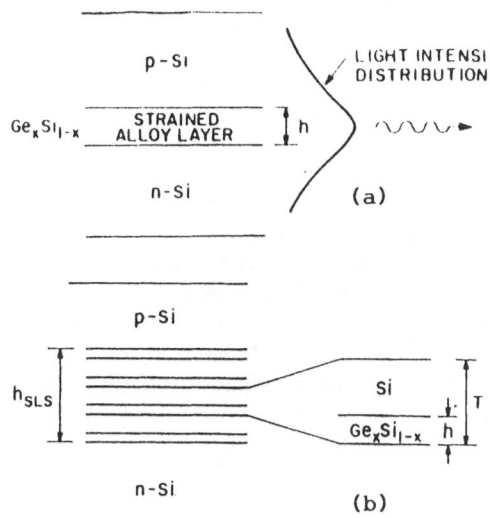

p–Si

Ge_xSi_{1-x} STRAINED
ALLOY LAYER $\updownarrow$ h

LIGHT INTENSITY
DISTRIBUTION

n–Si   (a)

p–Si

h_{SLS}

Si

Ge_xSi_{1-x} h

n–Si   (b)

FIGURE 12. Schematic illustration of strained-layer waveguide detectors (Luryi et al., 1986).
a) Single-layer core; poor confinement of light leads to unacceptably low absorption.
b) Superlattice core; calculated absorption coefficient is sufficient for high-speed operation. Waveguide detectors with SLS core have been successfully implemented by Temkin et al. (1986) and Pearsall et al. (1986).

decreasing r leads to less efficient absorption at the peak intensity, thus more than offsetting the gain. For $\lambda = 1.3\,\mu m$, we can expect an optimum value $\alpha_{eff} \lesssim 0.2\,\alpha \approx 20\,cm^{-1}$ in an SLS consisting of 12 periods of $60\,\text{Å}\ Ge_{0.6}Si_{0.4}\,/\,140\,\text{Å}$ Si. This means that the waveguide length must be of order 0.5 mm for high detector efficiency. If a *pin* detector represents a ridge waveguide of that length and the width $\leq 10\,\mu m$, then its capacitance is less than about 0.5 pF, assuming a typical depletion width of $1\,\mu m$. This value of the internal capacitance is acceptable and comparable to that of the conventional *pin* IR detectors. Note that the detector quantum efficiency grows with the optical path length, without degrading the speed of response.

3.2.3 Waveguide Avalanche Detectors

The detector sensitivity will be further improved by an avalanche gain in Si cladding layers. To reduce an excess noise, the APD design should be guided by the following principles: *i)* since $K \equiv \alpha_n/\alpha_p \gg 1$ in Si, the multiplication should be initiated by electrons rather than holes; *ii)* since K decreases sharply when the electric field much exceeds the ionization threshold E_i, the field in the avalanche layer should be near the threshold, $E \geq E_i \approx 3 \times 10^5$ V/cm, and the thickness of that layer should be well above $\alpha_n^{-1}\,(E_i) \approx 0.5\,\mu m$; *iii)* the field in the SLS layers should not exceed $\sim 10^5$ V/cm, the ionization threshold in Ge.

A possible waveguide APD structure is illustrated in Fig. 13. In addition to an undoped Ge_xSi_{1-x}/Si SLS of $x \geq 0.6$, $r \leq 0.3$, and thickness $h_{SLS} \geq 3\,000\,\text{Å}$, it contains an undoped Si avalanche layer of thickness $d \geq 2\,\mu m$ separated from the SLS by a thin ($\Delta \leq 10^{-6}$ cm) p-type Si layer. In the operating regime, the Δ layer must be depleted by an applied reverse bias. The total surface density of charge in this layer should, therefore, be of order $\kappa E_i \approx 2 \times 10^{12}$ e/cm². This will achieve the desirable hi-lo field separation of the absorption and multiplication layers and result in a low-noise SAM APD structure.

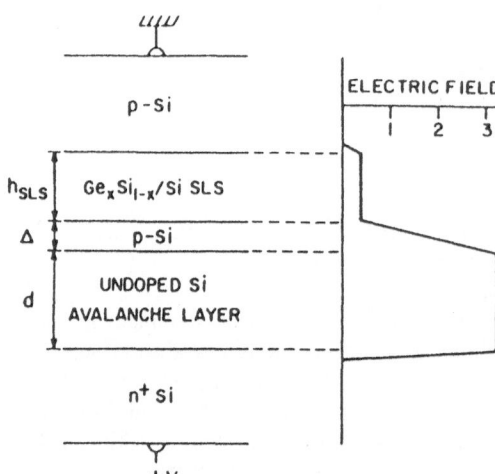

ELECTRIC FIELD (V/cm)

h_{SLS} Ge$_x$Si$_{1-x}$/Si SLS

p-Si

Δ p-Si

d UNDOPED Si
AVALANCHE LAYER

n$^+$ Si

+V

FIGURE 13. Schematic illustration of the proposed waveguide APD structure and the electric field profile (Luryi et al., 1986). The built-in acceptor charge sheet of surface density of order $\kappa E_i \approx 2 \times 10^{12}$ e/cm^2 will achieve the desirable hi-lo field separation of the absorption and multiplication layers and result in a low-noise SAM APD structure.

3.3 Infrared Detectors Based on the Photon Drag Effect and Intersubband Absorption by a 2DEG

Infrared absorption between quantum-well (QW) subbands was recently demonstrated by West and Eglash (1985) in a modulation-doped GaAs/AlGaAs superlattice and also by Levine et al. (1987) using conventionally-doped quantum wells. These dipole transitions are similar in nature to those used previously for intersubband spectroscopy of Si inversion layers and other charged-surface states. A number of novel infrared detectors have been proposed based on the intersubband transitions. Most of these utilize some sort of an internal photoemission mechanism, whereby electrons, photoexcited into a higher subband or a virtual-state resonance of a QW or a superlattice, then tunnel through or drift to a collector contact. A different detection scheme was recently proposed by Luryi (1987). The useful signal is derived not from the intersubband transition itself, but from the photon-drag current *in the plane* of the QW. The subbands are needed merely to enhance the absorption coefficient. Below we shall discuss the state of the motion of the QW electrons irradiated by an infrared light propagating parallel to the plane of a QW superlattice (situation quite feasible experimentally, since such superlattices usually form a dielectric waveguide). In general, the intersubband transitions will be accompanied by a measurable "photon-drag" current due to the momentum imparted by the absorbed photons. The frequency dependence of this current contains a wealth of information regarding the momentum-relaxation kinetics in 2D subbands.

3.3.1 Intersubband Absorption by a 2DEG

Consider a transverse-magnetic wave propagating along the waveguide formed by an AlAs/GaAs superlattice and two AlAs cladding layers, cf. Fig. 14. The superlattice core thickness is $D = p \times d$ and each of the p periods d consists of a d_W-thick QW and a d_B-thick barrier with a duty cycle $r \equiv d_W / d_B$. The superlattice is modulation doped, and each QW contains a 2DEG of areal density n. The fraction of light intensity contained within the core is then given by Eq. (3), viz. $\Gamma = (2\pi^2 D^2)/(\lambda_0^2) \, r \, \Delta\epsilon$, where λ_0 is the vacuum wavelength of the electromagnetic wave and $\Delta\epsilon \equiv \epsilon_{GaAs} - \epsilon_{AlAs}$ with $\epsilon_{GaAs} = 10.88$ and $\epsilon_{AlAs} = 8.16$ being the relative permittivities of the two materials. The absorption coefficient α can be calculated with the Golden rule for the transition probability per unit time

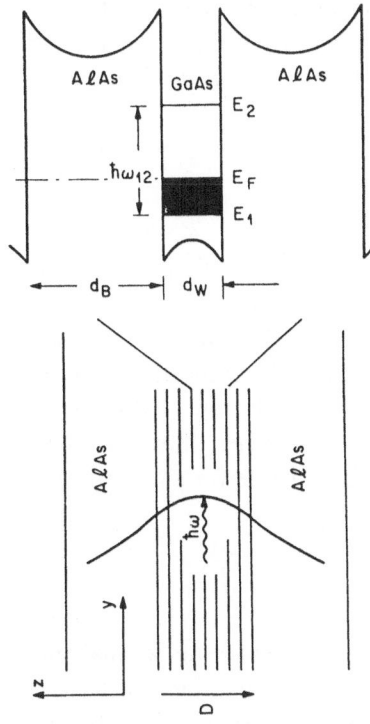

FIGURE 14. Schematic diagram of a photon-drag detector (Luryi, 1987b). We shall be using as an example an AlGaAs/GaAs superlattice waveguide with the following parameters: $n = 10^{12} \, \text{cm}^{-2}$, $d = 150 \, \text{Å}$ with $r = 1/3$, $D = 0.75 \, \mu\text{m}$ (i.e., $p = 50$ wells).

$$w \, (\vec{k}) = \frac{2\pi}{\hbar^2} \, e^2 \, (E/2)^2 \, | <1 \, |z| \, 2> |^2 \, \frac{\gamma / \pi}{[\omega - \omega_{12} - \hbar \, (\vec{k} \cdot \vec{q}) \, /m \,]^2 + \gamma^2} \, , \qquad (4)$$

where $\hbar\omega_{12}$ is the intersubband separation, $\vec{q}$ is the photon wavevector, m the electron effective mass (assumed, neglecting nonparabolicity, to be the same for all electrons in the 2DEG), E is the electric field amplitude in the direction normal to the QW, and $\gamma \equiv 1/2\tau_1 + 1/2\tau_2$ is the lifetime contribution to the width of the subband levels $|1>$ and $|2>$.

Let us ignore in first approximation the dependence of w on the electron momentum $\hbar\vec{k}$, i.e., neglect in (4) the term $\hbar \, (\vec{k} \cdot \vec{q}) \, /m$ which arises from the conservation of energy and 2D quasi-momentum during intersubband transitions. In this approximation, $w \, (\vec{k}) = \bar{w}$ and the total transition rate per unit volume equals $n \, \bar{w} \, /d$, giving

$$\bar{\alpha} \equiv \alpha \, [\, \bar{w} \, (\omega) \,] = \frac{n \, e^2 \, R_0 \, f \Gamma}{2m \, \bar{n} \, d} \, \frac{\gamma}{\Delta\omega^2 + \gamma^2} \, , \qquad (5)$$

where $f \equiv (2m\omega/\hbar) \, | <1|z|2> |^2$ is the oscillator strength ($f \approx 1$),[1] $\bar{n}$ is the average refractive index of the material, $\Delta\omega \equiv \omega - \omega_{12}$, and $R_0 \equiv \mu_0 c = 377 \, \Omega$ is the vacuum magnetic permeability constant. For $\lambda_0 \approx 10 \, \mu\text{m}$ and the superlattice parameters as in Fig. 1, one has $\Gamma \approx 0.1$ and with $\gamma = 10^{13} \, \text{s}^{-1}$ we find $\bar{\alpha} \approx 2 \cdot 10^3 \, \text{cm}^{-1}$.

In the above approximation, the entire photon-drag current is carried by electrons in the excited subband, since the cylindrical symmetry of the electron distribution in the lower subband is not disturbed by $\bar{w}$ transitions. Each excited electron receives an extra momentum $\hbar q = 2\pi\hbar / \lambda$, which, on the average, relaxes after the time τ_2. The photon-drag current is in the direction of the light propagation and its density J_0 is related to the radiation energy flux S in the waveguide as follows:

$$J_0 = -\frac{e\tau_2}{md}\, \hbar q\, n\overline{w} = -\frac{\overline{n}}{c} S\, \overline{\alpha}\, \mu_{\text{eff}} \;,\qquad (6)$$

where $\mu_{\text{eff}} \equiv (e/m)\,\tau_2$ is an effective mobility. A simple way of understanding (6) is to note that $\overline{\alpha}\,(\overline{n}/c)\,S$ represents the transferred momentum per unit volume per unit time, i.e., the drag-force density acting on electrons. Since $J_0 \propto \overline{\alpha}$, the spectrum $J_0(\omega)$ contains little additional information over the conventional absorption spectra. The current J_0 is present in the absorption of light by *any* conductor. However, as we shall now see, the intersubband photon-drag effect is much richer than that given by (6). The above approximation, although undoubtedly reasonable for calculating $\overline{\alpha}$, is totally inadequate for estimating the photon-drag current.

3.3.2 *The Enhanced Photon-Drag Current*

The gist of the matter is that the photon-drag current J represents a small net difference between two oppositely directed larger currents: one due to excited electrons in the upper subband, the other to remaining holes in the lower subband. Indeed, consider Fig. 15. If we ignore the level broadening, then the conservation laws of energy and momentum imply that the allowed transitions must obey

$$\hbar\omega + \frac{\hbar^2 k^2}{2m} = \hbar\omega_{12} + \frac{\hbar^2}{2m}\,(\vec{k} + \vec{q})^2 \;,\qquad (6)$$

i.e., to a good approximation, $\hbar\,(\vec{k}\cdot\vec{q})/m = \Delta\omega$. For example, if $\Delta\omega > 0$ then only those electrons can make a transition, whose momentum has a positive component in the direction of $\vec{q}$. Thus, in the presence of radiation *both* subbands acquire an electron distribution disturbed from cylindrical symmetry. Next point to note is that the relaxation times in the two subbands can be very different, $\tau_1 \gg \tau_2$, especially at lower temperatures. If we could follow the time evolution of J upon a short infrared pulse with $\Delta\omega > 0$, then we would observe the current to change sign after $\sim\tau_2$, with the magnitude $|J|$ rising, possibly, to a much higher value than the initial current and then decaying after $\sim\tau_1$ (during the time interval $\tau_2 < t < \tau_1$

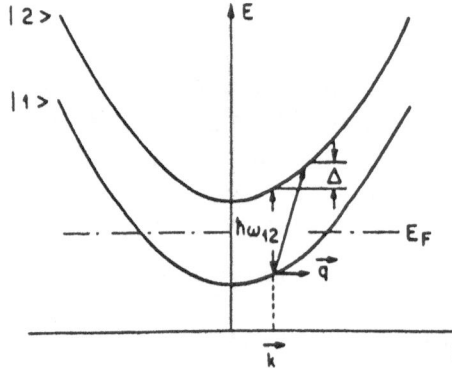

FIGURE 15. Intersubband transitions stimulated by an electromagnetic (TM) wave propagating along the plane of a 2DEG in the y-direction. The diagram illustrates the selection rule: $k_y > 0$ for $\Delta \equiv \hbar\Delta\omega > 0$. Similarly, $k_y < 0$ for $\Delta < 0$, as follows from the conservation laws for energy and momentum.

the characteristic drift velocity in the 2DEG will be of the order of its Fermi velocity, $\hbar k_F / m$, rather than $\hbar q / m$). In a steady state, the resultant current may be directed either along or against the "primary" current, depending on the sign of $\Delta \omega$, and its magnitude can be substantially higher than that given by (6). The effect is, of course, limited by the validity of the conservation laws (7), i.e., by the quality factor of the intersubband resonance. In the relaxation-time approximation this limitation is described by the width γ of the Lorentzian line (2).

Quantitatively, the total steady-state current is given by the difference of currents excited within each of the two subbands:

$$ J = \frac{e \tau_1 f_1}{md} - \frac{e \tau_2 f_2}{md} = \frac{e}{md} \left[(\tau_1 - \tau_2) f_1 - \tau_2 (f_2 - f_1) \right] , \qquad (7) $$

where f_i are the rates of momentum transfer in the ith subband per unit area of one 2DEG:

$$ f_1 = \int \frac{d^2 k}{2\pi^2} \, \hbar k_y \, w \, (\vec{k}) \approx \hbar q \, n \, \overline{w} \, \frac{\pi \hbar n}{m} \, \frac{\Delta \omega}{\Delta \omega^2 + \gamma^2} , \qquad (8a) $$

$$ f_2 = \int \frac{d^2 k}{2\pi^2} \, \hbar \, (k_y + q) \, w \, (\vec{k}) \approx f_1 + \hbar q \, n \, \overline{w} . \qquad (8b) $$

Evaluation of the integrals in (8) has been done to second order in the small parameter q/k_F. The result can be conveniently expressed in terms of the quantum efficiency of the drag-effect detector:

$$ \eta \equiv \frac{J/e}{S/\hbar \omega} = - \frac{2\pi \hbar \, \overline{\alpha}}{\lambda m} \left[\tau_2 - \frac{\pi \hbar n \, (\tau_1 - \tau_2)}{2\gamma m} \, \frac{2\gamma \Delta \omega}{\Delta \omega^2 + \gamma^2} \right] . \qquad (9) $$

The first term in the brackets corresponds to the primary drag current (6), whereas the second (which may be denoted by τ_1^{eff}) describes the novel enhanced effect resulting from the drag force (8a) and the difference in the relaxation times for the two subbands. Relative importance of the two terms depends on the frequency, the second term being maximized at $\Delta \omega \approx \gamma$ (at which point $\overline{\alpha}$ is reduced by a factor of 2 from its value at $\Delta \omega = 0$). The maximum ratio $[\tau_1^{\text{eff}} / \tau_2]_{\text{max}} \approx \pi \hbar n \tau_1 / m \equiv E_F \tau_1 / \hbar$ can substantially exceed unity (E_F is the 2DEG Fermi energy).

Varying the frequency of incident radiation in a narrow range near the intersubband resonance ω_{12} and measuring the photon-drag current $J(\omega)$, one obtains valuable information on the momentum-relaxation kinetics in 2D subbands. The predicted spectral curve is given by (9) or, for $\tau_1 \gg \tau_2$, by

$$ J(\omega) = J_0(\omega) \left[1 - \frac{E_F \tau_1}{\hbar} \, \frac{2\gamma \Delta \omega}{\Delta \omega^2 + \gamma^2} \right] , \qquad (10) $$

with J_0 given by (6). It is worth noting that the additional information is provided entirely by the enhanced effect, since $J_0(\omega)$ mimics the intersubband absorption spectra and depends only on $\tau_2 \approx (2\gamma)^{-1}$, whereas the enhanced photon-drag current spectrum is modulated by τ_1. Figure 16 displays the expected spectral curve. It may also be possible to probe the drug-current response at a fixed frequency, but varying carrier concentration.

267

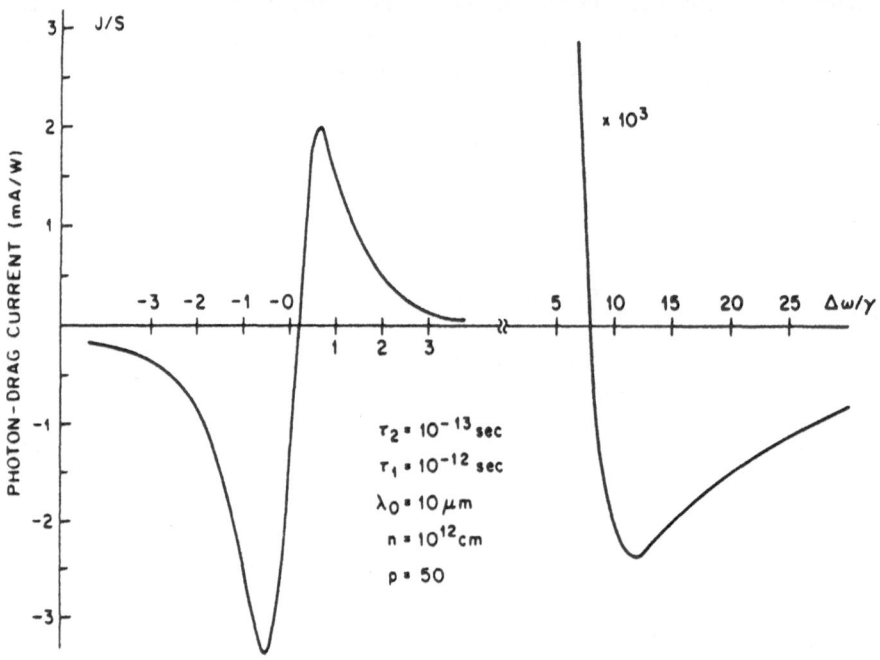

FIGURE 16. Plot of the expected photon-drag-current spectrum, Eq. (10), for an assumed set of realistic parameters. Note that the current changes sign *twice*.

3.3.3 Sensitivity of Photon-Drag Detectors

Let us estimate the detector sensitivity for the exemplary structure of Fig. 14 at temperatures well below the Debye Θ, say at $T = 77\,\mathrm{K}$. In this case, τ_2 is limited mainly by optical phonon emission and τ_1 by impurity scattering. Taking $\tau_2 \sim 7 \cdot 10^{-13}\,\mathrm{s}$, and $\tau_1 \sim 10\tau_2$, we find $[\tau_1^{\mathrm{eff}}/\tau_2]_{\mathrm{max}} \sim 300$ and $\eta \sim 6\%$. This corresponds to a sensitivity $J/S = 1.24\,\lambda_0\,[\mu\mathrm{m}] \times \eta \approx 0.7\,\mathrm{A/W}$ for $\lambda_0 = 10\,\mu\mathrm{m}$. This estimate should be compared with the thermal noise in the detector. In an ideal close-circuit current-measurement arrangement, the root-mean-square noise current in a bandwidth Δf is given by the Nyquist formula, $I_{\mathrm{noise}} = (4kT\,\Delta f / R)^{1/2}$, where the resistance R in our detector is given by $R^{-1} = e\,(pn)\,\mu\xi$ with $\mu = e\tau_1/m$ and ξ being a geometrical factor. In the above example, taking $R \sim 10\,\Omega$ and $\Delta f \sim 1\,\mathrm{GHz}$, we find $I_{\mathrm{noise}} \sim 6 \cdot 10^{-7}\,\mathrm{A}$, which implies that in order to beat the noise in a detector with 1 GHz response, we must couple into the waveguide at least $1\,\mu\mathrm{W}$ of infrared power.

REFERENCES

Ando, T., Fowler, A. B., and Stern, F. (1982) *Rev. Mod. Phys.* **54**, 832.

Bonnefoi, A. R., McGill, T. C., and Burnham, R. D. (1985a) *IEEE Electron Device Lett.* **EDL-6**, 636.

Bonnefoi, A. R., Chow, D. H., and McGill, T. C. (1985b) *Appl. Phys. Lett.* **47**, 888.

Brown, E. R., Sollner, T. C. L. G., Goodhue, and Parker, C. D. (1987) Appl. Phys. Lett. **50**, 83.

Büttiker, M. (1983) *Phys. Rev.* **B 27**, 6178.

Büttiker, M. and Landauer, R. (1982) *Phys. Rev. Lett.* **49**, 1739.

Capasso, F. and Kiehl, R. A. (1985) *J. Appl. Phys.* **58**, 1366.

Capasso, F., Mohammed, K., and Cho, A. Y. (1986) *IEEE J. Quant. Electronics* **QE-22**, 1853.

Chang, C. Y., Liu, W. C., Jame, M. S., Wang, Y. H., Luryi, S., and Sze, S. M. (1986) *IEEE Electron Device Lett.* **EDL-7**, 497.

Esipov, S. E. and Levinson, I. B. (1986) *Zh. Eksp. Teor. Fiz.* **90**, 330 [*Sov. Phys. - JETP* **63**, 191].

V. J. Goldman, D. C. Tsui, and J. E. Cunningham (1987) *Phys. Rev. Lett.* **58**, 1256.

Grinberg, A. A., Kastalsky, A., and Luryi, S. (1987) *IEEE Trans. Electron Devices* **ED-34**, 409.

Heiblum, M. (1981) *Solid State Electron.* **24**, 343.

Hess, K. (1983) *Physica* **117 B**, 723.

Hess, K., Morkoç, H., Shichijo, H., and Streetman, B. G. (1979) *Appl. Phys. Lett.* **35**, 469.

Honeisen, B. and Mead, C. A. (1972) *Solid State Electron.* **15**, 891.

Jogai, B. and Wang, K. L. (1985) *Appl. Phys. Lett.* **46**, 167.

Kastalsky, A. and Luryi, S. (1983) *IEEE Electron Device Lett.* **EDL-4**, 334.

Kastalsky, A., Luryi, S., Gossard, A. C., and Hendel, R., (1984a) *IEEE Electron Device Lett.* **EDL-5**, 57.

Kastalsky, A., Kiehl, R. A., Luryi, S., Gossard, A. C., and Hendel, R. H. (1984b) *IEEE Electron Device Lett.* **EDL-5**, 321.

Kastalsky, A., Luryi, S., Gossard, A. C., and Chan, W. K. (1985a) *IEEE Electron Device Lett.* **EDL-6**, 347.

Kastalsky, A., Abeles, J. H., Bhat, R., Chan, W. K., and Koza, M. (1986a) *Appl. Phys. Lett.* **48**, 71.

Kastalsky, A., Bhat, R., Chan, W. K., and Koza, M. (1986b) *Solid-State Electron.* **29**, 1073.

Kazarinov, R. F. and Luryi, S. (1981) *Appl. Phys. Lett.* **38**, 810.

Kazarinov, R. F. and Luryi, S. (1982) *Appl. Phys.* **A 28**, 151.

Kazarinov, R. F. and Suris, R. A. (1971) *Soviet. Phys. - Semicond.* **5**, 707.

Lang, D. V., People, R., Bean, J. C., and Sergent, A. M. (1985) *Appl. Phys. Lett.* **47**, 1333.

Lindmayer, J. (1964) *Proc. IEEE* **52**, 1751.

Levi, A. F. J., Hayes, J. R., and Bhat, R. (1986) *Appl. Phys. Lett.* **48**, 1609.

Levine, B. F., Malik, R. J., Walker, J., Choi, K. K., Bethea, C. G., Kleinman, D. A., and Vandenberg, J. M. (1987) *Appl. Phys. Lett.* **50**, 273.

Levine, B. F., Choi, K. K., Bethea, C. G., Walker, J., and Malik, R. J. (1987) *Appl. Phys. Lett.* **50**, 1092.

Luryi, S. (1985a) *Appl. Phys. Lett.* **47**, 490.

Luryi, S. (1985b) *IEDM-85 Tech. Digest*, 666.

Luryi, S. (1985c) *IEEE Electron Device Lett.* **EDL-6**, 347.

Luryi, S. (1985d) *Physica* **134 B**, 466.

Luryi, S. (1987a) in *Heterojunctions: a Modern View of Band Discontinuities and Device Applications*, ed. by F. Capasso and G. Margaritondo (Elsevier Science) Chap. 12.

Luryi, S. (1987b) *Phys. Rev. Lett.* **58**, 2263.

Luryi, S. and Capasso, F. (1985) *Appl. Phys. Lett.* **47**, 1347.

Luryi, S. and Kastalsky, A. (1985a) *Superlattices and Microstructures* **1**, 389.

Luryi, S. and Kastalsky, A. (1985b) *Physica* **134 B**, 453.

Luryi, S., Kastalsky, A., Gossard, A. C., and Hendel, R. H. (1984a) *IEEE Trans. Electron Devices* **ED-31**, 832.

Luryi, S., Kastalsky, A., Gossard, A. C., and Hendel, R. H. (1984b) *Appl. Phys. Lett.* **45**, 1294.

Luryi, S., Kastalsky, A., and Bean, J. C. (1984c) *IEEE Trans. Electron Devices* **ED-31**, 1135.

Luryi, S., Pearsal, T. P., Temkin, H., and Bean, J. C. (1986) *IEEE Electron Device Lett.* **EDL-7**, 104.

Luryi, S. and Sze, S. M. (1987) in *Silicon Molecular Beam Eptitaxy*, ed. by E. Kasper and J. C. Bean (CRC Uniscience Press, inc.), to be published.

Malik R. J., Hollis, M. A., Eastman, L. F., Woodard, D. W., Wood, C. E. C., and AuCoin, T. R. (1981) *Proc. 8th Biennial Conf. on Active Microwave Semicond. Devices and Circuits*, Cornell University.

Morkoç, H., Chen, J., Reddy, U. K., Henderson, T., and Luryi, S. (1986) *Appl. Phys. Lett.* **49**, 70.

Pearsall, T. P., Temkin, H., Bean, J. C., and Luryi, S. (1986) *IEEE Electron Device Lett.* **EDL-7**, 330.

People, R. (1985) *Phys. Rev.* **B32**, 1405.

People, R. and Bean, J. C. (1985) *Appl. Phys. Lett.* **47**, 322.

Rezek, E. A., Holonyak, Jr., N., Vojak, B. A., and Shichijo, H. (1977) *Appl. Phys. Lett.* **31**, 703.

Ricco, B. and Azbel, M. Ya. (1984) *Phys. Rev.* **B 29**, 1970.

Sollner, T. C. L. G., Tannenwald, P. E., Peck, D. D., and Goodhue, W. D. (1984) *Appl. Phys. Lett.* **45**, 1319.

Sollner, T. C. L. G., Le, H. Q., Correa, C. A., and Goodhue, W. D. (1985) *Appl. Phys. Lett.* **47**, 36.

Stevens, K. W. H. (1983) *J. Phys.* **C 16**, 3649.

Sze, S. M. (1969) *Physics of Semiconductor Devices*, 1st edition (Wiley, New York), Chap. 11.

Sze, S. M. and Gummel, H. K. (1966) *Solid-State Electron.* **9**, 751.

Temkin, H., Pearsall, T. P., Bean, J. C., Logan, R. A., and Luryi, S. (1986) *Appl. Phys. Lett.* **48**, 963.

Thornber, K. K., McGill, T. C., and Mead, C. A. (1967) *J. Appl. Phys.* **38**, 2384.

Tsuchiya, M. and Sakaki, H. (1986a) *Japan. J. Appl. Phys.* **25**, L185.

Tsuchiya, M. and Sakaki, H. (1986b) *Appl. Phys. Lett.* **49**, 88.

West, L. C. and Eglash, S. J. (1985) *Appl. Phys. Lett.* **46**, 1156.

Yokoyama N., Imamura, K., Muto, S., Hiyamizu, S., and Nishi, H. (1985) *Japan. J. Appl. Phys.* **24**, L853.

MULTILAYER OPTICS FOR X-RAYS

Eberhard Spiller

IBM T. J. Watson Research Center,
Yorktown Heights, NY 10598, U.S.A.

1. INTRODUCTION

The best conventional mirror materials have normal incidence reflectivities $R < < 10^{-4}$ for soft x-rays ($\lambda \approx 50\text{Å}$). The reflectivity decreases to still smaller values proportional to λ^4 at shorter wavelengths. These reflectivities are far too low to be of use for x-ray optics; only at very grazing angles are substantial reflectivities obtained with standard mirrors; grazing incidence instruments are presently the main optical components in the x-ray region. Normal incidence optics would have many advantages as easier fabrication, smaller size or larger collection area, smaller aberrations, and larger field size at improved resolution.

Useful normal incidence reflectivities in the soft x-ray region can be obtained by adding coherently and in phase the small reflections from many boundaries. The resulting multilayer x-ray mirrors can be seen as an extension of natural crystals to larger lattice spacings, of optical coatings to shorter wavelengths, as special reflection holograms for x-rays, or as superlattices with the largest possible contrast for soft x-rays. While Maxwells' equations are sufficient to describe the performance of a multilayer completely, the different fields have developed quite independently. For example, the full dynamical theory of x-ray diffraction by crystals was developed between 1912 and 1920 by Laue, Ewald and Bragg [1.1,1.2]. Optical coatings on the other hand were invented after 1930, and the matrix theory for optical multilayers, which is equivalent to the dynamical theory of x-ray diffraction, was published by Abeles [1.3] in 1950. The first generation of workers in x-ray diffraction had a strong background in optics and their publications use the language of classical optics. Later the fields separated and each developed its own terminology. The progress in the optics of soft x-rays during the last decade is bridging this gap again and most of the theoretical discoveries can simply be seen as a transfer of established knowledge from one side of the gap to the other.

For optical coatings in the visible, materials with different refractive indices are available that are completely absorption free. In this case it is possible to design coatings which in principle have any desired shape of the reflectivity versus wavelength curve or reflectivity versus angle-of-incidence curve. For high reflectivity mirrors, the most widely used design is the quarter wave stack of two materials with a different refractive index. In the quarter wave stack (optical thickness of each layer is $\lambda/4$), all boundaries add in phase to the reflected wave; therefore, this design gives the highest reflectivities with the fewest number of layers. Other designs give a better performance if one or both materials have absorption. An awareness of the standing wave field in a coating provides the intuitive understanding of the modified coating designs. By positioning critical components of a coating either into the nodes or antinodes of the standing wave generated inside the coating, interactions between the field and a material can be either enhanced or suppressed; examples are the reduction of the absorption losses in soft x-ray mirrors, the increase of the damage threshold in laser mirrors, or the enhancement of electro-optical or magneto-optical effects [1.4-1.9].

In x-ray diffraction, the reduction of the absorption for the case that the atomic planes are located in the nodes of the standing wave field was already known to Ewald and was experimentally demonstrated by Borrmann [1.10]. The possibility that non-periodic coatings with specifically tailored reflection curves can be fabricated is an option which was not previously available with crystals.

2. OPTICAL CONSTANTS FOR SOFT X-RAYS

In the x-ray diffraction theory, each atom in a crystal is described by the complex atomic scattering factor: $f = f_1 - if_2$. For an exact solution all waves scattered from each atom (including multiple scattering) are summed. In the optical theory, a multilayer consists of many single films with sharp boundaries and the reflections from all boundaries are added up. Each film is considered homogeneous with a complex refractive index $\tilde{n}$, which is obtained by adding the contributions of all atoms in the film:

$$\tilde{n} = 1 - \delta + ik = 1 - \frac{r_0 \lambda^2}{2\pi} N_{at}(f_1 - if_2). \qquad (2.1)$$

Here $r_0 = e^2/mc^2 = 2.82 \times 10^{-13}$ cm is the classical electron radius and N_{at} is the number of atoms of a specific type per cm^3 of material. Interactions between atoms which might influence the electronic structure in a solid are neglected in the summation of eq.(2.1). For photon energies above 100 eV and away from absorption edges the resulting errors are usually small, even if the scattering factors from isolated atoms in a gas are used. Within this approximation the optical constants of every compound can be obtained from the scattering factors of the elements by summing the contributions of each element in eq. (2.1). The best method to calculate the performance of a multilayer film is to use optical constants of the individual films that have been measured directly. Values of the atomic scattering factors for all elements and photon energies between 100 and 2000 eV have been tabulated by Henke et al. and are plotted in Fig. 2.1. The most prominent features in Fig. 2.1 are the K, L and M edges of the elements with the corresponding jump in f_2 or the absorption. At a fixed wavelength the available values for the cross sections vary by about a factor of 100 throughout the periodic table. For the softest x-rays in the plot (E=100 eV), f_1 and f_2 are of the same order; and the resulting ranges for δ and k are between 0.005 and 0.5 (hydrogen excluded). Outside the neighborhood of a resonance the value

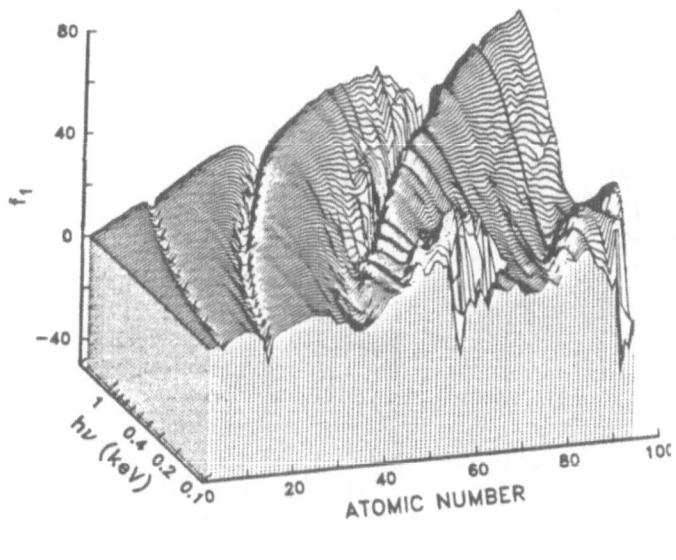

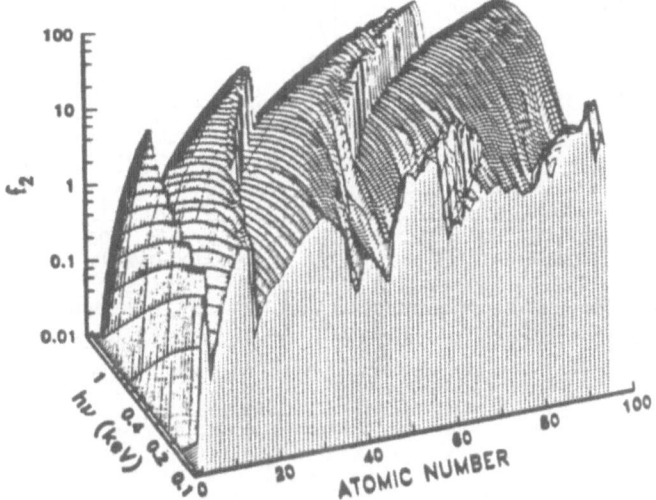

Fig. 2.1. Atomic scattering factors f_1 and f_2 for the elements in the soft x-ray range, data courtesy of B. Henke [2.1].

of f_1 is approximately equal to the number of "free electrons" (electrons with binding energy smaller than photon energy) per atom and approaches the total number of electrons per atom for large photon energies. In contrast, the values for f_2 decrease with increasing photon energies in regions well above an absorption (linearly in first order with wavelength); therefore at shorter wavelengths f_1 and δ are larger than f_2 and k; materials become more and more transparent for harder x-rays.

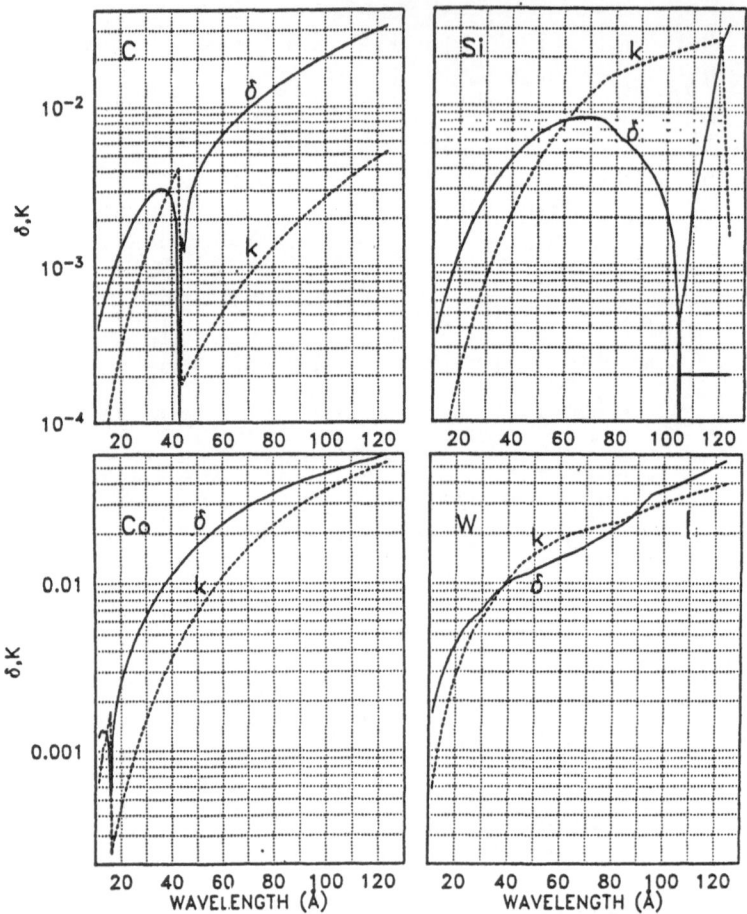

Fig. 2.2 $|\delta|$ and k for carbon, silicon, cobalt and tungsten obtained from the atomic scattering cross-sections. The values of δ can be negative close to an absorption edge.

The slow variation in f_1 with increasing photon energy (decreasing wavelength) leads to a quadratic decrease of δ with wavelength ($\delta \propto \lambda^2$) while the decrease of f_2 with λ leads to a faster drop in the absorption index k (k $\propto \lambda^3$, see eq. (2.1)). Figure 2.2 is a plot of δ and k for selected materials. The linear absorption coefficient

$$\alpha = 4\pi k/\lambda \qquad (2.2)$$

describes the decrease of the intensity of a wave propagating through thickness z of material

$$I/I_0 = e^{-\alpha z}. \qquad (2.3)$$

Figure 2.3 gives the linear absorption coefficient for some of the strongest and some of the weakest absorbers in the soft x-ray region.

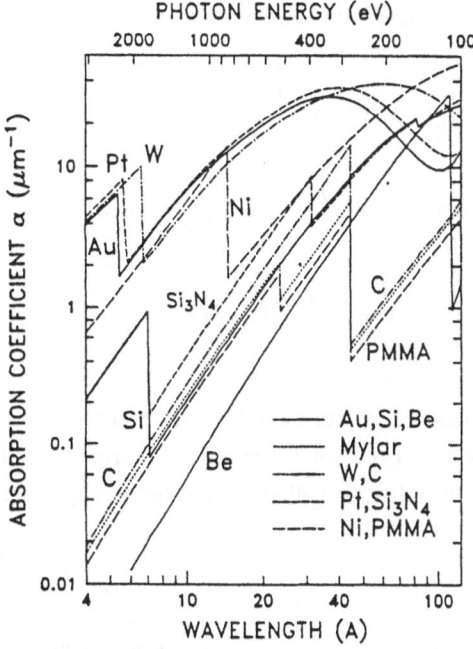

Fig.2.3. Linear absorption coefficient for some of the strongest and some of the weakest absorbers in the soft x-ray region. The absorption edges are approximated by jumps without structure.

3. REFLECTION AT A SINGLE BOUNDARY

The reflected amplitude r_{12} from a single boundary between two materials of complex refractive indices n_1 and n_2 and propagation angles α_1, α_2, is given by the Fresnel equations

$$\overset{s}{r}_{12} = \frac{n_1 \cos\alpha_1 - n_2 \cos\alpha_2}{n_1 \cos\alpha_1 + n_2 \cos\alpha_2}, \quad \text{s-polarization} \qquad (3.1a)$$

$$\overset{p}{r}_{12} = \frac{n_1 \cos\alpha_2 - n_2 \cos\alpha_1}{n_1 \cos\alpha_2 + n_2 \cos\alpha_1}, \quad \text{p-polarization,} \qquad (3.1b)$$

where the propagation angles α are calculated from the propagation angle α_0 in the incident medium (index n_0) using Snell's law:

$$n \sin\alpha = n_0 \sin\alpha_0. \qquad (3.2a)$$

$$\cos\alpha = \sqrt{1 - (n_0/n)^2 \sin^2\alpha_0}. \qquad (3.2b)$$

In the soft x-ray region, the refractive indices of all materials (except vacuum) are complex leading also to complex values of the propagation angle in eqs. (3.2). The incident angle α_0 is measured from the normal and can be replaced by the grazing angle $\theta = 90 - \alpha_0$.

For normal incidence reflectivity the Fresnel equations simplify to

$$R = rr^* = \left| \frac{\tilde{n}_1 - \tilde{n}_2}{\tilde{n}_1 + \tilde{n}_2} \right|^2 . \tag{3.3}$$

With $\tilde{n} = 1 - \delta + ik$ and assuming $\delta, k \ll 1$ we obtain for the reflectivity between vacuum ($\tilde{n} = 1$) and a material

$$R = \frac{1}{4}(\delta^2 + k^2) . \tag{3.4}$$

For $\lambda = 50\text{Å}$ the largest available values for δ and k are around 10^{-2}, which leads to a maximum reflectivity $R < 10^{-4}$, a value much too small for usable instruments. For still shorter wavelengths, the reflectivity drops very fast to still smaller values, roughly proportional to λ^4 (eqs. (3.4) and (1.1)).

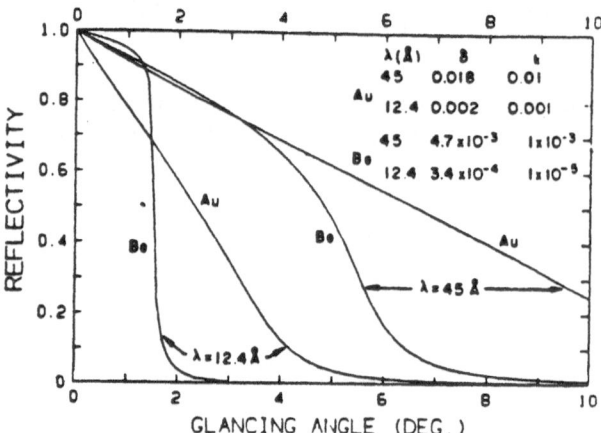

Fig. 3.1. Reflectivity near grazing incidence for different sets of optical constants for s-polarization and soft x rays.

Substantial reflectivity for x rays is only obtained near grazing incidence and most optical instruments use grazing incidence reflection. Figure 3.1 shows calculated reflectivity curves typical for the soft x-ray region. For absorption free materials, there exists a critical angle below which the reflectivity is 100% (total reflection). The critical angle is defined by

$$\sin\Theta_c = \sqrt{2\delta} . \tag{3.5}$$

We can use eq. (2.1) to obtain an approximate expression for all materials, if we equate f_1 with the number of free electrons per atom

$$\sin\theta_c = \lambda \sqrt{r_0 N_0 / \pi} , \tag{3.6}$$

where N_0 is now the free electron density of the material. From eq. (3.6) we see that the theoretical resolution limit of grazing incidence optical instruments is independent of wavelength, due to the fact that resolution is proportional to the wavelength/aperture angle.

For absorbing materials the critical angle is complex and defined by $\cos\alpha = 0$ in eq. (3.2b):

$$\cos\Theta_c = \sin\alpha_c = \tilde{n}. \tag{3.7}$$

We can rewrite the Fresnel equations by replacing the refractive indices with the respective critical angles and obtain

$$r_{12}^s = \frac{\sqrt{\sin^2\Theta_0 - \sin^2\Theta_{c1}} - \sqrt{\sin^2\Theta_0 - \sin^2\Theta_{c2}}}{\sqrt{\sin^2\Theta_0 - \sin^2\Theta_{c1}} + \sqrt{\sin^2\Theta_0 - \sin^2\Theta_{c2}}}, \tag{3.8a}$$

$$r_{12}^p = \frac{\dfrac{\cos\Theta_{c1}}{\cos\Theta_{c2}}\sqrt{\sin^2\Theta_0 - \sin^2\Theta_{c1}} - \dfrac{\cos\Theta_{c2}}{\cos\Theta_{c1}}\sqrt{\sin^2\Theta_0 - \sin^2\Theta_{c2}}}{\dfrac{\cos\Theta_{c1}}{\cos\Theta_{c2}}\sqrt{\sin^2\Theta_0 - \sin^2\Theta_{c1}} + \dfrac{\cos\Theta_{c2}}{\cos\Theta_{c1}}\sqrt{\sin^2\Theta_0 - \sin^2\Theta_{c2}}}. \tag{3.8b}$$

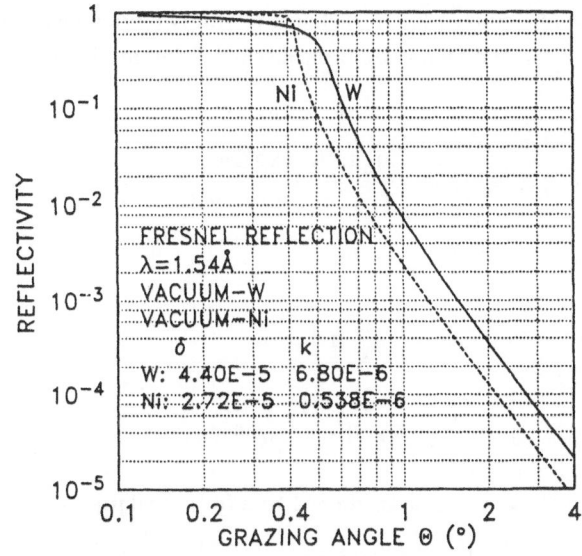

Fig. 3.2. Typical grazing incidence reflectivity at $\lambda = 1.54\text{Å}$

For $\Theta_c \ll \theta_0$ the reflected intensity for both polarizations can be approximated by

$$R = \frac{|\sin^2\Theta_{c2} - \sin^2\Theta_{c1}|^2}{16\sin^4\Theta_0}, \tag{3.9a}$$

$$R = \frac{(\delta_2 - \delta_1)^2 + (k_2 - k_1)^2}{4\sin^4\Theta_0}, \text{ for } \delta, k \ll 1. \tag{3.9b}$$

Above the critical angle the reflectivity drops very fast proportional to $1/\sin^4\Theta$ to very small values with increasing grazing angle.

The Fresnel equation have always the form

$$r_{12} = \frac{q_1 - q_2}{q_1 + q_2},$$ (3.10)

where q is called the admittance in electrical engineering and optics, and the normal component of the momentum transfer in x-ray diffraction.

4. MULTILAYER THEORY

The methods for the calculation of the performance of a multilayer in the spectral region of visible light can also be used for x rays and can be found in any textbook on thin films [4.1]. As a reference for our discussion we give in the following section the formulas for the recursive method, where the parameters for a multilayer are obtained by repeated use of the formula for a single film.

The reflected amplitude of a single film of thickness d with two boundaries (top and bottom) with amplitude reflectivity r_t, r_b and amplitude transmission of the top boundary t_t^+, t_t^- is given by

$$r_f = r_t + \frac{r_b \, t_t^+ \, t_t^- \, \exp(2i\varphi)}{1 + r_t \, r_b \, \exp(2i\varphi)},$$ (4.1)

with

$$\varphi = \frac{2\pi}{\lambda} nd \cos\alpha,$$ (4.2a)

$$\varphi = \frac{2\pi}{\lambda} d\sqrt{n^2 - n_0^2 \sin^2\alpha_0}.$$ (4.2b)

The amplitude transmission $t^+ = t_{12}$ from medium 1 to 2 for s- and p-polarization is given by

$$t_{12}^s = \frac{2n_1 \cos\alpha_1}{n_1 \cos\alpha_1 + n_2 \cos\alpha_2},$$ (4.3a)

$$t_{12}^p = \frac{2n_1 \cos\alpha_1}{n_2 \cos\alpha_1 + n_1 \cos\alpha_2},$$ (4.3b)

and the corresponding value $t^- = t_{21}$, by exchange of the indices 1 and 2 in eqs. (4.3).

One can easily verify from the Fresnel equation that

$$t_{12} \, t_{21} + r_{12}^2 = 1$$ (4.4)

and use this expression to rewrite eq. (4.4) to obtain

$$r_f = \frac{r_t + r_b \exp(2i\varphi)}{1 + r_t \, r_b \exp(2i\varphi)}.$$ (4.5)

278

Equations (4.1) and (4.5) are valid for any thin film, even if it is bounded on either side by multiple layers. Recursive application of eq. (4.5) gives the reflectivity of any coating, for example by starting with r_b using the Fresnel coefficient between the first film and the substrate and then using the calculated value of r_f as a new value for r_b when the next film is added.

Imperfect boundaries can easily be incorporated into the calculation by describing the reduction of the reflected amplitude at a boundary by a Debye-Waller factor

$$DW = \exp - 2\left(\frac{2\pi\sigma\cos\alpha}{\lambda} \right)^2 \tag{4.6}$$

and by multiplying the Fresnel coefficients (eq.(3. 1)) with DW before using them in the recursive formulas eqs.(4.5) or (4.1). By using eq. (4.5) for the calculation, we assume the boundary consists of a smooth transition layer of width σ without any scattering losses. The transition layer reduces the reflectivity but increases the transmission by an equivalent amount (see eq. (4.4)).

Boundaries with scattering losses can be described by eq. (4.1). By using the values for the transmission coefficients t, as given by eqs. (4.3), we assume a reduced reflectivity without increase in transmission at each boundary, equivalent to a loss at each boundary corresponding to scattering away from the specular beam.

High reflectivity is obtained by a multilayer, when all periods (period length p) in the multilayer add in phase. The phase factor in eqs. (4.1) and (4.5) should be 2π (or $2\pi m$ for reflectivity in m^{th} order). From eqs. (4.2) with $n_0 = 1$ and $n = 1 - \delta + ik$ and neglecting quadratic terms in δ and k we obtain

$$m\lambda = 2 p \sin\Theta_0 \sqrt{1 - \frac{2(\delta - ik)}{\sin^2\Theta_0}}. \tag{4.7}$$

This expression becomes the Bragg equation for $\delta = k = 0$. For small grazing angles the correction to the Bragg condition can become quite large. One can deduce the effective index of a coating from a measurement of the wavelength and the angle of the reflectivity maximum [4.2]. Assuming $k < < \delta$ we obtain

$$\delta_{eff} = \sin\theta_0 (\sin\theta_0 - \frac{m\lambda}{2p}). \tag{4.8}$$

where the effective index is defined by

$$\delta_{eff} = \frac{d_1\delta_1 + d_2\delta_2}{d_1 + d_2} ; \tag{4.9}$$

d_1, d_2, $(d_1 + d_2 = p)$ are the individual thicknesses and δ_1, δ_2 the indices of the two materials that make up one period of the multilayer coating.

Vector Model and Fourier Transform

Multiple reflection and the depletion of the incident beam are neglected in the kinematical theory of X-ray diffraction or the vector model of the optical theory. The reflectivity is simply calculated as the vector sum of the reflected amplitudes of the individual boundaries:

$$r = \sum_{j=0}^{N} r_j e^{2i\varphi_j}, \tag{4.10}$$

where φ_j is now the phase retardation of the jth boundary measured from the top surface r_0. Introducing a variable q which is proportional to the momentum transfer at the reflection and has the unit of reciprocal space

$$q = \frac{4\pi}{\lambda} n\cos\alpha = \frac{4\pi}{\lambda} n\sin\theta, \tag{4.11}$$

we can rewrite eq. (4.10) as a Fourier series

$$r = \sum_{j=0}^{N} r_j(z_j) e^{iqz_j}, \tag{4.12}$$

where z is the depth coordinate in the multilayer and r_j is the reflected amplitude of the boundary at coordinate z_j.

For the case that r_j is a continuous function r(z) eq. (4.12) becomes a Fourier integral

$$r(q) = \int_0^{\infty} r(z) e^{iqz} dz \tag{4.13}$$

with the inversion

$$r(z) = \frac{1}{2\pi} \int_0^{\infty} r(q) e^{-iqz} dq. \tag{4.14}$$

The Fresnel reflection coefficient eq. (3.1) can be expressed in the variable q as

$$r_{12} = \frac{q_1 - q_2}{q_1 + q_2}, \tag{4.15}$$

which for δ, $k << 1$ can be approximated by

$$r_{12} = \frac{\Delta(\delta - ik)}{2\sin^2\theta_0} . \tag{4.16}$$

The calculation of a multilayer reflectivity by Fourier transform takes considerably less computer time than the full theory. It is also possible to approximate the depletion of the incident beam by multiplying r(z) in eqs. (4.12) or (4.13) with a weighting factor $e^{-z/z_{max}}$, where z_{max} represents an estimate of the penetration length of the radiation into the multilayer. The inverse Fourier relation eq. (4.14) can be useful as a first step in the design of a multilayer with a desired reflectivity curve r(q). The main difficulty to derive the structure of a multilayer from a measured reflectivity curve r(q) is the fact that usually only the amplitude but not the phase of r(q) is known.

Atomic Planes Within a Thin Film

The optical theory assumes homogeneous materials and has to be modified, when films are only a few atoms sizes thick. We can still use the optical theory, if we restrict ourselves to specular reflection from planes parallel to the surface of a mirror. For this case the atomic structure of a material is of importance only in the z-direction normal to the surface and we can assume homogenous properties in the x,y-directions. (However, in general we will have to know the 3-dimensional distribution of the atoms to derive the density of atoms and the resulting refractive index along the z-direction.) We divide each film into subfilms, which represent the atomic planes within each film and apply the optical theory to this structure in the usual way.

It is instructive to consider first the influence of such a subdivision on the reflectivity of a single film surrounded by vacuum in the approximation of the vector model or the kinematical theory (no multiple reflections (Fig. 4.1)).

We obtain for a single film the reflected amplitude

$$r_f = r_{12} + r_{21} e^{2i\varphi} \tag{4.17}$$

where φ is defined in eq. (4.2). and r_{12}, r_{21} are the amplitude reflections at the top and bottom of the film. We have $r_{21} = -r_{12}$, and neglecting absorption, we obtain a maximum reflectivity of the film for $\varphi = \frac{\pi}{2}$. This value corresponds for normal incidence to the quarter wave film with a reflected amplitude $r_f = 2r_{12}$ and reflectivity $R_f = 4R_{12}$. The reflected amplitude at the boundary is obtained from the Fresnel formula and for normal incidence we obtain

$$r_{12} = \frac{\delta - i k}{2 - \delta + i k} \simeq \frac{1}{2} (\delta - i k) \tag{4.18a}$$

$$r_{12} \simeq \frac{r_0 \lambda^2}{4\pi} N_{at} (f_1 - i f_2). \tag{4.18b}$$

The approximations $\delta, k \ll 1$ are used at the right side of eq. (4a) and in eq. (4b).

We now replace the quarter wave film by N subfilms filling a fraction γ of the volume of the single film with the same amount of matter. Each term in eq. (4.17) is replaced by N terms, representing the boundaries of the N subfilms and the Fresnel coefficient at each boundary is replaced by r_{12}/γ, due to the higher density of the subfilm. We increase the density of the subfilms in order to keep the average density equal to that of the homogeneous film. After summation, we obtain for the reflected amplitude of the stack

$$\frac{r_f}{r_{12}} = \frac{2}{\gamma} \frac{1 - e^{i\frac{\pi}{N}\gamma}}{1 - e^{-i\frac{\pi}{N}}}. \tag{4.20}$$

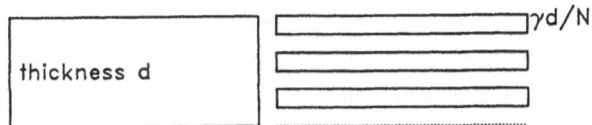

Fig. 4.1 A single film of thickness d is replaced by N subfilms (atomic planes) of thickness $\gamma d/N$ with a spacing of d/N.

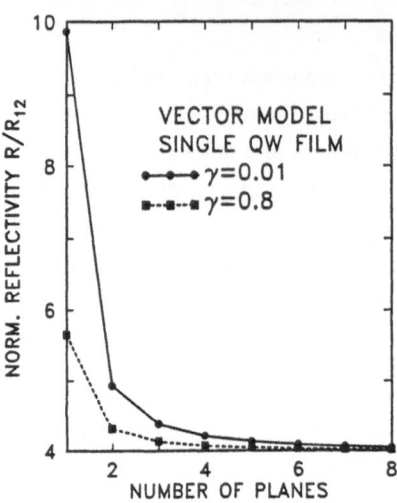

Fig. 4.2 Reflectivity of a single quarter wave film which has been replaced by N subfilms for values of the fill factor $\gamma=0.01$ and 0.8 obtained with the vector model (eq. (4.9)). R_{12} is the reflectivity of one boundary of the quarter wave film.

Figure 4.2 is a plot of the normalized reflected intensity for values for up to 8 atomic planes and 2 values of γ. We see that the reflectivity of the N-plane system becomes higher than that of a homogeneous film, for $N < 5$.

The enhancement of the reflectivity is due to the fact that near the quarter wave thickness the reflectivity of a film has a maximum and does in first order not change with thickness, while the reflectivity of the single boundary increases as r_{12}/γ, if we maintain the average density over the volume of the quarter wave film. In the limit $\gamma \to 0$ and for $N=1$, we obtain $r_f/r_{12} = \pi$ from eq. (4.20). The reflectivity increases from $4\,r_{12}$ to a value $\pi^2 r_{12}$, when the material in a single quarter wave film is concentrated into a thin plane. Most of the increase in eq. (4.20) is already obtained when γ is still of the order 1, a value $\gamma=0.1$ is for all practical purposes equivalent to $\gamma=0$. As a consequence the approximations δ, $k \ll 1$ used in eq. (4.18) remain valid for all practical cases in the x-ray region.

The approximations which lead to eq. (4.9) are most useful for the understanding of the physical principle and for an estimate of the maximum possible enhancement. The full, rigorous theory gives similar results. Figure 4.3 gives one example. The dashed curve is obtained for homogeneous films of thickness $d_1 = 2a$ and $d_2 = 3a$ (a=3.33 Å), $\delta_1 = 5.41\text{x}10^{-3}$, $k_1 = 3.97\text{x}10^{-3}$, $\delta_2 = 1.54\text{xx}10^{-3}$, $k_2 = 4.63\text{x}10^{-4}$; the value a corresponds to the spacing of atomic planes. For the full curve each subfilm of thickness a is replaced by two films: one film of 1/10 the previous thickness with 10 times larger optical constants and one film of vacuum extending for the remaining 90% of the volume. The curve is identical with calculations obtained in ref. [4.3] from Ewald's dynamical theory of x-ray diffraction [4.4]. (On the scale of the plot $N=1024$, and $\gamma=0.1$ gives the same results as $N=\infty$ and $\gamma=0$.)

The description of a film by its atomic planes is suitable for epitaxially grown films, where atomic planes are well defined. No such system with sufficient contrast in the optical constants have been fabricated up to now. All good multilayer mirrors produced up to now are better described by the optical theory with homogeneous layers. This is due to the fact that film materials are in most cases amorphous (for best boundary sharpness) and in the remaining cases polycrystalline.

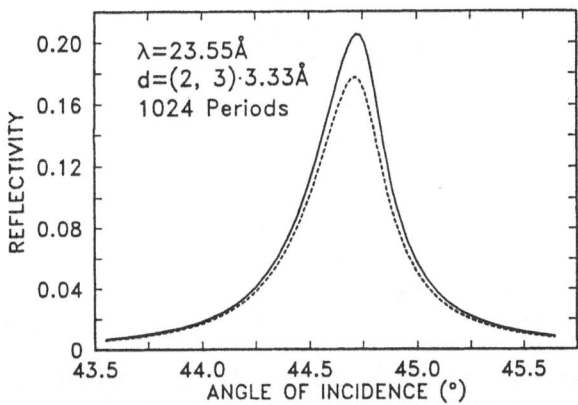

Fig. 4.3 Calculated reflectivity of a multilayer with 1024 periods using the optical multilayer theory. Each period consists of 2 layers with $d_1 = 6.66$Å. $d_2 = 9.99$Å , $\delta_1 = 5.41 \times 10^{-3}$, $k_1 = 3.97 \times 10^{-3}$, $\delta_2 = 1.54 \times 10^{-3}$, $k_2 = 4.63 \times 10^{-4}$ for the dashed curve. The full curve is obtained by replacing the films with 2 and 3 thinner films ($\gamma = 0.1$) and is identical to that obtained in Ref. [4.3] from Ewald's theory.

5. MULTILAYER DESIGN

There are two well known geometries which utilize interference effects to produce high reflectivity mirrors: the quarter wave stack, and the ideal Bragg crystal (Fig.5.1). The quarter wave stack consists of alternating layers of high (H) and low (L) refractive index, each of the same optical thickness nd $= \lambda /4$ (for normal incidence), such that all boundaries add with equal phase to the reflected wave. For the case that both film materials are completely absorption free, the quarter wave stack gives the highest reflectivity with the fewest number of layers and approaches a reflectivity R = 100%. In an ideal Bragg crystal the atomic planes are usually much thinner than a quarter wave, different planes are spaced $\lambda /2$ apart (for normal incidence) and contribute in phase to the reflected wave. For the case that the space between the atomic plane is absorption free, the reflectivity approaches R = 100% for a large number of layers even if the thin atomic planes are absorbing. The elimination of the absorption losses is due to the fact that the atomic planes are located at the nodes of the standing wave field generated by the superposition of the incidence and reflected wave (see Fig.5.1). This behavior is made plausible in Fig. 5.2. In front and within the top layers of a perfect reflector with R $\rightarrow$ 1 exists a perfect standing wave with zero intensity in its nodes. The intensity increases quadratically with distance from the node. This leads to an absorption that decreases as the third power of the thickness for a very thin film. The reflectivity of a very thin film decreases only quadratically with thickness. Therefore, the absorption losses can be made arbitrarily small compared to the reflectivity if thinner and thinner films are used. In the limit, a very thin film of absorbing material is equivalent to an absorption-free film of low reflectivity. In the quarter wave stack, on the other hand, each layer extends from a node to an antinode; if one or both layers are only slightly absorbing the performance of the quarter wave stack deteriorates fast due to the large absorption losses at the antinodes of the standing wave field.

Figure 5.3 demonstrates that even for the case that no differences in the refractive index are available, very high reflectivities can be obtained by alternating an absorbing material with an absorption-free spacer. The reflectivity approaches 100% for a stack with many layers of very thin absorbers. If no absorption free spacer material is available the penetration depth and the maximum possible reflectivity are limited by the absorption (Fig. 5.4).

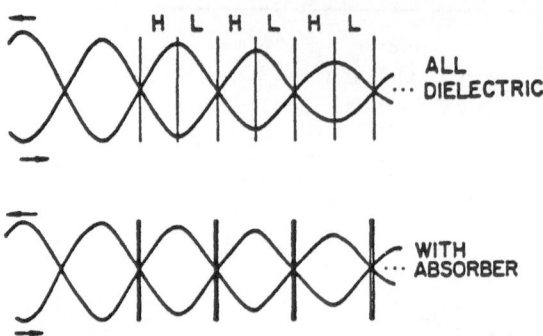

Fig.5.1. Two designs for high reflectivity mirrors. The quarter-wave stack (top) gives the fastest increase in reflectivity with increasing number of layers, but deteriorates fast in performance if one of the layers is absorbing. The ideal Bragg crystal (bottom) minimizes absorption by positioning the layers (atomic planes) into the standing wave produced by the superposition of the incident and reflected wave.

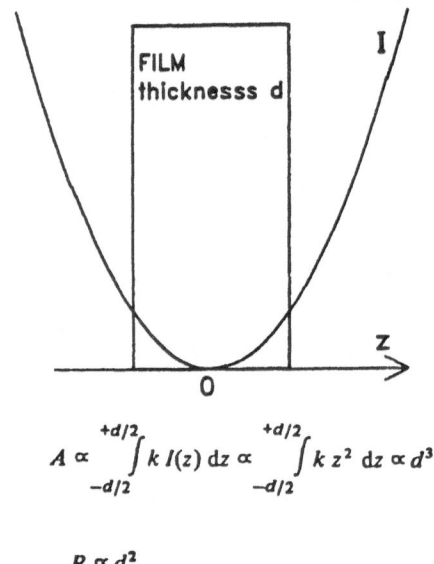

$$A \propto \int_{-d/2}^{+d/2} k\, I(z)\, \mathrm{d}z \propto \int_{-d/2}^{+d/2} k\, z^2\, \mathrm{d}z \propto d^3$$

$$R \propto d^2$$

Fig. 5.2. A thin film centered at the node of a standing wave.

Material Selection Rules

For the highest possible reflectivity materials should be selected by the following rules:

1. Select a first material with the lowest possible absorption constant k_{sp} as a spacer material.

2. Find a second material with the largest possible reflection coefficient at the boundary with material 1.

3. If several materials give similar reflection coefficients, choose the one with the smaller absorption coefficient.

4. Make sure that the material can be deposited with sharp, smooth boundaries, such that $\sigma < 0.1p$ for the desired multilayer period p.

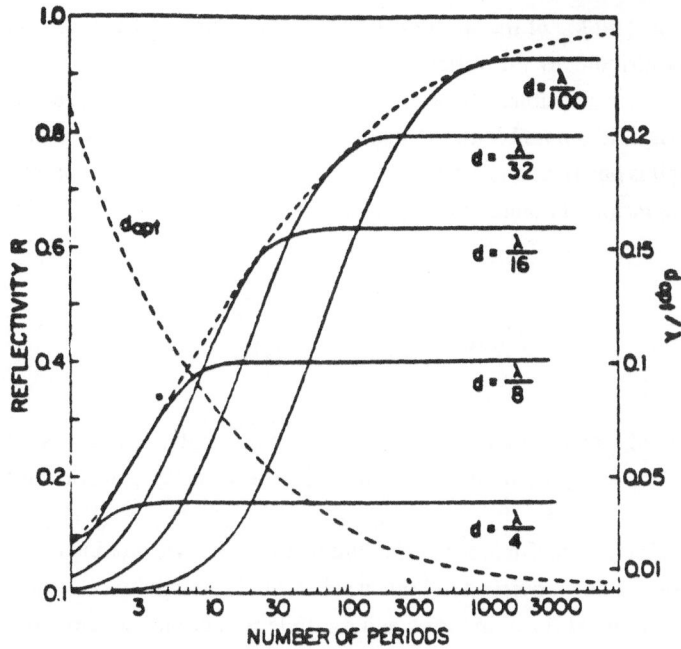

Fig. 5.3. Normal incidence reflectivity for a periodic structure of period $\lambda/2$, where an absorbing material with $k = 0.5$ alternates with a non-absorbing material vs. the number of periods in the structure. The parameter d is the thickness of the absorbing material. Dashed curves: Optimal thickness d_{opt} of the absorber and highest reflectivity obtainable with a periodic structure. The dot represents the reflectivity obtainable with an optimum aperiodic structure of nine layers (4.5 periods). The refractive index is assumed to be 1 for all layers (from Ref.1.4, 1972).

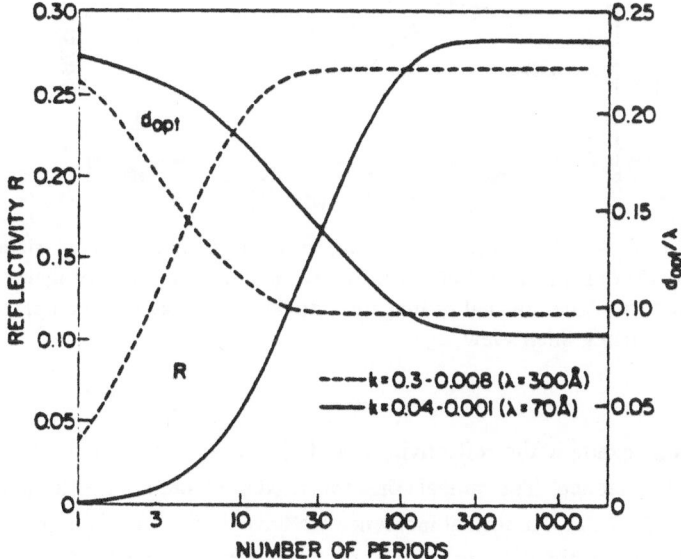

Fig.5.4 Normal incidence reflectivity for periodic structures of period $\lambda/2$, where a material with high absorption alternates with a material of lower absorption. The thickness d_{opt} of the strong absorber is chosen to give the highest peak reflectivity. The chosen k values are possible near $\lambda = 300$ Å and 70 Å.

285

The absorption constant k_{sp} of the spacer material determines the maximum number of periods N_{max} which can contribute to the reflectivity. For $k_{sp} = 0$ there is no limit to N_{max} and a reflectivity R = 100% is possible. Figure 5.3 demonstrates this fact. Even for the case that δ = 0 for both materials, it is possible to get high reflectivities in the limit that the absorbing material is very thin and is positioned in the nodes of the standing wave in the coating. For absorbing spacer materials the maximum number of layers is

$$N_{max} = \frac{\cos^2\alpha}{2\pi k_{sp}} = \frac{\sin^2\theta}{2\pi k_{sp}} , \qquad (5.1)$$

where α is the angle of incidence and $\theta = 90 - \alpha$ the grazing angle. The limitation in the number of periods defines also an upper limit for the maximum reflectivity and for the highest spectral resolution $\lambda / \Delta \lambda$.

The Fresnel reflection coefficient r at the boundary of the two coating materials defines the minimum number of periods required to obtain a substantial reflectivity. Considering the quarter wave stack and neglecting absorption losses and multiple reflections, we can estimate N_{min} as

$$N_{min} = 1/(2 \bullet |r|) . \qquad (5.2)$$

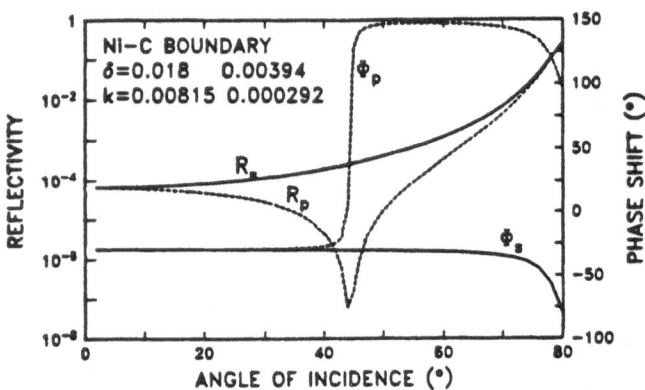

Fig.5.5. Reflectivities R_s and R_p and phase shifts on reflections for s and p polarization at a Ni-C boundary using optical constant from Ref.2.1 for λ = 50.7 Å. The angle of incidence is the external angle of incidence. The internal propagation angle in each layer is complex and obtained from the plotted angle using Snell's law.

Figure 5.5 gives as an example the reflectivity R = $|r|^2$. at a Ni-C boundary as a function of the external angle of incidence. The optical constants used are from Henke [2.1] and correspond to a wavelength λ = 50.7 Å . At normal incidence we have $|r| = 10^{-2}$ and we conclude from (5.2) that we need at least about 50 periods to obtain a substantial reflectivity. We see from Fig.5.4 and eq.(5.1) that both N_{min} and N_{max} decrease for more grazing angles of incidence. In a first approximation, the maximum reflectivity changes only little with changes in the angle of incidence. However, because fewer layers are required at more grazing incidence angles, the integrated reflectivity increases with decreasing grazing angles. The period length in a multilayer increases with decreasing grazing angle for a fixed wavelength; therefore, the influence of roughness decreases for multilayers designed for more grazing angles of incidence. For normal incidence and using the fact that all refractive indices are close to 1, we can express

$$N_{min} = 1 / \sqrt{\Delta n^2 + \Delta k^2} \tag{5.3}$$

by the differences Δn and Δk in the optical constants of the coating materials. Designs for the highest possible resolution are similar to a Bragg crystal while designs for largest integrated reflectivity are more similar to the quarter wave stack. However, due to the fact that all materials are absorbing in the soft x-ray region, the two design limits never represent an optimum design in the soft x-ray region.

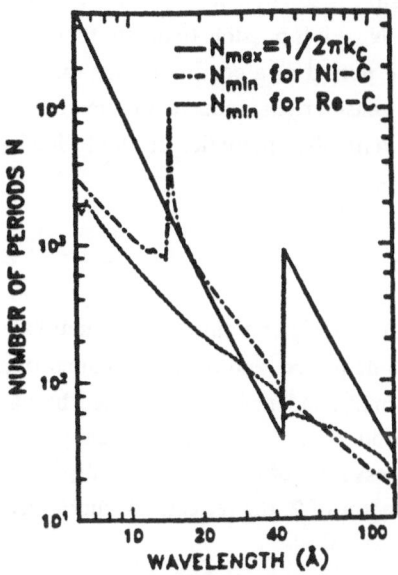

Fig.5.6. The maximum number N_{max} of periods possible in a multilayer with a carbon spacer layer due to the absorption of carbon, and the minimum number N_{min} in a Ni-C and Re-C system required for reflectivities approaching 1. Optical constants from Ref.2.1, normal incidence.

Figure 5.6 is a plot for normal incidence of N_{max} and N_{min} for Ni-C and Re-C multilayers. We see that for wavelengths $\lambda > 44$ Å the values for N_{max} are considerably larger than those for N_{min} such that we can expect good reflectivities from both material combinations. The values for N_{min} are comparable for both material combinations in this wavelength region. However, the absorption coefficient of Ni is smaller than that of Re and, therefore, rule 3 will let us prefer the NiC system. Figure 5.7 confirms this prediction; here, the maximum reflectivity R_{max} obtainable with a large number of periods is plotted for both systems and the Ni-C clearly gives higher reflectivities.

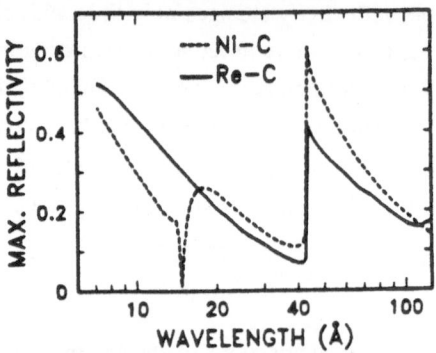

Fig. 5.7. Maximum reflectivity obtainable with periodic Ni-C and Re-C multilayers for normal incidence, a large number of periods and smooth, sharp boundaries. Optical constants from Ref.2.1.

The ratio N_{max}/N_{min} gives us a measure on the variety of possible coating designs. For spectroscopic applications the possible resolution $\lambda/\Delta\lambda$ can be between the values of N_{min} and N_{max}. For the case that $N_{max} >> N_{min}$ it is possible to obtain relative spectral bandwidths larger than $1/N_{min}$ by changing the base period of a coating throughout its depth. Coatings can be classified by being either periodic, quasiperiodic (constant period but changing distribution of the two materials within a period) or generally aperiodic. In the following, we will give some performance curves for each case.

Periodic Designs

Periodic designs have only three design parameters: the multilayer base period, the distribution of the two materials within one period, and the total number of periods. Figure 5.8 shows how the reflectivity curves for normal incidence are influenced by changing the thickness ratio of the two materials within one period. The curves in Fig. 5.8 show that the performance is rather insensitive to changes in the thickness ratio of the two films; despite the fact that the thicknesses of the Re films differ by nearly a factor of 2, the reflectivity curves are very similar. The dashed

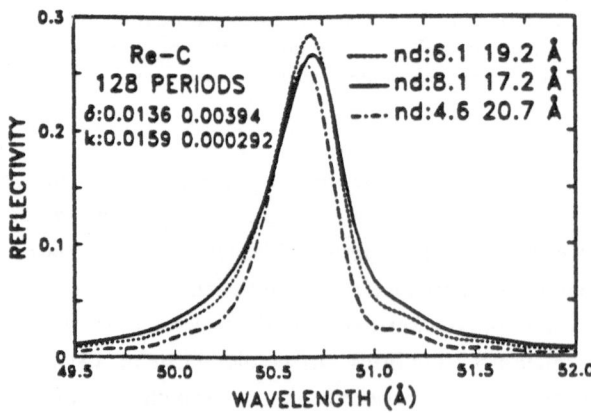

Fig.5.8. Calculated normal incidence reflectivity versus wavelength for 128 period coatings of Re-C for different thickness ratios of the two materials within one period.

curve represents the design for the highest peak reflectivity; the full curve is the largest integrated reflectivity. The dash-dotted curve has a higher spectral resolution than the other designs. The spectral resolution, in this case, is limited by the number of periods and can be increased further to a value close to N_{max} by using a larger number of periods. To obtain higher resolution, more layers have to contribute to the reflectivity and the contribution of each individual period has to be reduced in order to reach the deeper layers. This is achieved by making the heavy material films very thin, so that the reflections from its front and back side cancel to a large extent. It is not required to choose a material combination with low values of the Fresnel coefficient $|r|$ in order to achieve the maximum possible spectral resolution.

Figure 5.9 gives the reflectivity curve of the design with the highest peak reflectivity for the Ni-C system film. The peak reflectivity is higher than for the corresponding Re-C multilayer. Due to the smaller absorption in the Ni, all Ni films can be made thicker for similar absorption losses. This results in a more in-phase contribution between the waves reflected from the front and back of each film, giving higher reflectivity per period.

Analytical formulas for the optimum design and the performance of periodic multilayers with a large number of periods have been derived by Vinogradov and Zeldovich [5.1]. The optimum ratio β_{opt} of the thickness of the heavy material (index $n_1 + ik_1$) divided by the period of the multilayer is obtained from a solution of the equation

$$\tan(\pi\beta_{opt}) = \pi\left(\beta_{opt} + \frac{n_2 k_2}{n_1 k_1 - n_2 k_2}\right). \tag{5.4}$$

For $k_2 << k_1$ an approximate solution for small values of β_{opt} is

$$\beta_{opt} \simeq \frac{1}{\pi}\left(\frac{3\pi n_2 k_2}{n_1 k_1}\right)^{\frac{1}{3}}. \tag{5.5}$$

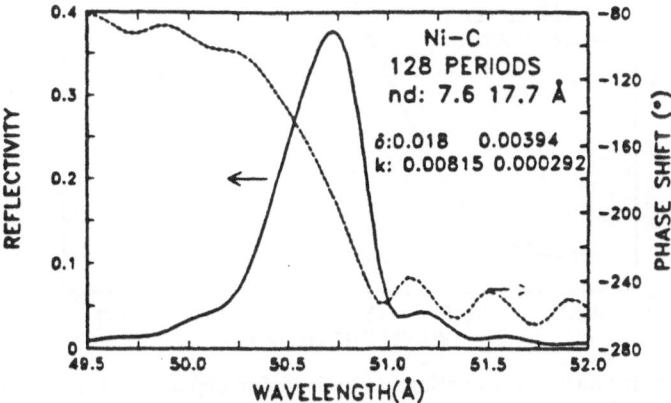

Fig.5.9. Calculated normal incidence reflectivity for a Ni-C multilayer with 128 periods and phase shift on reflection.

289

Figures 5.3 and 5.4 showed that the optimum thickness of the heavy film in a periodic multilayer decreases if the number of periods is increased. The optimum design which maximizes the peak reflectivity with the fewest number of layers is quasiperiodic; the thickness of the heavy material decreases from the bottom to the top of the multilayer in a similar way as the curve for d_{dopt} in Fig. 5.3. In contrast to Fig. 5.3, for real materials in the x-ray range d_{dopt} does not go to zero for a large number of layers but settles at a finite thickness which can be obtained from eq. (5.5) and is mainly determined by the absorption in the materials (see Fig. 5.4). One can obtain the optimum design quite easily on a computer by searching during the deposition of each period for the thickness ratio that gives the largest increase in reflectivity. The result is a coating that resembles a quarter wave stack at the bottom with the thickness of the heavy material gradually decreasing towards the top. The reflectivity of this design is always higher than that of the optimum periodic design, especially for multilayers which contain only few periods ($N < 10$). For multilayers with a large number of periods the differences between the optimum quasiperiodic and the optimum periodic design become very small due to the fact that the designs differ only in the bottom layers, while the top layers give the largest contribution to the reflectivity. Figure 5.10 shows a comparison of the reflectivity versus wavelength curve for a 127 layer quasiperiodic and a 128 layer periodic design both optimized for maximum reflectivity for $\lambda = 50.75$ Å and an angle of incidence of $45°$. The optical constants used are those from Hagemann et al. [5.2] for Au and C. In general, the optical constants of Ref.[5.2] give a slightly inferior performance than the more recent compilation of Henke et al.[2.1].

The thickness of the heavy film in each period determines the weight of the contribution of this period to the total reflectivity. The weight is maximum for the case that the film covers 50% of the period and goes to zero if the film thickness goes to zero. For a periodic structure and a large number of periods the weight of each period decreases exponentially with depth from the top of a coating and the line shape approaches a Lorentzian (= Fourier transform of the exponential). The possibility of adjusting the weight of each period in a quasiperiodic design make it

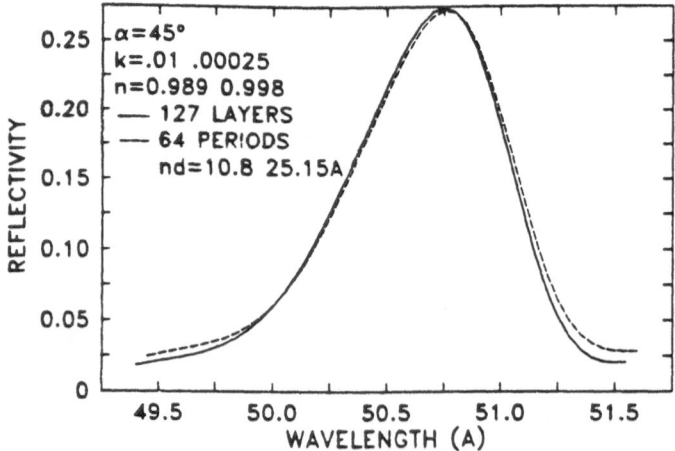

Fig.5.10. Reflectivity curves for coatings optimized for highest peak reflectivity. Full curve: quasiperiodic design with 127 layers; dashed curve: periodic design with 128 layers.

possible to produce coatings with modified line shapes. Quasiperiodic designs are most important for $\lambda > 150$Å, where only few layers are needed. In addition, they are preferred for *in-situ* monitoring in order to increase the reflectivity as fast as possible in the first few layers for a stronger monitor signal.

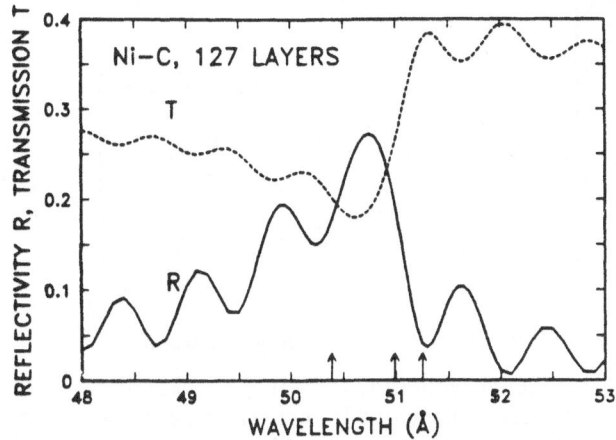

Fig. 5.11. Normal incidence reflectivity and transmission versus wavelength for a multilayer which contains three different periods. The peak wavelengths for each period are marked by arrows. The smaller periods are located at the top of the stack.

Aperiodic Designs

In the wavelength range of visible light where absorption-free materials are available, it is in principle possible to design coatings which have any predetermined reflectivity curve. The calculations are much more involved than the straightforward determination of the performance of a given design. Computer algorithms to synthesize a coating, which have been developed for visible light, can be transferred to the design of x-ray coatings, however, the design freedom is much more limited due to the absorption in all materials.

The most important application of aperiodic designs for soft x rays will probably be the increase of the bandwidth and integrated reflectivity over that which is possible with periodic designs. Figure 5.11 gives the reflectivity and transmission curves for a design, where three coatings with slightly different periods are deposited on top of each other. The integrated reflectivity is more than a factor of 2 larger than that possible with a periodic design like that given in Fig.5.8. Please note that the total number of layers in Fig.5.11 is 127 versus 256 in Fig.5.9. With the optical constants of Ni and C and for wavelengths around $\lambda = 50$ Å near normal incidence an increase in the integrated reflectivity by about a factor of 4 is possible with aperiodic designs.

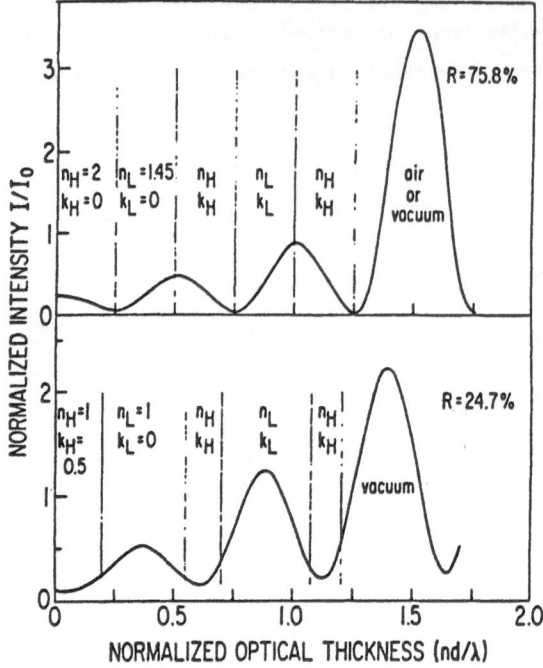

Fig.5.12. Intensity distribution within and in front of 5-layer reflectors. a) lossless quarter wave stack of 2 materials with different refractive indices. b) optimized coating of one absorbing and one non-absorbing material with the same real refractive index. Light is incident from the right.

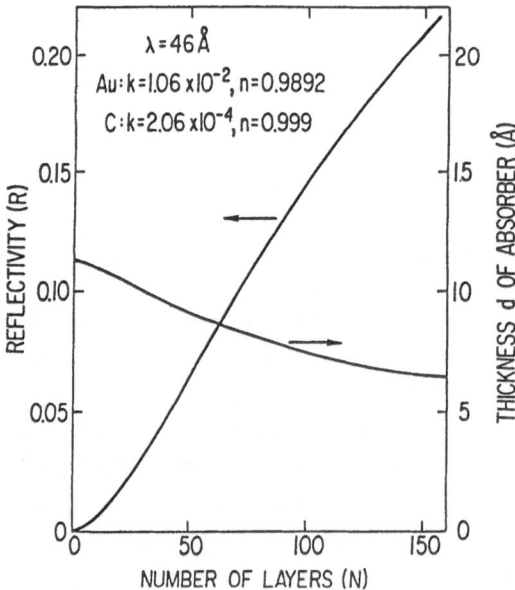

Fig. 5.13. Calculated normal incidence reflectivity for an optimized quasi-periodic mirror versus the layer number and the required thickness for the heavy material.

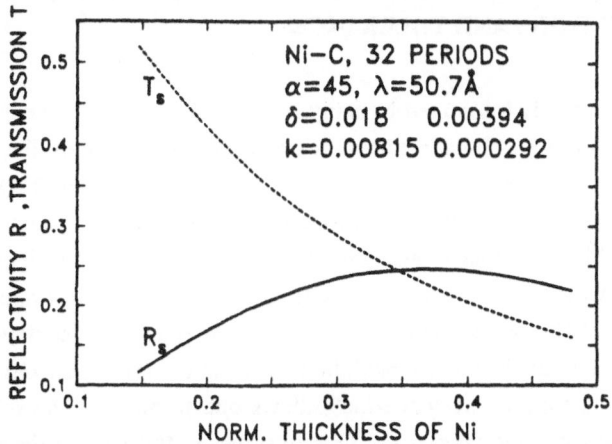

Fig. 5.14. Reflectivity and transmission of a 45°beamsplitter of Ni-C for lambda = 50.7Å as a function of the nickel thickness normalized to the period of the multilayer. The reflectivity for p-polarization is about 1000 times smaller

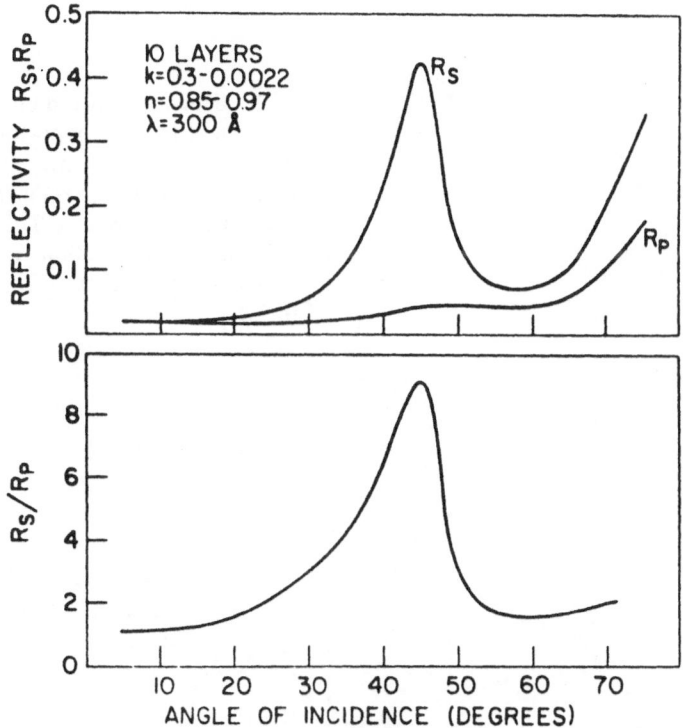

Fig. 5.15. Reflectivities for s- and p-polarization (top) and degree of polarization (bottom) versus angle of incidence for a polarizing mirror at $\lambda = 300$Å. Optical constants of substrate: n=0.89, k=0.092. Thickness of each layer counted from the substrate in Å :157.7, 72.8, 168.1, 57.7, 178.9, 45.6, 186.9, 38.9, 191.1, 37.1.

6. FABRICATION METHODS AND TOLERANCES

X-ray mirrors are presently fabricated in many laboratories and have become commercially available in 1984. Sputtering [6.1] and electron beam evaporation [6.2] are the most commonly used deposition methods, and satisfactory results have been obtained by both. In the sputtering system the thickness of each layer is simply controlled by timing. Sputter systems (Fig. 6.1) have shown a remarkable stability; multilayer systems with over 200 layers and insignificant thickness errors have been obtained. In the evaporation method, the thickness was originally only monitored with a quartz microbalance and thickness errors become significant for depositions of more than 10 layers. In order to avoid the accumulation of thickness errors, an x-ray reflectometer was installed into the evaporation system which allows one to monitor the reflectivity *in-situ* during the deposition of a coating (Fig. 6.2). One observes in this system the interference between the amplitude reflected from the top boundary of a growing film with that reflected from the stack underneath, and one period in the oscillations of this interference corresponds to one

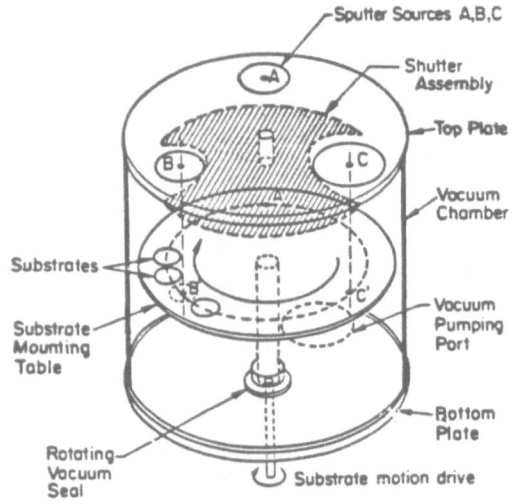

Fig. 6.1. Schematic of a multisource sputtering system for the deposition of multilayer films. A,B and C are sputter source positions. Substrates are on a rotating table and pass by each source. Thicknesses are determined by deposition rate, rotation speed and source substrate spacing (from ref.[6.1]).

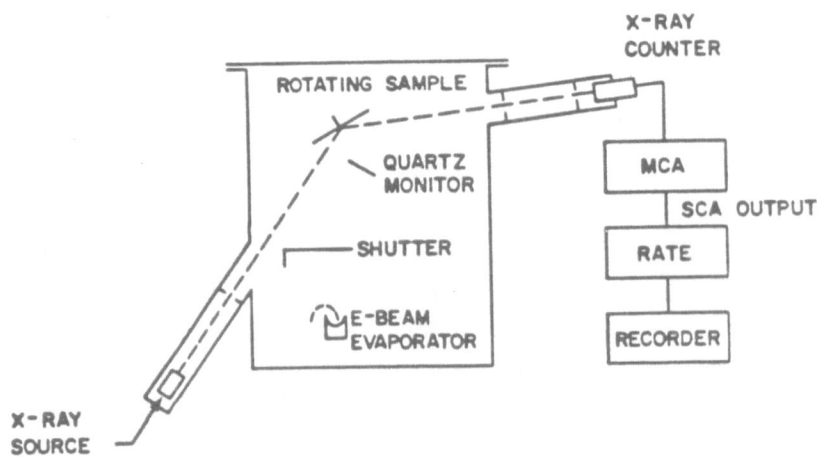

Fig. 6.2. Multilayer deposition system with *in-situ* monitoring of the reflectivity during deposition.

period (2 films) of the desired multilayer. This monitoring system is error compensating, i.e., an error in the thickness in one layer is automatically compensated for in the next layer, and a large number of layers can be deposited without the accumulation of thickness errors.

In-Situ Monitoring

Single Film Deposition

The in-situ monitoring system uses the interference between waves reflected from the top of a growing film (r_t) with those reflected from the structure beneath (r_b). An interference maximum is obtained for thicknesses when r_t adds in phase with r_b, a minimum for out of phase addition. The amplitude reflected from the film can be represented by

$$r = \frac{r_t + r_b \exp(2i\varphi)}{1 + r_t r_b \exp(2i\varphi)}, \tag{6.1}$$

with $\varphi = \frac{2\pi}{\lambda} nd \cos\alpha$; $n = 1 - \delta + ik$ is the refractive index, d the thickness of the film; α is the angle of propagation.

For the case that $r_t r_b \ll 1$ we can simplify eq. (6.1)

$$r = r_t + r_b \exp(2i\varphi), \tag{6.2}$$

$$r = r_t + r_b \exp\left(\frac{4\pi i}{\lambda} d \, Re(n \cos \alpha)\right) \exp\left(-\frac{4\pi}{\lambda} d \, Im(n \cos\alpha)\right), \tag{6.3a}$$

which can be approximated to

$$r = r_t + r_b \exp(2\pi i \, d/p) \exp((-2\pi \, d/p)(k/\cos^2\alpha_0)); \tag{6.3b}$$

$p = \lambda/2 \, Re(n \cos \alpha)$ is the period of the oscillation. We have used the approximation $Im(n \cos\alpha) = (1 - \delta) k / \cos\alpha_0$ in eq. (6.3b) and α_0 is the angle of incidence in vacuum. The amplitude vector r in eq. (6.3b) can be considered as the sum between the vector r_t and a vector which rotates around r_t with a slowly decreasing amplitude, and k can be determined from the decrement of this amplitude. When a single film is deposited on top of a multilayer structure, r_b represents the reflectivity of this structure and usually we have $r_b > r_t$. In this case it is advantageous to shift the reference plane for the phase to r_b by multiplying eq. (6.3a) with the factor $\exp\left(-\frac{4\pi i}{\lambda} d \, Re(n \cos alpha)\right)$. We obtain

$$|r| = |r_t \exp\left(\frac{-4\pi i}{\lambda} d \, Re(n \cos\alpha)\right) + r_b \exp\left(\frac{-4\pi}{\lambda} d \, Im(n \cos\alpha)\right)|, \tag{6.4a}$$

$$|r| = |r_t \exp(-2\pi i \, d/p) + r_b \exp((-2\pi \, d/p)(k/\cos^2\alpha_0))|. \tag{6.4b}$$

For the case that the second term is larger than the first, eq.(6.4) represents an oscillation with amplitude r_t around a slowly decreasing average value. The absorption index k can be obtained from the decrease of this average value.

A nice side feature of this monitoring system is that it becomes possible to judge the quality of films *in-situ* during the deposition. As an example, Fig. 6.3 shows the monitor signal (actually the square root of the reflectivity) obtained during the deposition of a film of Ni on a glass substrate. For a film which perfectly replicates the surface topography of the substrate, we expect an oscillation of the reflected amplitude around a constant average value -- the amplitude reflectivity of the top surface of the film. The damping of the oscillations is determined by the absorption index k of the film. A film that becomes rougher during the deposition will have a decreasing reflectivity at its top interface resulting in a reduction in the average between the maxima and minima of the monitor curve. Under the assumption that the decrease in reflectivity is described by eq. (6.3b), we can deduce from the measurement (dotted curve) the change in the roughness of the film during growth. For the example in Fig. 6.3 the experimental curve is obtained by theory if we assume that the roughness increases linearly with thickness to $\sigma = 3$ Å for 100Å thickness. Smooth boundaries are very important for most multilayer films and the *in-situ* reflectometer gives an easy evaluation of possible materials suitable for multilayers. Roughness values for different materials obtained by this method are given in Table 1. The data are in qualitative agreement with SEM micrographs of the same films. [6.3]. The smoothest boundaries are obtained with amorphous carbon films; these films can smooth the roughness produced by the growth of metal films, such that in many cases the roughness of a multilayer does not increase for increasing number of layers.

TABLE 1. Increase in the surface roughness σ of the top surface and absorption index k of evaporated films of thickness d, measured during the deposition by the method of Fig.6.3. Negative values indicate smoothing. (flgl.=float glass)

material	substrate	λ(Å)	$k(10^{-3})$	σ(Å)	d(Å)
B	ReW	44.8	2.8	1	200
C	ReW	44.8	0.12	0	200
C	ReW	67.6	1	0	150
C	Fe	23.6	0.3	0	100
C	Fe(Zr)	44.8	0.08	-11	465
Si	ReW-C	44.8	2.2	6	100
Si	Co	44.8	2.0	3	175
Fe(Zr)	ReW-C	44.8	3.3	11	175
Fe(Zr)	ReW-C	23.6	0.6	4	40
FeCoNiCr	flgl.	44.8	4.6	-3	175
Fe	Fe-C	44.8	7	4	100
Co	Co-Si	44.8	2.7	3	175
LiF	ReW	44.8	6.5	17	100
Ta	Si	44.8	5.8	5.2	100
W	Si	44.8		3	100
ReW	Si	44.8	10	0 to 2.5	100
ReW	flgl.	44.8		-3 to +2	100
Re	Si	44.8	10	5.2	100
Mo	flgl.	67.6	4.7	8.5	475
Mo	Fe-C	44.8	3.5	5	100
Os	Si	44.8	9	5.3	100
Ir	Si	44.8	7	4.2	100
Pt	Si, flgl.	44.8	8	2.8 to 6	100
Au	Si	44.8	9	8	100
AuPd	Si	44.8	7.5	3.3	100
PtIr	Si	44.8		7	100
Ni	Si	44.8	6	0-3	100

During the deposition of a multilayer the reflectivity oscillates, and one can keep the period constant and avoid the accumulation of thickness errors by locking the switch-over from material 1 to material 2 and vice versa to this oscillation. The method is widely used for the fabrication of dielectric interference coatings for visible light; in this region the highest reflectivity is obtained by switching to the high index material at a minimum and to the low index material at a maximum of the reflectivity curve. The resulting coating is the quarter wave stack with equal optical thickness nd for both materials. Due to the absorption of all materials, the thickness ratio is not as easily determined from the monitor signal in the soft x-ray region. The top boundary r_t is a boundary against vacuum during the deposition and becomes an internal boundary with a different phaseshift after the deposition of the next layer. A switch-over between materials at the extrema produces coatings where the heavy material usually is too thick for maximum reflectivity (see Fig.6.4). Furthermore, monitor wavelength and angle of incidence are usually different from those at which the coating will later be used.

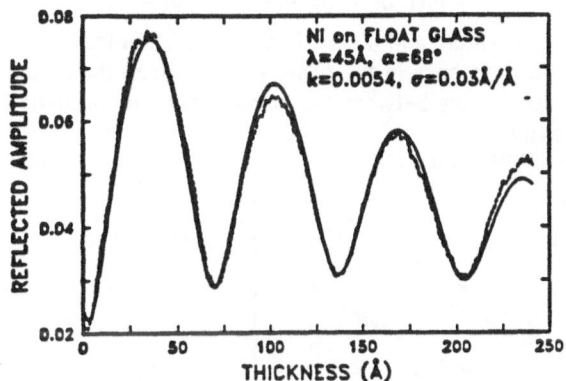

Fig. 6.3. Measured reflected amplitude ($\sqrt{R}$) obtained during the deposition of a Ni film (dotted) and the theoretical curves for a film, where the roughness of the top boundary increases linearly with thickness to $\sigma = 3$ Å for 100 Å thickness.

We determine in a first step the optimum coating parameters of a mirror at the intended wavelength and angle of incidence with a computer. In a second step we determine the optimum monitor wavelength and geometry using the calculated Fresnel coefficients versus vacuum for the two materials and the possible monitor wavelengths as a guide (Fig. 6.5). Figures. 6.6 and 6.7 are examples of calculated monitor curves. Figure 6.7 demonstrates a limitation of the *in-situ* monitoring system for multilayer periods p < 30Å. The signal is very small at the beginning of the deposition and the contrast of the modulation becomes small at the end. These problems can be reduced somewhat by the use of precoated monitor mirrors and by using different deposition geometries or times for monitor mirror and workpiece [6.4, 6.5].

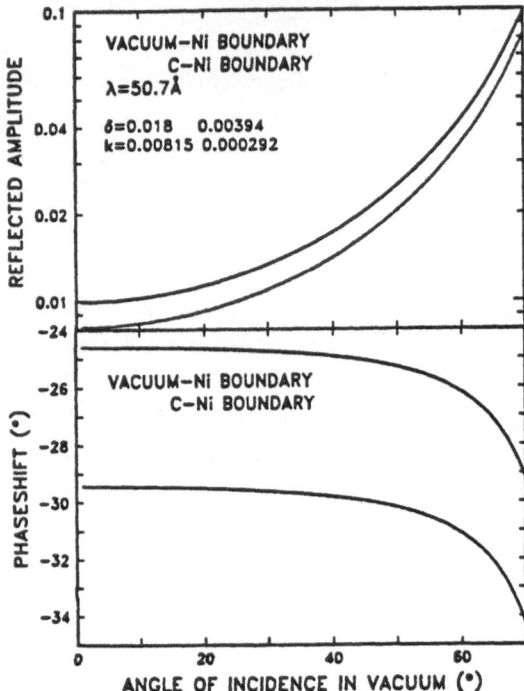

Fig. 6.4. Reflected amplitude and phase shift on reflection for a vacuum-Ni and a C-Ni boundary. The difference in the phase shift between the 2 boundaries is the reason that *in-situ* monitoring between the extrema does not produce the design for the highest reflectivity.

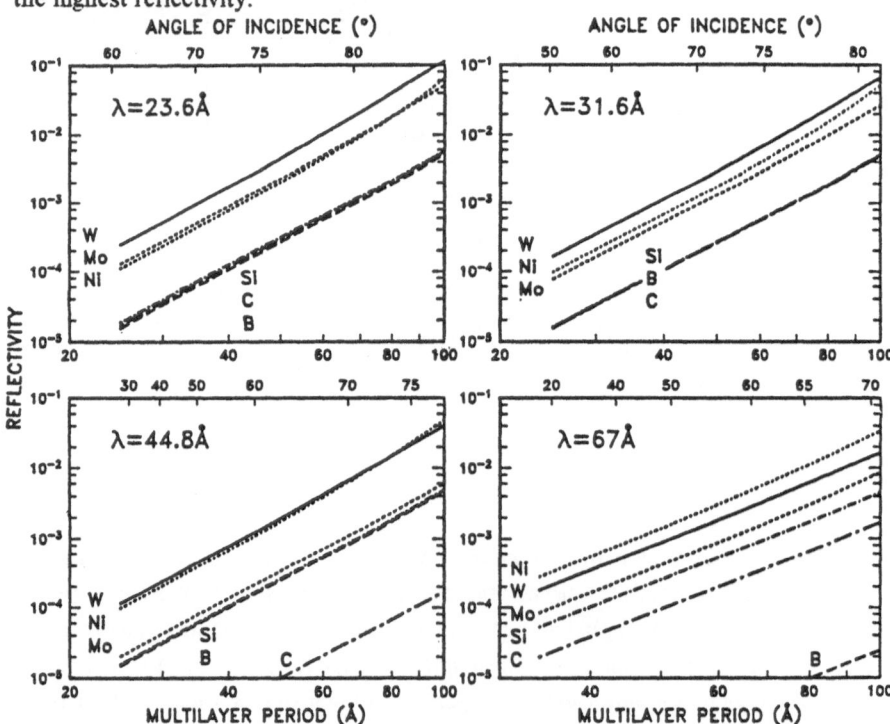

Fig. 6.5. Reflectivity of the vacuum boundary of six materials versus angle of incidence (top scale) or versus period of a multilayer (bottom scale) monitored at that angle in first order for 4 characteristic soft x-ray lines. For short wavelength the reflectivity decreases as the fourth power of the period.

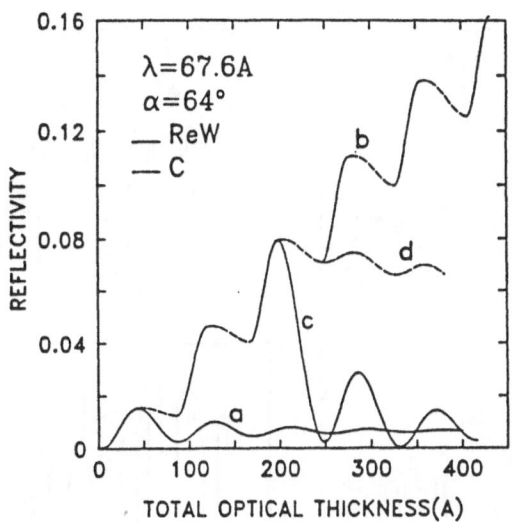

Fig. 6.6. Calculated reflectivity for $\lambda = 67$ Å , an angle of incidence $\alpha = 64°$, and s-polarization obtained during the deposition of various films. Optical constants n=0.98+ 0.09i for full curve sections (ReW) and n=0.998+0.001i for dashed sections (carbon). a: single film of ReW; b: a multilayer optimized for maximum reflectivity with the following layer thicknesses starting from the substrate (42.9, 43, 35, 45.3, 32.4, 48, 29.1, 50.4, 27.6, 51.4, and 26.6 Å); c: a ReW film deposited on top of an optimized 5-layer coating, and d: a carbon film on top of an optimized 5-layer coating.

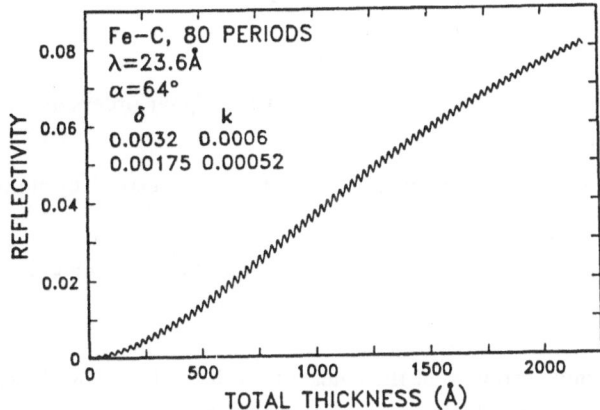

Fig. 6.7 Calculated reflectivity for the deposition of a Fe-C multilayer with period p = 27 Å.

Thickness Errors in Multilayer Mirrors

The influence of thickness errors can be studied numerically on a computer by comparing the calculated reflectivity of coatings, where errors of different magnitude have been introduced into the thickness of each layer. Depending on the deposition method errors can accumulate randomly from layer to layer or errors in one layer can be compensated in subsequent layers (*in-situ* monitoring of the reflectivity produces error compensation). Examples of reflectivity curves for multilayers with random thickness error are given in Fig. 6.8. The curves demonstrate the following behavior:

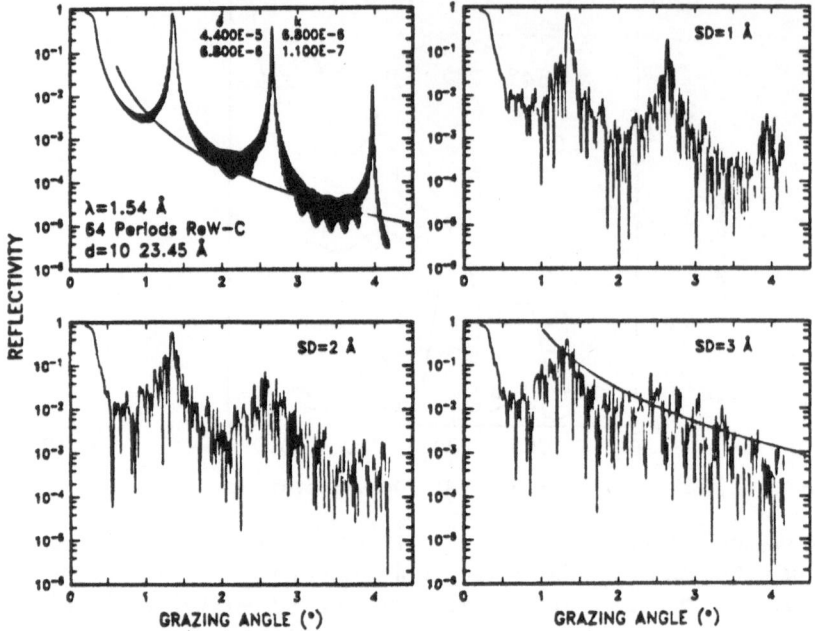

Fig. 6.8. Calculated reflectivities at λ = 1.54 Å for 128 layer periodic ReW-C multilayers with random thickness errors of standard deviations SD = 1, 2 and 3 Å. Thickness errors are obtained by a random walk process. The errors for SD = 2 and 3 are obtained by multiplying the errors for SD = 1 with 2 and 3. The reflectivity of the single ReW-C boundary R_{12} is also plotted in (a) while (d) gives $N\,R_{12}$.

1. Thickness errors reduce the peak reflectivities and the higher orders are stronger affected than the first order peak.

2. Thickness errors increase the reflectivities between the peaks, especially at the 1/2, 3/2, 5/2 order.

3. The regular spacings between the side maximum becomes irregular when thickness errors are introduced.

In order to obtain a simple estimate of the change in reflectivity due to thickness errors, we will first consider the case in which the reflectivity of each boundary $R_{12} << 1$, the total reflectivity $R << 1$, and absorption can be neglected. This will generally be true at sufficiently high order of reflection. In this case, the multilayer can be described by a simple vector model, where the reflected amplitude vectors of each boundary are added to obtain the total reflected amplitude. Thickness errors modify the contribution of the j th boundary in the vector sum by a factor exp $(2i\Delta\varphi_j)$, where the phase error for this boundary is given by the accumulation of the errors from all the boundaries beneath: $\Delta\varphi_j = \dfrac{2\pi}{\lambda} \sum\limits_{l=1}^{j} n_l\,\Delta d_l\,\cos\alpha_l$. Perfect error compensation without accumulation is equivalent to $< \Delta\varphi_j\Delta\varphi_k > \propto \delta_{jk}$, i.e. the summed phase errors of the various boundaries are independent.

As shown in Ref. 6.6 the expectation value for the reflectivity R for an N layer system in the vector model is given by

300

$$\langle R \rangle = N R_{12}(1 - \exp(-4 \langle \Delta\varphi^2 \rangle)) + \sin^2\gamma\, N^2 R_{12}\exp(-4 \langle \Delta\varphi^2 \rangle), \quad (6.6)$$

where $\langle \Delta\varphi^2 \rangle$ is the variance in φ due to random thickness errors and $\gamma = \pi d_1/p$ describes the nominal phase separation between the top and bottom of each film. We have $\sin \gamma = 1$ for the quarter wave stack, where $d_1 = d_2 = p/2$, with $p = \dfrac{\lambda_{max}}{2n\, \cos\alpha_{max}}$ being the period in the multilayer. The value of $\sin \gamma$ is close to 1 for practical x-ray mirrors ($\sin \gamma = 0.87$ for $d_1 = p/3$). We can use eq. (6.6) to express the error in φ by a thickness error:

$$\Delta d = (p/\pi m)\, \Delta\varphi, \quad (6.7)$$

where Δd and $\Delta\varphi$ denote the standard deviation, $\Delta\varphi = \sqrt{\langle \Delta\varphi^2 \rangle}$, and m is the order of the reflection. Equation (6.6) can be inverted to express the variance as a function of the peak reflectivity:

$$\langle \Delta\varphi^2 \rangle = \frac{1}{4}\, \ln \frac{N \sin^2\gamma - 1}{(R/NR_{12}) - 1}. \quad (6.8)$$

Rough boundaries will not alter the validity of eq. (6.8), since the corresponding Debye-Waller factor cancels in the ratio (R/NR_{12}). Because the ratio (R/NR_{12}) can be directly determined from the R vs Θ curve eq. (6.8) provides a means to obtain thickness errors separately from roughness.

For very small thickness errors ($\langle \Delta\varphi^2 \rangle \to 0$) the first term in eq. (6.6) vanishes and the reflectivity is close to $N^2 R_{12}$ and decreases with increasing errors in the same way as described by the Debye-Waller factor. For very large thickness errors, the first term becomes important and gives a reflectivity of NR_{12}.

Small thickness errors can be most easily recognized near the half-order points (m = 1/2, 3/2, 5/2 ...) of the reflectivity curve. At half orders the reflectivity of four adjacent layers adds up to zero and the average reflectivity stays near R_{12} for an error-free coating. The second term in eq. (6.6) has to be replaced by a value close to R_{12} and the first term indicates the increase in reflectivity due to random errors. We can obtain the variance $\langle \Delta\varphi^2 \rangle$ from the increase of the reflectivity $\Delta R = \langle R \rangle - R_{12}$,

$$\langle \Delta\varphi^2 \rangle = -\frac{1}{4}\, \ln\left(1 - \frac{\Delta R}{N R_{12}}\right) \approx \frac{\Delta R}{4 N R_{12}} \quad (6.9)$$

for m = 1/2, 3/2, 5/2 ...

The curves in Fig. 6.8 are in reasonable agreement with eq. (6.9).

Rough Boundaries and Transition Layers

The simplest method to describe a transition layer of width σ in a mirror is to multiply the reflectivity curve with a Debye-Waller factor (DW^2 in eq. (4.6)). This method produces the same reflectivity loss for a given coating independently, if it is tested at short wavelengths near grazing incidence or at long wavelengths near normal incidence. However, tests of actual coatings showed a larger loss in reflectivity at long wavelengths than at $\lambda = 1.54\text{Å}$ [6.6]. This behavior can be explained if a Debye-Waller factor is incorporated at each boundary during the multilayer calculations. The method can easily be extended to multilayers where different films have different values for σ.

Figure 6.9 gives the calculated reflectivity curves for ReW-C multilayers for different widths σ of the transition layers for $\lambda = 1.54$ Å and grazing incidence and around $\lambda = 66$ Å at normal incidence. The drop in reflectivity as a function of σ (for the case that all boundaries have the same σ) is plotted in Fig. 6.10 for the two wavelengths and is compared to that given by a simple Debye-Waller factor for a single boundary. We see from Figs. 6.9 and 6.10 that for small values of σ and increasing σ the reflectivity decrease at $\lambda = 1.54$ Å is smaller than that at $\lambda = 66$ Å or that given by the Debye-Waller factor.

This behavior can be understood by noting that the number of layers in the coating is larger than is required for high reflectivity at $\lambda = 1.54$ Å. A transition layer reduces the reflectivity and increases the transmission at each boundary. As a consequence, the incident light penetrates

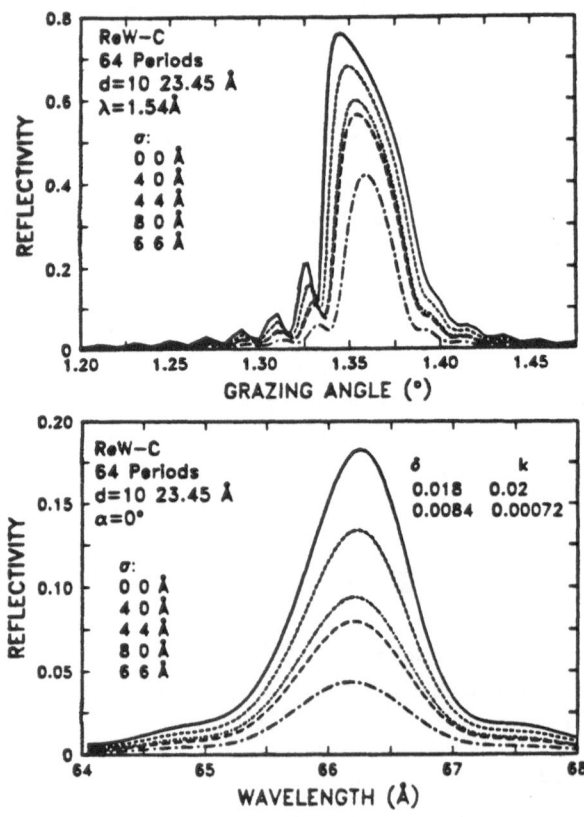

Fig. 6.9. Reflectivity of a multilayer mirror of ReW-C with 64 periods and different values of the σ of the boundary between the 2 materials. The first value of σ represents the ReW boundary coated with carbon, the second the carbon ReW boundary. Top curve for $\lambda = 1.54$ Å and grazing incidence, bottom for normal incidence and soft x-rays.

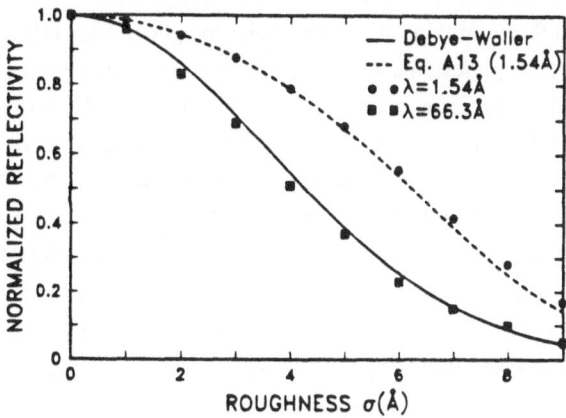

Fig. 6.10. Reflectivity loss due to a transition layer or a rough boundary as a function σ for the first order peak at $\lambda = 1.54$ Å and $\lambda = 66.3$ Å (points) using the multilayer calculation (eqs. (4.5) and (4.6)). The dashed curve is obtained from eq. (A13) in ref. [6.6] and the full curve is the Debye-Waller factor. All boundaries are assumed to have the same value of σ.

deeper into the coating and is reflected from deeper layers. This reflection compensates partly for the reflection losses at the top boundaries. (For a dielectric coating without absorption and a sufficient number of layers, the reflectivity remains at R = 100% if transition layers are introduced, only the linewidth would decrease with increasing σ.) The compensation process ceases to be effective when the number of layers contributing to the reflectivity is limited either by absorption (as for $\lambda = 66$ Å) or by the actual number of layers present.

A necessary condition for the compensation as described is R$\approx$1 . For the case R < 1, a Debye-Waller factor for the entire reflectivity curve gives practically the same result as the same Debye-Waller factor applied to each boundary during the calculation.

A multilayer can still have very good performance, even if only one of the two boundaries is smooth. In the limit of one very rough boundary, we have only one reflecting surface per period and by doubling the number of periods the same peak reflectivity is obtained if absorption can be neglected. This compensation is strongly evident in Fig. 6.9 for $\lambda = 1.54$ Å and very much reduced at $\lambda = 66$ Å. While roughness or a transition layer described by Debye-Waller factor always decreases the reflectivity of a single boundary, an unequal roughness distribution can produce an increase in the higher order peaks of a multilayer. For example, the coating with $\sigma = 8$ Å at the ReW-C boundaries and $\sigma = 0$ at the C-ReW boundaries in Fig. 6.9 has a higher reflectivity in the second-order peak than a coating where both boundaries are smooth.

Distinction Between Smooth Transition Layers and Rough, Scattering Boundaries

The reflectivity curves in Figs. 6.9 and 6.10 have been obtained using eqs. (4.5) and (4.6), i.e., they represent transition layers without any scattering. Equation (4.1) with a Debye-Waller factor only applied to the reflected amplitude and unmodified values for the transmission in eq. (4.3) represents the case, where the Debye-Waller factor reduces the reflectivity by scattering.

For our examples in Fig. 6.9, we obtain only very small changes in the reflectivity curves; even for the largest values of the roughness the difference for scattering or non-scattering boundaries is less than 1%. This result appears surprising at first sight but can easily be understood. For the example in Fig. 6.9, the reflectivity from a single boundary is about $R_{12} = 10^{-4}$ and with a roughness $\sigma = 4$ Å the scattered intensity is 50% of the reflected intensity. The multilayer with N layers can scatter not more than N times the amount of the single boundary (i.e., about 0.5% for our example), while the specular reflectivity is increased by a factor close to N^2 over that of the single boundary. We expect that the reduction of the background intensity due to scattered light by a factor N will be very important to the application of multilayers to x-ray imaging especially for astronomy where often large dynamic intensity ranges occur in the images.

7. MULTILAYER CHARACTERIZATION

The final test of an x-ray mirror is its reflectivity curve around the design wavelength as given in Fig. 7.1. However, this curve represents the integrated effect of all the film parameters, and different sets of parameters for the structure can usually be found that give agreement with the measured curve. An independent determination of each parameter, like optical constants, thickness errors, boundary and film structure is therefore desirable.

We can divide the characterization methods into two groups, those that analyze the finished coatings and those that monitor the growth of the films during the deposition (Fig. 7.2). We discussed the characterization of a coating during the deposition by the *in-situ* x-ray reflectivity in chapter 6. The *in-situ* monitoring method has also been used to obtain more detailed information on film properties; in principle density, an estimate of the optical constants and boundary roughness can all be obtained during the growth of a film [7.1]. The nature of the chemical bonds and the detailed structure of a multilayer can be obtained by applying the methods used in surface science to a multilayer. Auger spectroscopy has been used to obtain depth profiles in the composition of a multilayer [7.2]. The conventional methods of depth profiling using sputtering to remove material are limited in their resolution by the nonuniformity and the mixing produced by the sputter process. Higher resolution becomes possible when these processes are used *in-situ* during the growth of the film. Chauvineau et al. have obtained Auger spectra during film growth [7.1], and there is no doubt that atomic scale depth resolution could be obtained with the various surface sensitive analytical methods applied *in-situ* (photoemission, SEXAF, RHEED, etc.). Excellent depth resolution has also been obtained with ellipsometry of a growing film [7.3]. Figure 7.3 shows the ellipsometric parameters obtained with visible light during the deposition of a W film. The smooth curve represents the calculated values for a W film with bulk values for the optical constants and density. The measured points show an interesting periodicity, every 9 Å the bulk values are obtained, in between the films deviate and seem to be rougher and less dense.

The x-ray reflectivity curve at a short x-ray wavelength (Fig. 7.4) can be used to estimate the width of the boundaries and the thickness errors in a coating. For this characterization one tilts the mirror to grazing angles of incidence such that the reflectivity peak shifts to short wavelengths (e.g. $\lambda = 1.54$ Å) and measures the reflectivity as a function of grazing angle Θ over the largest possible range of Θ. Thickness errors are obtained by comparing the reflectivities between the

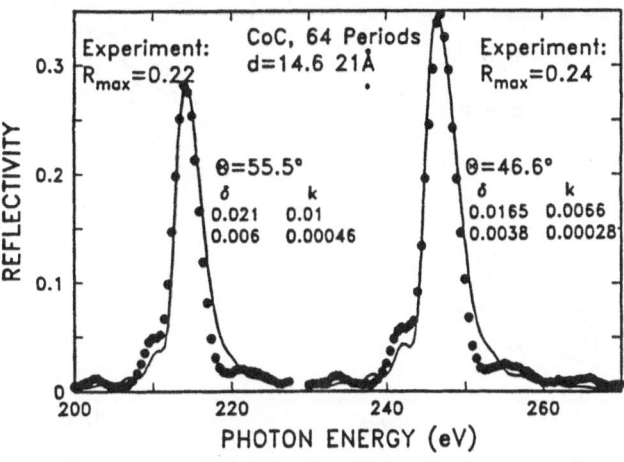

Fig. 7.1. Comparison between measured (dots) and calculated (solid lines) reflectivity curves for a Co-C multilayer mirror with 128 layers.

1. in-situ

- soft x-ray reflectivity
- Auger spectroscopy
- ellipsometry
- ......

2. finished mirror
- soft x-ray reflectivity near normal incidence
- x-ray reflectivity at 1.54 Å near grazing incidence
- electron microscopy
- tunnel microscopy
- Auger spectroscopy
- .....

Fig. 7.2 Characterization Methods.

main peaks to that of the main peaks using eq. (6.8). while roughness is obtained by comparing the measured reflectivities to their theoretical value. We estimate from the 2nd order peak in Fig. 7.4 a value of $R/NR_{12} \simeq 10$, and obtain from eqs (6.8) and (6.7) a standard deviation of 3.5 Å for the thickness error (p=31 Å). We can estimate the width or roughness σ of the boundaries by comparing the average background reflectivity around the 2nd order peak with the calculated

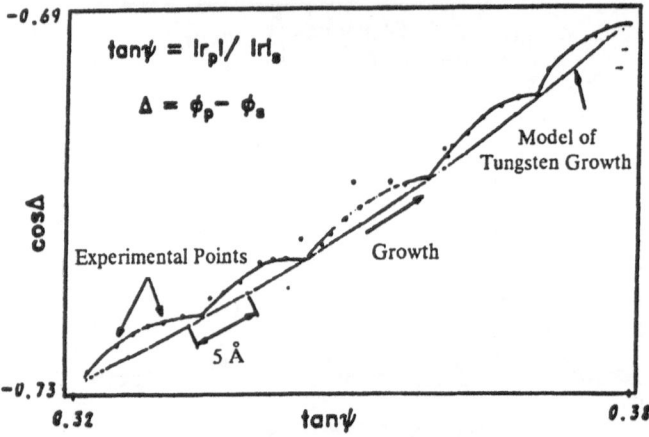

Fig. 7.3 Characterization of a W film during deposition with visible light ellipsometry. Experimental points are spaced by 1.25 Å , the model curve is calculated for a W film with bulk properties (from Ref. [7.3]).

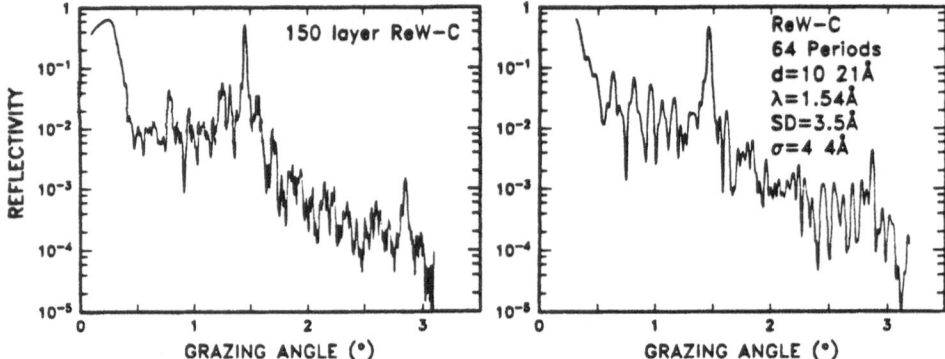

Fig. 7.4. Measured reflectivity curve at λ = 1.54 Å of a 150 layer ReW-C mirror (a) and calculated reflectivity for a 128 layer coating with values for thickness error and roughness as obtained from the measured curve (b). The finite resolution of the reflectometer is included in (b) by replacing the calculated reflectivity at each point by the average of 5 points spaced by 0.005°.

value $N\,R_{12}$ for a fully randomized coating (see Fig.6.8). Around θ = 2.8° we have a value of R_{12} = 2×10^{-5} and a measured average reflectivity of 2×10^{-4}. With N = 150 the measured reflectivity is a factor of 15 smaller than the expected value for a smooth boundary giving us a value of σ = 4 Å from eq. 4.6.

Electron microscopy has been used to look sideways at a coating [7.4]. Beautiful pictures of multilayer structures have been obtained (Fig. 7.5), and observation of the structure during heating gives the most direct visualization of diffusion, roughening due to phase transition from amorphous to crystalline, and melting of the structure [7.5]. The limited spatial resolution does not allow to obtain thickness errors as accurate as from the x-ray reflectivity curve.

The highest spatial resolution is presently obtainable with the tunnel microscope [7.6, 7.7]. Figure 7.6 shows the surface contour of the top carbon surface, a 5-layer Fe-C coating. The

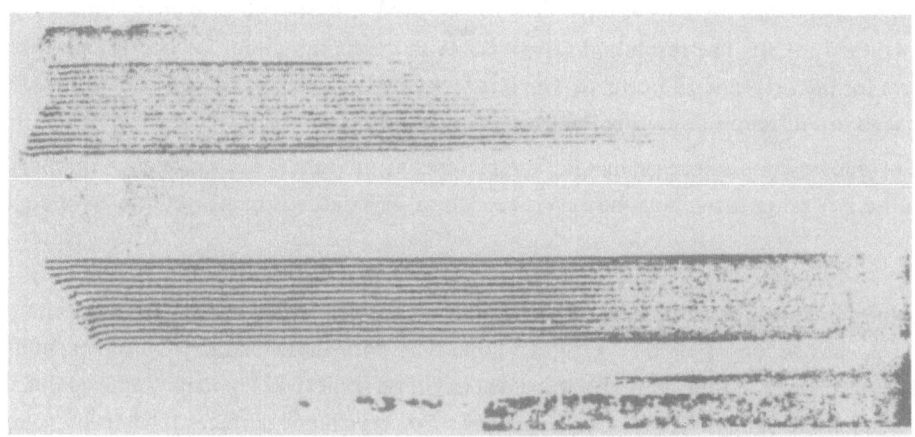

Fig. 7.5. Electron micrograph of the cross-section of a Fabry-Perot. Period in the two mirrors is 32.5 Å and the carbon spacer is 480 Å thick (from Ref.7.4), coating by T. Barbee).

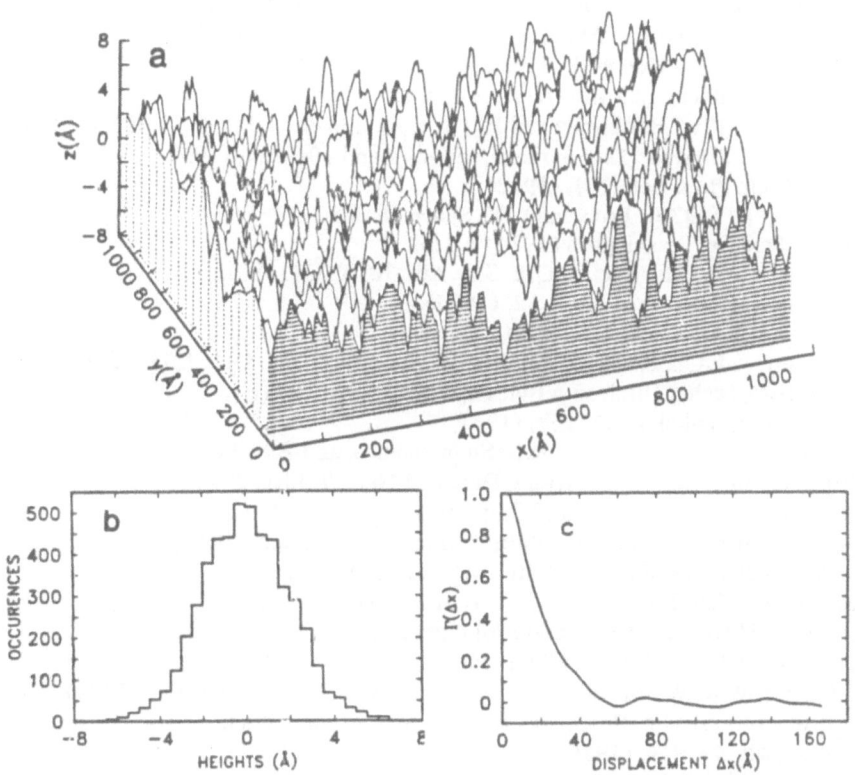

Fig. 7.6. Topography (a) of the top carbon surface of a 5-layer (C-Fe-C-Fe-C, thicknesses 30,14.10,10, 30 Å) coating obtained with a tunnel microscope; heights distribution (b) and autocorrelation function (c) of the surface. Tunnel micrograph courtesy of Mark Welland.

surface map is obtained digitally and the data can be used to derive the complete statistics of the surface (Figs. 7.6b,c). The resolution (about 0.2 Å in depth and about 2Å lateral) is sufficient to remove for the first time all fitting parameters from the theoretical description of a surface and will allow to test different scattering theories quantitatively.

Optical profilometers produce similar surface pictures as the tunnel microscope. A depth resolution in the 1 Å range is possible, however, the lateral resolution is limited by the wavelength of light.

The measurements show that characterization of the properties of thin films on the scale around 1 Å has become possible. One can hope that, with these characterization capabilities, control of the growth of films at the atomic level might be achieved. Up to now amorphous films have produced the smoothest surfaces (Fig. 7.6). Polycrystalline surfaces tend to be rougher; amorphous multilayer films that are made polycrystalline by heating, generally lose their high reflectivities. Epitaxial films can be atomically smooth; however, no epitaxial multilayer system is presently known where the two components have sufficient contrast to produce high reflectivity for x rays.

References

1.1. P.P. Ewald, Ann. Phys. (IV) **49**, 1 and 117, (1916).
1.2. A.H. Compton and S.K. Allison, X-ray in Theory and Experiment, (van Nostrand, New York 1935); R.W. James, The Optical Principles of Diffraction of X Rays (Cornell U. Press, Ithaca NY 1965); M. von Laue, Röntgenstrahl-Interferenzen (Akademischer Verlag, Frankfurt 1960); B.W. Batterman and H. Cole, Review Mod. Phys. **36**, 681, (1964).
1.3. F. Abeles, Ann. Physique 12th Series **5**, 596-640 and 706-784, (1950).
1.4. E. Spiller, Appl. Phys. Lett. **20**), (1972); Appl. Opt. **16**, 89 (1976); Proc. ICO-IX, Space Optics, Natl. Acad. Science, Washington (1974) p.525.
1.5. C.K. Carniglia and J.H. Apfel, J. Opt. Soc. Am. **70**, 523, (1980).
1.6. R.H. Miller, Opt. Spectra **9**, No.7, 32, (1975).
1.7. B.E. Newnam, J. Opt. Soc. Am. **70**, 1051, (1980).
1.8. H.E. Bennett, A.J. Glass, A.H. Guenther, B. Newnam, Appl. Opt. 2375, (1980).
1.9. E. Spiller, IBM Techn. Disclosure Bull.**16**, 3789, (April 1974).
1.10. G. Borrmann, Physikal. Z. **42**, 157, (1941).
2.1. B.L. Henke, P. Lee, T.J. Tanaka, R.L. Shimabukuro, and B.K. Fujikawa, AIP Proc.Vol. 75, 340, (1981); Atom. Data and Nucl. Tables, 27,1, (1982).
4.1. P.S. Heavens, Optical Properties of Thin Films, (Dover, NY 1965); A. Vasicek, Optics of Thin Films (North Holland, Amsterdam, 1960); H.A. Macleod, Thin-Film Optical Filters, (Elsevier, NY, 1969); M. Born and E. Wolf, Principles of Optics, 5th ed. (Pergamon Press, 1975); P.H. Berning, Theory and Calculations of Optical Thin Films, in **Physics of Thin Films**, ed. by G. Hass, **1**, 69 (Academic Press, NY, 1963).
4.2. T. W. Barbee, W.K. Warburton, and J.H. Underwood, J.O.S.A. B1, 691,(1984); H. van Brug, M.P. Bruijn, R. van der Pol, and M.J. van der Wiel, Appl. Phys. Lett. 49, 914, (1986).
4.3. O. Litzman and I.Sebelová, Optica Acta, **32**, 839, (1985).
4.4. P.P. Ewald, Ann. Phys., **49**,1, 117, (1916); ibid. **54**, 519, 557 (1917); Rev. Mod. Phys. **37**, 46, (1963).
5.1. A.V. Vinogradov and B.Ya. Zeldovich, Appl. Optics **16**, 89, (1977).

5.2. H.J. Hagemann, W. Gudat and C. Kunz, J. Opt. Soc. Am. **65**,
      742, (1975) and DESY Report SR-74/17.
6.1. T.W. Barbee, Jr., Proc. SPIE Vol.**563**, 3, (1985).
6.2. E. Spiller, AIP Proc. Vol. <u>75</u>, 125, (1981).
6.3. A. N. Broers and E. Spiller, in <u>Scanning Electron</u>
      <u>Microscopy</u>, (SEM Inc., AMF O'Hara 1980) p.201, (1980).
6.4. E. Spiller, Proc. SPIE Vol.**563**, 367, (1985).
6.5. M.P. Brujn, P. Chakraborty, H. van Essen, J. Verhoeven, M.J. van der Wiel,
      Proc. SPIE **563**, 36, (1985).
6.6. E. Spiller and A.E. Rosenbluth, Proc. SPIE **563**, 221, (1985);
      Optical Engineering **25**, 954-963, (1986).
7.1. J. P. Chauvineau, J. Corno, D. Decanini, L. Nevot, and B. Pardo, Proc.
      SPIE Vol **563**, 245, (1985).
7.2. K.D. Rachocki, D.R. Brown, R.W. Springer, and P.N.Arendt, Applications of
      Surface Science **18**,165, (1984).
7.3. E. Ziegler, P. Houdy, and L. Nevot, Proc. SPIE Vol. **563**, 306, (1985).
      M.Yamamoto, A.A. Arai, H. Shibata, and T. Namioka, Conference
      Digest ICO-13, 626, (1984).
7.4. Y. Lepetre, I.K. Schuller, G. Rasigni, R. Rivoira and R. Philip,
      Proc. SPIE Vol. **563**, 258, (1985);
      A.K. Petford-Long, M.B. Stearns, C.H. Chang, S.R. Nutt, D.G. Stearns,
      N.M. Ceglio, and A.M. Hawryluk, J. Appl. Phys. 61, 1422, (1987).
      H.W. Deckman, J.H. Dunsmuir, and B. Abeles, Appl. Phys. Lett.**46**,171, (1985).
7.5. K. Schuller, Y. Lepetre, E. Ziegler and E. Spiller, Appl. Phys. Lett. **48**,1354, (1986).
7.6   G. Binning, H. Rohrer, C. Gerber, and W. Weibel, Phys. Rev. Lett. 50, 120, (1983).
7.7   M.E. Welland and R.H. Koch, Appl.Phys. Lett.**48**, 724, (1986).

Additional References

Multilayer theory

P. Lee, Optics Commun. **37**, 159 (1981).
J.H. Underwood and T.W. Barbee Jr., AIP Proc. **75**, Low Energy
X-Ray Diagnostics, 1981, ed. by D.T. Attwood and B.L. Henke, p.176;
Appl. Opt. **20**, 3027 (1981).

Scattering

J.M. Eastman, in **Physics of Thin Films**, eds. G. Hass and M.H. Francombe,
10,(1978);
C.K. Carniglia, Opt. Engin. **18**, 104 (1979);
J.M. Elson, J. Opt. Soc. Am. **69**, 48 (1979).

General Overviews

G.F. Marshall, editor, Applications of Thin Film Multilayered
Structures to Figured X-Ray Optics, Proc. SPIE Vol. **563** , (1985).
T. W. Barbee, Jr., AIP Proc. Vol. 75, Low Energy X-Ray
Diagnostics, 1981, ed. D. T. Attwood and B. L. Henke, p. 131, (1981).
E. Spiller, AIP Proc. Vol. 75, 125, (1981).

Tests

L. Golub, E. Spiller, R. J. Bartlett, M. P. Hockaday,
D. R. Kania, W. J. Trela, R. Tatchyn, Appl. Opt. **23**, 3529, (1984).
T.W.Barbee, S. Mrowka and M.C. Hettrick, Appl. Opt. **24**, 883, (1985).

MULTILAYER NEUTRON OPTICAL DEVICES

F. Mezei

Hahn-Meitner-Institut, Pf. 390128
D-1000 Berlin 39

INTRODUCTION

Multilayer neutron mirrors represent those multilayer structures which have been produced in by far the largest quantity - over 100 m^2 since 1977. Many multilayer devices have been used on various neutron spectrometers for several years. The oldest device in continuous use since 1978 is the neutron beam polarizer on the Neutron Spin Echo (NSE) spectrometer (IN11) at the Institut Laue-Langevin (ILL) in Grenoble. There are four vacuum deposition units in operation dedicated to multilayer neutron mirror production: two electron beam gun evaporators at ILL (about 1 m^2/week production capacity each), another similar instrument at the Leningrad Institute of Nuclear Physics, and a diode sputtering apparatus at Brookhaven National Laboratory. A fifth machine (a triode sputtering unit with a capacity of 1 - 2 m^2/week) is being installed at the Hahn-Meitner-Institute in Berlin. It is already apparent that the need for neutron mirrors amounts to some 200 m^2 at least within the next 2 - 4 years.

The whole state-of-the-art with multilayer neutron mirrors is determined by the single most important specific feature that large amounts are required in any real application. This is why up to now only relatively simple production methods have been explored, which do not require too sophisticated and lengthy procedures and which can be fully automated and run by a technician. Another aspect of the same problem is that with quite advantageously usable mirrors being available since 1977 (when few other multilayer devices were available) much of the effort in recent years has been directed to the mass production and relatively little to further research and developement. It is my personal feeling, however, that in the near future we shall witness a spectacular progress in the field of neutron optical multilayer mirrors, namely by adapting the successful techniques and knowledge which emerged in the last decade in other areas of multilayer research. This basically requires the implementation of these techniques on scaled-up, fast turn-over equipment. This is by no means a simple task, and beyond just adding another field of applications of multilayers, I see a special significance of neutron mirror developement in its "quasi-industrial" character.

Of course, similarly to X-rays, neutrons can also be used for the characterization of multilayers. Here we face the same situation as in neutron scattering research in general: since available neutron beam intensities are many orders of magnitude less than those of X-ray and light beams, neutrons can only be profitably used in order to probe features inaccessible to these other kinds of radiation. For multilayers there are two particular areas of this kind: the study of the ferromagnetic behaviour and distinction between atomic species with atomic numbers close to each other (such as Al and Mg, or Ni and Cu).

In what follows I shall first introduce the elements of neutron optics in a way which focusses on the basic mechanism and limitations. In several respects this presentation is complementary to other contributions in this volume. Subsequently, we shall examine the experimentally observed neutron reflectivities of various multilayer structures. This will be followed by a description of practical applications of multilayer neutron mirrors, such as monochromators, polarizers and neutron guide tubes. Various points during the discussion will illustrate the opportunities offered by neutron reflectometry for the study of multilayer structures.

ELEMENTS OF NEUTRON OPTICS

The neutron beams available for the investigation of condensed matter consist of particles with energies typically between 1 meV and 1 eV, corresponding to neutron wavelengths of about 9 Å and 0.28 Å, respectively. These wavelengths are very large compared to the size of atomic nuclei and the range of nuclear forces. Therefore, at these low energies the interaction between neutrons and nuclei can be described by a delta function Fermi pseudopotential

$$V(\mathbf{r}) = \frac{2\pi\hbar^2}{m} \, b \, \delta(\mathbf{r-R}) \quad , \tag{1}$$

where m is the neutron mass, **R** the position of the nucleus, and b is the so called scattering length of the nucleus.[1] This potential has been defined in a way that b is positive and equal to the radius of an inpenetrable sphere. In principle b can be a complex number, and its real part can either be positive or negative. The imaginary part corresponds to the absorption of the neutron via a nuclear reaction. It turns out that with very few exceptions, b can be taken as real for all isotopes and it is constant over the energy range of our interest. The values of b are tabulated for all elements and most isotopes of the periodic table,[1] and these values range from $-0.87 \cdot 10^{-12}$ cm (for ^{62}Ni) to $4.95 \cdot 10^{-12}$ cm (for ^{164}Dy). For most isotopes b lies between 0 and 10^{-12} cm, and the variation is fully irregular for the different atomic species (for example ^{58}Ni shows a b value of $1.44 \cdot 10^{-12}$ cm, very different from ^{62}Ni). Although some materials absorb neutrons to a substantial degree (penetration depth of about 0.01 mm) the absorption contributes to b by a generally small imaginary part, which is undectably small for most elements. Note that the absorption varies inversely proportional to the neutron velocity.

Particle scattering on the pseudopotential (1) can be described by the following relation:

$$e^{ikr} \implies \frac{b}{|\mathbf{r-R}|} e^{ik|\mathbf{r-R}|} + e^{ikr} , \qquad (2)$$

i.e. an incoming plane wave of unit amplitude and wavenumber **k** will continue basically unperturbed and the scattering initiates an additional spherical wave component around **R**; k is the absolute value of vector **k**. It can be shown that particle current carried by the spherical wave corresponds to the particle current in the initial plane wave across an area $4\pi b^2$ perpendicular to **k**. This is why $4\pi b^2$ is called the scattering cross section, and it amounts to 10^{-23}-10^{-24} cm^2 for most nuclei. This number illustrates the weakness of neutron interaction with matter of usual density: the physical cross section of an atom (about 10^{-15} cm^2) is much bigger than its scattering cross section.

The interaction of a neutron beam with an extended body is the result of the interference of all the spherical waves scattered by the nucleus in that body. (Note that if the atoms are in motion, the scattered wavenumber will be changed due to the Doppler effect). As long as we can neglect the interference between the scattered spherical waves and the initial plane wave (i.e. the total scattered amplitude is small) this interference pattern is readily evaluated (so called kinematic theory). Often the more complicated situation with multiple scattered-transmitted beam interaction applies (dynamic theory). For the understanding of the fundamentals here we may remain within the framework of kinematic theory, but we shall also discuss results obtained by dynamic theory.

Refractive Index

The familiar optical quantity, the refractive index n is defined by the ratio of the <u>phase</u> <u>velocities</u> of the neutron wave in vacuum (o) and a medium (m)

$$n = \frac{v_o^{(ph)}}{v_m^{(ph)}} \quad . \tag{3}$$

Note that due to the finite mass of the neutron the phase velocity is not equal to the particle velocity, and is in fact an ambiguous quantity, because it depends on the zero of our energy scale. Taking the potential energy zero in vacuum, i.e.

$$E = \hbar\omega = \frac{1}{2} mv_0^2 \quad ,$$

the phase velocity of the corresponding de Broglie wave $\exp[i(kr-\omega t)]$ becomes

$$v_0^{(ph)} = \frac{\omega}{k} = \frac{1}{2} v_0 \quad , \tag{4}$$

since $\hbar k = mv_0$. The phase velocity of the particle inside a medium can be evaluated by considering the interference of the transmitted plane wave and the scattered spherical waves.[2] We shall, however, choose a simpler argument based on eq.(1). Since the interaction is weak with individual atoms, we can consider the particle motion in the average pseudopotential of the medium defined as

$$U = \frac{2\pi\hbar^2}{m} \bar{b}\rho \quad , \tag{5}$$

where $\bar{b} = \Sigma\, b_i/N$ is the average scattering length per atom in the medium, and ρ is the atomic density (number of atoms per unit volume). Therefore inside the medium we have

$$E = \hbar\omega = \frac{1}{2} mv_m^2 + U \quad . \tag{6}$$

Thus by energy conservation

314

$$v_m = \sqrt{v_0^2 - \frac{2U}{m}}$$

and

$$v_m^{(ph)} = \frac{\omega}{k_m} = \frac{E}{mv_m} = \frac{1}{2}\frac{v_0^2}{v_m}$$

which gives

$$n = \frac{v_0^{(ph)}}{v_m^{(ph)}} = \frac{v_m}{v_0} = \sqrt{1 - \frac{2U}{mv_0^2}} \quad . \tag{7}$$

Noting that with $b \sim 10^{-12}$ cm and $\rho \sim 10^{23}$ cm^{-3}, U is typically of the order of 10^{-7} eV, i.e. much smaller than the neutron energies E of our interest, we can finally write

$$n = 1 - \frac{U}{2E} = 1 - \frac{m}{h^2}U\lambda^2 \quad , \tag{8}$$

where $\lambda = 2\pi/k = h/mv$ is the neutron wavelength. Thus we find that n is even closer to unity than for X-rays of comparable wavelength: $|1-n|$ typically is 10^{-4}. For most materials, except for a few natural elements (H, Li, Ti, V, Mn), μ is smaller than 1 (corresponding to repulsive potential).

At this point, we have to note that in the presence of a magnetic field a Zeeman term has to be added to the energy equation due to the magnetic moment μ of the neutron:

$$E = \frac{1}{2}mv^2 + U \pm \mu B \quad , \tag{9}$$

where the + and - signs apply to neutron spin orientations parallel or anti-parallel to the field. (The neutron spin direction is opposite to that of the magnetic moment.) Thus, in eqs.(7) and (8) U has to be replaced by $U \pm \mu B$. In practical units the numerical value of μ is $6.1 \cdot 10^{-8}$ eV/Tesla. In ferromagnetic materials B can be substantial (e.g. the saturation magnetization of Fe is 2.2 Tesla) which makes μB to be of the same order of magnitude as U. Equation (9) also implies that a magnetic field in vacuum has to be considered as a refractive medium, and we can describe neutron beam propagation in an inhomogeneous field with the help of varying refractive indices n_+ and n_- for the two spin states (Stein-Gerlach effect).

Total Reflection

The refractive index describes the change of the beam direction on entering a material via Snell's law. With n very close to 1 measurable effects only occur at small grazing angles θ.

Total reflection occurs if Snell's law cannot be satisfied, i.e. for $\theta \leqslant \theta_c$ where the critical angle θ_c is defined as

$$\cos\theta_c = n \quad . \tag{10}$$

In this definition n stands for the relative refractive index of the two media on both sides of the reflecting surface, since for neutrons $n \simeq 1$ we can express the relative refractive index n_{12} between media 1 and 2 as

$$n_{12} = 1 - n_1 + n_2 \quad , \tag{11}$$

where n_1 and n_2 are the refractive indices of the two media relative to vacuum.

Equation (10) can easily be solved in terms of $\sin\theta_c$, viz. in view of eqs.(7) to (9):

$$\sin\theta_c = \sqrt{1 - n^2} = \lambda\sqrt{\frac{1}{\pi}\overline{b}\rho \pm mh^{-2}\mu B} \quad .$$

If B corresponds to the magnetization of a medium $4\pi M$ only, we can express it in terms of the average magnetic moment per atom $\overline{s} = M/\rho$ and rewrite this equation as

$$2d_c\sin\theta_0 = \lambda \quad , \tag{12}$$

where we introduced the definitions of $2d_c = \sqrt{\pi/\rho(\overline{b} \pm \overline{p})}$ and $\overline{p} = g\overline{s}$ with $g = 0.270 \cdot 10^{-12}$ cm/Bohr magneton.

Equation (12) has the familiar form of Bragg's law with d_c playing the role of a hypothetical lattice spacing. In reality d_c characterizes the length scale which separates "thin" and "thick" films in neutron optics. At grazing angles $\theta \ll \theta_c$ it can be shown that the radiation field of the totally reflected beam penetrates to a depth of d_c in the reflecting medium. A thin layer will reflect like the bulk material if its thickness is large compared to d_c. On the other hand multilayer structures with a repeat distance much smaller than d_c will have the total reflection properties of a homogeneous mixture of the components. The value of d for various materials is typically a few hundred Å. (For example, for iron d_c equals 245 Å and 500 Å, respectively, for the two neutron spin orientations relative to the Fe magnetization direction, provided that the magnetization is parallel to the surface. Namely, when the field is perpendicular to the surface, B is the same in the vacuum and in the ferromagnetic medium and there is no magnetic contribution to the relative refractive index.)

Note that for a surface between two media the total reflection condition [eq.(12)] holds with d_c replaced by $d_c^{(1,2)}$ defined by the equation

$$\left(\frac{1}{d_c^{(1,2)}}\right)^2 = \left(\frac{1}{d_c^{(2)}}\right)^2 - \left(\frac{1}{d_c^{(1)}}\right)^2 \quad , \tag{13}$$

where $d_c^{(1)}$ and $d_c^{(2)}$ characterize the two media with respect to the vacuum.

A final remark concerns the role of absorption. The absorption losses in a total reflection situation are due to the penetration of the evanescent wave in the reflecting medium. Since the penetration depth d_c is small compared to the absorption range of even the most absorbing materials (for example about 3 μm in Gd for $\lambda = 5$ Å), the reflection is not affected. Therefore, a neutron beam impinging on a flat surface at a small grazing angle can only be absorbed if no total reflection takes place. This is the case for all grazing angles for media with $n \geq 1$, i.e. with b < 0. None of the elements with b < 0 have a high absorption cross section, but the alloy of a b < 0 and a highly absorbing material can give a suitable so called "antireflecting" layer, which absorbs neutrons at small grazing angles and e.g. can make a totally reflecting surface opaque. Antireflecting layers have been introduced by Drabkin,[3] and Ti(20 at % Gd) is the most often used alloy composition.

Diffraction Theory

Let us consider the reflection of neutron beams on a periodic stack of flat layers (which can also be atomic layers in a crystal) in the kinematical limit, i.e. assuming that we can neglect the attenuation and refraction of the incident beam. This means that we have to consider the spherical scattered waves originating at each atom as described by eq.(2) and evaluate the interference of these individual and independent spherical waves. There is one fundamental question we have to answer first: How big is the coherence volume within which interference can take place in a given beam geometry? The introduction of the classical notion of Fresnel zones answers this question.

Let us consider various optical paths around the ideal reflection path from point A to B (Fig.1). The boundary of the first Fresnel zone is defined by the locus of points X for which

$$\overline{AX} + \overline{XB} - (R_A + R_B) = \frac{\lambda}{2} \quad ,$$

where $\overline{AX}$ and $\overline{XB}$ stand for the distances from A to X and from X to B, R_A and R_B are the distances between A and B, respectively, and the point of ideal geometrical reflection on the surface (which is the centre of the Fresnel zones). Obviously, the Fresnel zone is an ellipsoid with axes x and y, where x is determined by the equation

$$\frac{\lambda}{2} = \sqrt{R_A^2 + x^2} + \sqrt{R_B^2 + x^2} - (R_A + R_B) \simeq \frac{x^2}{2R_A} + \frac{x^2}{2R_B} \quad .$$

This gives

$$x = \sqrt{\frac{\lambda R_A R_B}{R_A + R_B}} \tag{14}$$

and it can similarly be shown that

$$y = x/\sin\theta \quad . \tag{15}$$

With $\lambda \sim 4$ Å and $R_A \sim R_B \sim 5$ m (a rather typical experimental situation) we get x = 0.03 mm and y = 3 mm for a grazing angle $\theta \sim 0.5°$. These are rather large numbers and we see that the Fresnel zone is of macroscopic size. This will be an important element in our later discussion of surface qualities.

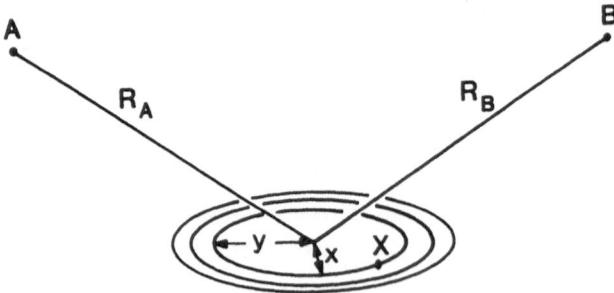

Fig. 1. Fresnel zones for mirror reflection from point A to point B

The reflected beam amplitude can be deduced as follows: A spherical wave emitted at point A will have an amplitude $1/R_A$ at reflecting surface, and thus the amplitude of the spherical wave scattered from an atom i will be $b_i/R_A R_B$ at point B, and the total amplitude at B is

$$\psi_{AB} = \frac{1}{R_A R_B} \sum_i b_i e^{ik\left(r_i^{(A)} + r_i^{(B)}\right)} \quad ,$$

where $r_i^{(A)}$ and $r_i^{(B)}$ are the distances between atom i and A and B, respectively. In the spirit of the Fresnel zone construction we assume that all atoms in a plane parallel to the surface scatter in phases within an area of xy, and the scattering by atoms outside this area averages to zero due to the rapid variation of the phase. We are thus left with the variation of the phase between parallel layers at various depths which is given by e^{iqr_i} where q is the momentum transfer vector, perpendicularly pointing to the surface and having a magnitude of $q = 2k\sin\theta$ and r_i is the position vector of the atom. Hence

$$\Psi \simeq \frac{y}{R_A R_B} \sum_i b_i e^{iqr_i} \quad ,$$

where the sum refers to unit surface area. In order to calculate the reflectivity R we have to relate this amplitude to the one we would obtain by direct propagation between A and B, i.e. $1/(R_A + R_B)$. In view of eqs.(14) and (15) we find that

$$R \simeq \frac{\lambda^2}{\sin^2\theta} \left| \sum_i b_i e^{iqr_i} \right|^2 \quad . \tag{16}$$

We shall evaluate this general equation for a periodic multilayer consisting of alternating layers of two materials with a thickness of d/2 each. By Bragg's law we get maxima of the reflectivity when

$$2d\sin\theta = \ell\lambda \quad , \tag{17}$$

where ℓ is an integral. The sum in the bracket in eq.(16) is the Fourier transform of the distribution of b_i density perpendicular to the surface, and if the momentum transfer q corresponds to distances large compared to atomic spacing, i.e.

$$\frac{1}{q} = \frac{\lambda}{4\pi\sin\theta} = \frac{d}{2\pi\ell} \gg 1 \text{ Å} \quad ,$$

we can express the sum in eq.(16) in terms of the average scattering length densities of the two materials. The result for odd orders $\ell=1,3,5...$ is (for ℓ even the sum is 0)

$$\sum_i b_i e^{iqr_i} = \frac{dN}{\pi\ell} (\rho_1 b_1 - \rho_2 b_2) = \frac{dN}{\ell} \left(\frac{1}{2d_c^{(1,2)}}\right)^2 \quad , \tag{18}$$

where N is the number of periods in the multilayer and in the last transformation we made use of the definition of d_c and eq.(13). For simplicity, in what follows we will drop the superscript (1,2) in d_c. Finally, eqs. (16), (17) and (18) lead to our final result

$$R = \frac{1}{4} \frac{N^2}{\ell^4} \left(\frac{d}{d_c}\right)^4 \quad . \tag{19}$$

The main feature of this relation is that R drops very rapidly as the layers get thinner.

We have to remember that this relation is the kinematic approximation. It is, however, relatively simple to take into account the main dynamic effects in a satisfactory fashion: (a) If the reflectivity approaches unity, the incoming beam intensity will attenuate across the stack; consequently, R approaches unity slower than predicted by eq.(19) and it turns out that this saturation can rather well be described by replacing R by

$$R_s = 1 - \exp(-R) \quad . \tag{20}$$

(b) The refraction effects make q deviate in a medium from the value calculated from Bragg's law with λ taken in the vacuum. This means that in order to maintain the Bragg condition with an apparent lattice spacing d, the actual layer thickness of material j has to be

$$\frac{d/2}{\sqrt{1-\left(\frac{d}{\ell d_c^{(j)}}\right)^2}} \tag{21}$$

instead of d/2. This refraction correction can readily be derived from Snell's law and eq.(12) and it becomes very substantial if d is close to $d_c^{(j)}$. This means that if d is close to one of $d_c^{(j)}$'s in the stack, the apparent d values will be different for various orders ℓ. In practical applications we only concentrate on $\ell = 1$, which gives the highest reflectivity.

An apparent $d > d_c^{(j)}$ means that the layer structure becomes totally reflecting if one of the j layers is sufficiently thick ($\gtrsim 2d_c^{(j)}$).

The Role of Imperfections

The reflectivity of a non-ideal layer structure can be evaluated by considering eq.(16) in the light of a Fresnel zone construction. This implies
that we have to regard this sum over the substantial Fresnel zone area with
the direction of q perpendicular to the surface as being defined as an average plane over this zone area. Thus, if the interfaces between the two materials are not flat or not parallel on a scale smaller than the Fresnel
zone dimensions, the effective scattering length density profile of the multilayer stack becomes washed out, and it changes from the ideal square wave
shape to a more or less deeply oscillating smooth pattern. The interdiffusion of the two materials across the interface results in the same kind of
pattern (see Fig.2). The sum in eq.(16), essentially a Fourier transform,
becomes considerably reduced for such a contrast pattern.

Deviation of the interfaces from ideal parallel planes on a length scale
large compared to the Fresnel zone dimensions x and y do not effect the
reflectivity itself, but they result in so called optical figure errors,
i.e. deviations of the reflected beam from the ideal direction. Since for
the small grazing angles in question the Fresnel zone is a very much elongated cigar-shaped ellipsoid, anisotropic surface irregularities may lead
to different reflectivities for different reflection directions. A further
possible Fresnel zone effect is a wavelength dependence of the reflectivity:
at longer wavelengths the grazing angle θ becomes bigger, whereby the y
dimension of the Fresnel zone gets smaller. Averaging over a smaller area
may lead to an effective profile with more pronounced modulation.

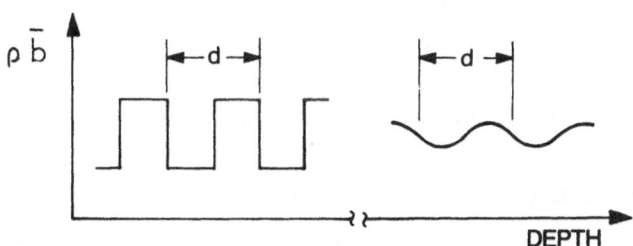

Fig. 2. Effective neutron scattering-length density variation in an ideal
        multilayer (left) and in a multilayer structure with rough inter-
        faces (right)

Another kind of imperfection is the scatter of individual layer thicknesses. It is easy to verify by inspecting the Fourier sum in eq.(16) that moving a single interface around its ideal position by 10 - 20% of the single layer thickness $d/2$ only makes negligible changes. On the other hand, the inversion of two layers (putting them in opposite phases) reduces the effective number of periods by 2, which is a substantial effect. This observation stresses that what matters in this respect are the cumulative errors which are defined as the actual distance of a given interface from its ideal position. It does not matter if individual layer thicknesses fluctuate by as much as 30 - 40% as long as these fluctuations compensate each other in the sense that each interface stays within some 20% of $d/2$ at its ideal position defined by perfect periodicity.

Exact Computational Methods

The problem of plane wave propagation in a stack of homogeneous layers with ideally plane and parallel interfaces can be solved exactly for any layer sequence by considering the propagation of the incoming and reflected beams within each layer and matching them at the interfaces. It turns out that each layer can be represented by a wavelength-dependent so called transfer matrix, and the matrix representing the whole stack is the product of the easily calculable individual matrices for each layer. The roughness of the interfaces and absorption can also be built into these kinds of algorithms.[4]

These computational methods provide little insight into the mechanism of building up reflectivities, but they allow trial-and-error type approaches. Furthermore, they are indispensable for checking the conclusions and predictions obtained by using approximate ideas such as the above mentioned kinematic theory. Actually, the exact calculations prove that our results, primarily eqs.(19), (20) and (21), provide very good approximations.

NEUTRON REFLECTION PROPERTIES OF MULTILAYER MIRRORS

In this chapter we shall review the neutron reflectivity performance of various kinds of multilayer structures and compare them to theoretical predictions. The period length of neutron mirrors is of the order of 100 Å, and this is basically determined by the magnitude of d_c values. Indeed, in order to reach high reflectivities in Bragg reflection the number of bilayers required is about $2(d_c/d)^2$, cf. eq.(19), which becomes a very large number if $d \ll d_c$, and d_c's are typically of the order of 300 Å. Since in neutron scattering we are primarily interested in wavelengths below 10 Å, the grazing angles θ are always rather small, typically 0.5 - 5°. (Only first order ℓ = 1 reflections are of practical interest.)

Periodic Structures: Monochromators

Periodic multilayer neutron mirrors were investigated first by Schoenborn et al. in 1974.[5] Figure 3 shows the measured and calculated ideal reflectivity of N = 10 bilayer Ge-Mn structure with d = 180 Å periodicity as a function of the grazing angle. The pronounced peak at θ = 0.65° corresponds to the expected Bragg reflection, whose rather large width of about 10% also corresponds to what is to be expected for a small number of periodics N, viz. relative FWHM of about 1/N. The peak reflectivity of the order of about 20% is considerably smaller than we would expect from eqs. (19) and (20), viz. about 80% (for the Ge-Mn combination d_c is 347 Å). There are also other substantial differences between the shape of the mea-

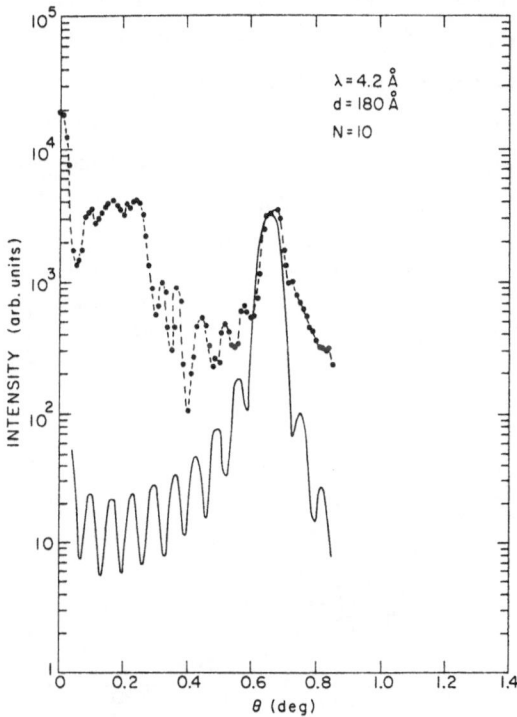

Fig. 3. Reflectivity vs. grazing angle plot for an N=10 bilayer stack consisting of alternating nominally 90 Å thick Ge and Mn layers, deposited by vacuum evaporation. The measured points have been determined by a finely collimated neutron beam of a wavelength of 4.2 Å. The continuous line is the calculated curve with the peak reflectivity adjusted to the measured one at the ℓ=1 Bragg reflection at θ=0.67°. (After Ref.6)

sured and the numerically evaluated exact reflection curves, although there are general similarities. (Note that the calculated line was adjusted to match the peak of the measured Bragg reflection.) These discrepancies are certainly due to irregularities in the layer thickness, and some surface roughness effects might also be involved.

Figure 4 shows similar measured reflectivity curves for an N = 10 and d = 108 Å Fe-Ge multilayer structure saturated in a strong magnetic field, so that for Fe we are concerned with two scattering length densities, viz. for "up" and "down" moments. The d_c values are 245 Å and 500 Å, respectively. The "up" reflectivity of the first-order Bragg peak is 23% at least, the measured apparent value being diminished by instrumental resolution effects, which also broaden the reflection. Thus the reflectivity is rather close to the expected 39%. This difference can also be experessed by noting that the reflectivity is at least equivalent to the ideal one of N_{eff}= 8 bilayers, compared with the deposited N = 10. This shows that this structure is rather close to ideal, in contrast to the one of Fig.3. This is confirmed by the ratio of the ℓ = 1 and ℓ = 3 "up" spin Bragg reflectivities, namely about 200 : 1 compared to the theoretical 81 : 1. Higher order reflections are more sensitive to imperfections, since they have to be referred to a shorter period d/ℓ. Thus we have further evidence for interface roughness and/or thickness irregularities, but $N_{eff} \simeq$ 5 for ℓ = 3 still indicates a rather good structure even on the scale of $d/3 \simeq$ 36 Å.

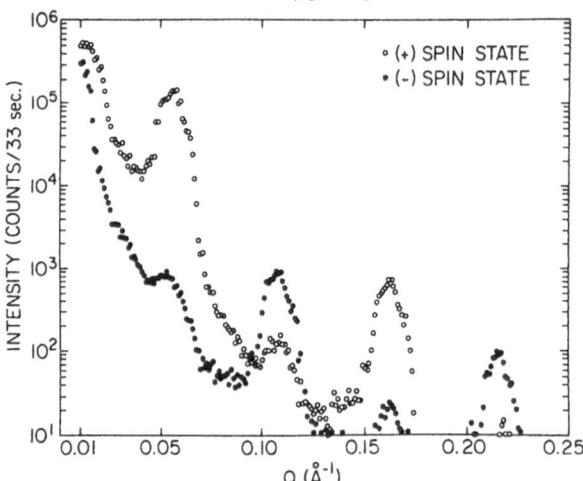

Fig. 4. Reflectivity curves for spin "up" (open circles) and "down" (dots) neutrons measured in an N=10 bilayer Fe-Ge multilayer structure with d=108 Å average period length. The horizontal axis shows the neutron momentum transfer $Q=4\pi\sin\theta/\lambda$. The sample has been prepared by diode sputtering. (After Ref. 7)

There are some very interesting details to be learned about the magnetic behaviour of the Ge-Fe interface from the spin "down" (-) reflectivity curve. The first-order reflection (around $q = 0.06$ Å^{-1}) is about 200 times weaker than the "up" reflection, exactly as expected. (This illustrates the power of these mirrors as neutron polarizers: with an unpolarized incoming beam the reflected beam contains to 99.5% "up" spin states, only.) The $\ell = 2$ Bragg reflection on the other hand is as strong as the $\ell = 1$ reflection, while it is ideally expected to be zero, as it is very weak indeed in the "up" reflectivity. The only possible explanation for this kind of behaviour is that some of the Fe atoms are not (fully) magnetized (or that some of the Ge atoms became magnetic, but on physical grounds we can discard this latter assumption.) Considering eq.(16) it can be shown that this anomalous $\ell = 2$ spin "down" reflectivity can be explained by the equivalent of 2.5 atomic layers of non-magnetic Fe atoms at the Fe-Ge interfaces. On the other hand, we can say nothing about the exact nature of this reduction in Fe magnetism. As far as the neutron reflectivity is concerned, it is equivalent if (a) there is a magnetically "dead" Fe layer at the sharp Fe-Ge interface, or (b) there is some interdiffusion between Fe and Ge atoms at the interface and an equivalent number of Fe atoms become non-magnetic in a Ge environment or (c) a larger number of Fe atoms display reduced moments in the interdiffusion range or (d) due to magnetic anisotropies or other surface anomalies some Fe atoms are not magnetically aligned parallel to the magnetization of the bulk of the Fe layers (e.g. antiferromagnetic coupling at some special positions at the interface). In fact, a combination of some or all of these effects can contribute to the observed anomalous reflectivity. This example indicates that polarized neutron reflectivity measurements can reveal the existence of magnetic anomalies at the interfaces in a multilayer structure with a very high sensitivity (less than one atomic layer would be well visible in this example), but in themselves do not allow the identification of the detailed nature of these anomalies.

Periodic multilayer neutron mirrors have been occasionally used as spin polarizers for monochromatic neutron beams and their use has been considered for producing long wavelength monochromatic beams. There is, however, a major problem with these kinds of applications: for the typical divergencies of neutron beams (0.2 - 0.5°) the uncertainty of the Bragg scattering angle is much too big (cf. Fig.3) to match the inherent wavelength resolution of the multilayer, $1/N$. This results in poor neutron economy.

Aperiodic Structures: Supermirrors

In neutron beam optics the most frequently used devices are neutron mirrors. As we have seen above, media with n < 1 display total reflection for a beam coming from vacuum at grazing angles $\theta < \theta_c$. The critical angles θ_c are rather small for the neutron wavelength range we are interested in, cf. eq.(12). For example for the most often used total reflecting material, $d_c \simeq 285$ Å, which gives approximately $\theta_c/\lambda \simeq 0.1$ °/Å for $\lambda \ll d_c$. In practice Ni mirrors are produced by depositing an Ni layer 1000 - 2000 Å (i.e. a few times d_c) thick on optically polished glass substrates. More recently Ni has been replaced with isotopic ^{58}Ni, for which $\theta_c/\lambda \simeq 0.12$°/Å. The small quantity of ^{58}Ni isotopic material needed costs almost as much as the polishing of the glass. By far the most frequent type of application of these mirrors is in the construction of neutron guide tubes. These guides consist of typically 10 - 50 cm^2 cross section hollow tubes with optically polished Ni coated walls and they work like optical fibers: Neutrons which propagate within $\pm\theta_c$ parallel to the axis of the guide tube are totally reflected on the walls and therefore can be transported to long distances without substantial beam intensity losses. This allows us to increase the efficiency of utilization of the very expensive neutron sources by placing instruments at various distances from the source. Without guides, the neutron flux drops with the distance from the source as $1/r^2$ and the only available location for instrumentation is the circumference of the about 10 m diameter biological shielding around the source. Compared to the about 10 beam positions conventionally available, modern neutron scattering facilities equipped with neutron guides offer 20 to 40 useful beams. Typically, neutron guide tubes measure 10 - 50 m in length and actually they cost about $3000 per meter.

Another classical application of total reflection neutron mirrors is for the production of polarized beams. As we have seen, n is different for the "up" (↑) and "down" (↓) neutron spin states in ferromagnets, and therefore the critical angles $\theta_c^{\uparrow}$ and $\theta_c^{\downarrow}$ are also different. Thus for grazing angles θ such that $\theta_c^{\downarrow} < \theta < \theta_c^{\uparrow}$ the reflected beam has a high intensity and it is also highly polarized (in practice 97% - 98% are typical polarization degrees.) It turns out that $\bar{b} - \bar{p}$ is negative for Co, i.e. there is no total reflection for ↓ spins, and $\theta_c^{\uparrow}/\lambda \simeq 0.09$°/Å. Until recently, bulk polished Co plates had been used as mirrors, since glass substrates have critical angles of about $\theta_c/\lambda \simeq 0.05$°/Å for both spin states, thus no advantage could be taken of the vanishing Co critical angle for ↓ spins. With the introduction of the antireflecting layers we have discussed above,[3] it has now become possible to utilize thin Co films deposited on the top of an antireflecting layer on a glass substrate. This technique has various advantages. Primarily it is much easier to fully magnetize thin films, which is crucial for obtaining a high degree of neutron polarization.

In both applications of neutron mirrors the neutron intensity is determined by the beam divergence accepted by the mirror device: for neutron guide tubes this is $2\theta_c$ both vertically and horizontally and for simple polarizing mirrors it is about $2/3\theta_c$ (grazing angles around zero are not very effective, since the probability of incidence at such small angles goes to zero). Therefore, it is highly desirable to obtain neutron mirrors with the highest critical angles possible. This can be achieved by the use of non-periodic multilayer structures, so called "supermirrors",[8] which are in fact the most frequently used multilayer neutron optical devices. (Some 100 m^2 is the total area in use by now, representing the total production since 1978).

Let us consider the reflectivity of a non-periodic multilayer stack with a monotonic change of the bilayer thickness. We assume that both layers in bilayer i have equal optical thicknesses $d(i)/2$, and we choose $\partial d(i)/\partial i$ to be negative. The reflectivity corresponding to the Bragg condition with $d(i)$ lattice spacing will be determined by the effective number of bilayers $N_{eff}(i)$ which act in constructive interference. As an approximate criterium for this we require that the interfaces of these bilayers lie within 45° phase error where their ideal position would be in a periodic stack around bilayer i with spacing $d(i)$. Thus the cumulative thickness deviation has to be less than $d(i)/8$, i.e. N_{eff} is determined by the condition

$$\sum_{j=0}^{N_{eff}(i)/2} d(i+j) - d(i) = \frac{d(i)}{8} \quad . \tag{22}$$

There are other $N_{eff}/2$ coherently reflecting layers at $j < 0$. The sum can approximately be evaluated as

$$\frac{\partial d(i)}{\partial i} \sum_{j=0}^{N_{eff}/2} j = \frac{1}{8} \frac{\partial d(i)}{\partial i} N_{eff}^2(i) \quad .$$

In order to simulate total reflecting mirrors, we require that R = 1 for every $d(i)$, i.e. by virtue of eq.(19) for first order reflection $\ell = 1$

$$N_{eff}^2(i) = 4\left(\frac{d_c}{d(i)}\right)^4 \quad ,$$

where d_c refers to the diffraction index difference between the two materials in the stack. Finally, this leads to the differential equation

$$\frac{\partial i}{\partial d(i)} = -4 \frac{d_c^4}{[d(i)]^5}$$

the solution of which is

$$d(i) = \frac{d_c}{\sqrt[4]{i}} \quad . \tag{23}$$

The actual layer thicknesses to be deposited have to be determined by the application of the diffraction correction [eq.(21)] to the $d(i)/2$ values given by eq.(23).

Since d_c characterizes the total reflection between the two materials we see that with for instance 16 bilayers we can obtain a "supermirror" with $2\theta_c$ cut-off angle, but with a reflectivity of about $1 - e^{-1} \simeq 65\%$ only [cf. eqs.(20)]. Exactly this has been achieved with an Fe-Ag bilayer structure,[9] in which the combination for spin ↑ neutrons $d_c = 280$ Å, i.e. virtually the same as for Ni (see Fig.5). The spin ↓ refractive index of Fe is roughly equal to that of Ag and also that of the glass substrate. Therefore for ↓ neutrons we have total reflection only. While with N = 17 bilayers the supermirror reflectivity agrees very well with the one expected, be routinely produced in a fully automated process. Nowadays supermirrors

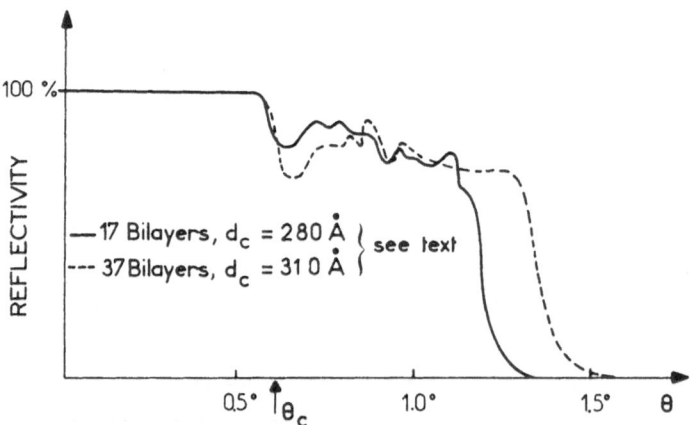

Fig. 5. Spin "up" neutron reflectivity curves of two Fe-Ag supermirrors with parameters indicated in the figure. θ_c shows the critical angle of total reflection for "up" spin neutrons between Fe and Ag, and it coincides with the one between vacuum and Ni. The supermirrors were produced by vacuum evaporation method. The measurement was performed by a $\lambda = 6.7$ Å monochromatic beam. The spin "down" reflectivity is <2% above a total reflection cut-off angle of $\theta_c^{\downarrow} \simeq \theta_c/2$. (After Ref. 9)

this is not the case for N > 17. Obviously, with an increasing number of layers, their quality starts to suffer due to increasing interface roughness. For a while this can be off-set by depositing more layers than ideally necessary by using a higher value of d_c in eq.(23). The dashed curve in Fig. 5 illustrates this: The reflectivity falls short of the calculated $1-e^{-2.3} \approx 90\%$. (Note that the oscillations of the reflectivity curves around an average are to be expected for interference in a structure with a small number of elements.) Up to now, no improvement could be achieved by depositing more bilayers, but with d_c = 340 Å and N = 37 the observed reflectivity becomes greater than 95% up to $2\theta_c$, i.e. in the whole angular range covered in the figure by the N = 17 curve. (The theoretical reflectivity would be about $1 - e^{-4.5} \simeq 98.5\%$.) Note that in these mirrors the smallest d spacing (about 140 Å) is still quite large compared with the ones used successfully in X-ray mirrors.

These Fe-Ag mirrors have been used as neutron polarizers and polarization analysers in full scale operation since 1978.[10] In order to eliminate the ↓ spin reflectivity below $\theta_c/2$ Schärpf has developed a method[11] to combine antireflecting layers[3] and Co-V supermirrors. These mirrors show a reflectivity of about 70% up to $1.75\theta_c$. The production of these Co-V mirrors ʰʸ electron beam gun evaporation is very easy, and large quantitites could

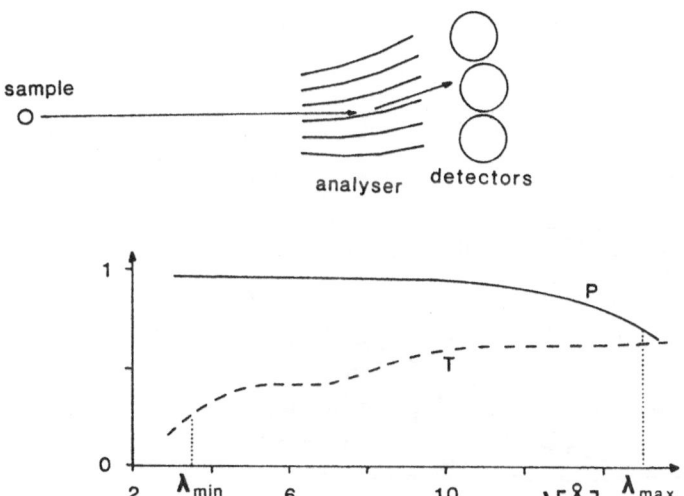

Fig. 6. Schematic lay-out (top) and performance (bottom) of a large solid angle supermirror polarization analyser system. The continuous line shows the wavelength dependence of the neutron polarization efficiency and the dashed line that of the transmission coefficient for spin "up" neutrons. The real device use on IN11 spectrometer at ILL contains 24 10x20 cm supermirror plates arranged in 8 channels. It can handle a 8,5x9,5 cm cross section beam coming from a 3 cm wide source at 3 m distance. (After Ref. 12)

provide the most powerful polarizers and analysers for polarized neutron work. A typical mirror arrangement is shown in Fig. 6.

In many instances it is more convenient to use a transmitted beam than a reflected one. Perfect silicon single crystals are practically transparent for neutron beams. With large diameter Si wafers and nearly 100% reflectivity supermirror coatings now available in newer designs, the ↓ polarized transmitted beam or both the ↑ polarized reflected and ↓ polarized transmitted beams, cf. Fig. 7, are often utilized.[12] (θ_c of Si roughly coincides with that of glass or Ag.)

Of course, supermirrors can also be used to improve the beam divergence, i.e. neutron intensity in neutron guides. Non-polarizing supermirror structures have been experimentally tested[13] (Ni-Ti combination with a cutoff angle of ~ $2\theta_c^{Ni}$) and theoretically evaluated[14] (e.g. Ni-Mn combination with $N = 202$ periods and $33\theta_c^{Ni}$ cut-off angle, Fig. 8). Such mirrors could either be used to cover the whole guide or to compress the beam of a large cross section of an ordinary Ni coated guide to a smaller sample area, by which the beam intensity per unit area is proportionally increased. Both solutions have been studied in detail and systems are being built along both lines. No full scale operational experience is yet available with either method.

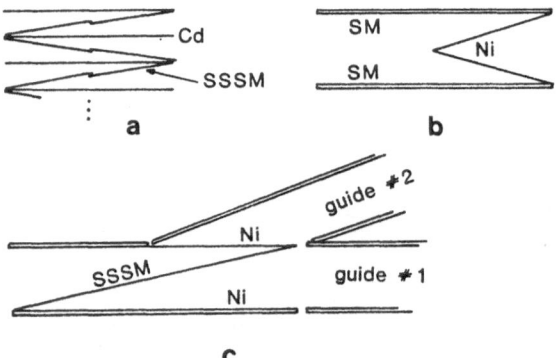

Fig. 7. Proposed polarizer arrangements using supermirrors evaporated on neutron-transparent Si substrates (SSSM) in transmission geometry; (a) polarizing collimator, (b) polarizing beam splitter: beam # 2 is "up" polarized and beam # 1 is "down" polarized, (c) polarizing neutron guide tube: the walls are supermirror coated (SM) and the V shaped insert consists of Ni coated perfect Si crystal plates. (After Ref. 12)

Over the past decade multilayer neutron optical devices, primarily supermirrors, have become a fundamental ingredient in neutron scattering instrumentation. Due to the large beam sizes and small grazing angles involved, full scale devices usually need several m^2 of coating. This is why neutron mirrors are those multilayer structures produced in by far the largest quantities up to now.

The necessity of producing large surface coatings fundamentally determines the deposition techniques. The deposition procedures had to be adapted to substrate sizes of the order of 10 - 50 cm, and therefore they had to be simple, easily reproducible and fully automated. Thus they represent a further developement challenge compared to the ultra-sophisticated techniques used in order to produce the finest and most precisely grown structures. Understandably the structural qualities of large area neutron optical multilayers (primarily the smoothness of the interfaces) are way below those achieved with X-ray mirrors and other advanced structures. Adapting features of the sophisticated techniques in large deposition area, fast turn-over, easy to handle, coating facilities should lead in the near future to a substantial improvement in the optical properties of neutron mirrors. In particular, we should achieve a reflection performance near to the value theoreti-

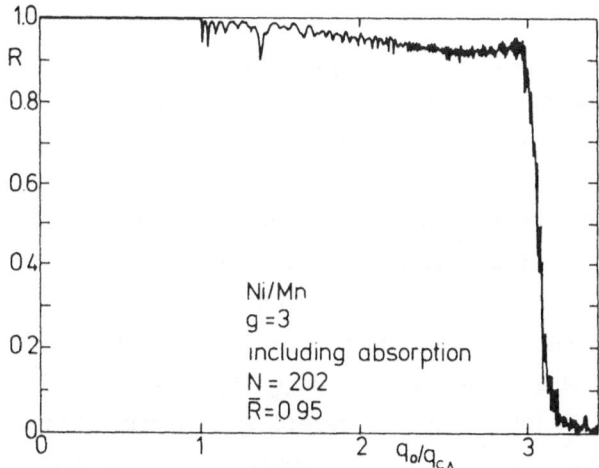

Fig. 8. Calculated reflectivity curve of an Ni-Mn non-polarizing supermirror with N = 202 bilayers. The abscissa corresponds to the normalized grazing angle θ/θ_c^{Ni}. No interface roughness or layer thickness errors have been assumed. (After Ref. 14)

cally predicted for structures with 50 - 200 bilayers and with bilayer thicknesses less than 100 Å. The possibilities of straightforward evaporation techniques are probably exhausted by the existing machines, but there is certainly a lot of room for development in the application of sputtering methods.

On the down-to-earth practical side, a major neutron project requires the shear out-put capability of a \$ 400.000 - 800.000 evaporation or sputtering apparatus for 1 or 2 years. This means that, with usual industrial overhead and mark-up (in particular when specifications have to be rigorously met) the commercial production of multilayer mirrors in these quantities is likely to cost several times the value of the deposition equipment. Thus, while for small quantities ($< 5 - 10 \, m^2$ total surface) it is certainly advisable to try to purchase custom-made multilayer neutron mirrors, beyond a requirement of some 50 m^2 the development of one's own supermirror production capability is basically the only affordable option.

REFERENCES

1. See S.W. Lovesey, "Theory of Neutron Scattering from Condensed Matter", Clarendon Press, Oxford (1984).
2. I.I. Gurevich and L.V. Tarasov, "Low-energy Neutron Physics", North Holland, Amsterdam (1968), p. 46.
3. G.M. Drabkin, A.I. Okorokov, A.F. Schebetov, N.V. Borovikova, A.G. Gukasov, V.A. Kurdriashov, V.V. Runov, Nucl. Instr. and Meth. 133:453 (1976).
4. P. Croce and B. Prado, Nouv. Rev. Opt. Appl. 1:229 (1970); J. Schelten and K. Mika, Nucl. Instr. and Meth. 160:287 (1979).
5. B.P. Schoenborn, D.L.D. Caspar and O.F. Kammerer, J. Appl. Cryst. 7:508 (1974).
6. A.M. Saxena and B.P. Schoenborn, Acta Cryst. A33:805 (1977).
7. C.F. Majkrzak and L. Passell, Acta Cryst. A41:41 (1985).
8. F. Mezei, Commun. Phys. 1:81 (1976); a partially similar suggestion has been forwarded by V.F. Turchin in 1967 (unpublished).
9. F. Mezei and P.A. Dagleish, Commun. Phys. 2:41 (1977).
10. P.A. Dagleish, J.B. Hayter and F. Mezei, in: "Neutron Spin Echo", F. Mezei, ed., Springer Verlag, Heidelberg (1980), p. 66.
11. O. Schärpf, AIP Conf. Proc. No. 89, J. Faber, ed., American Institute of Physics, New York (1982) p. 182.

12. F. Mezei, in: "Use and Development of Low and Medium Flux Research Reactors", O.K. Harling, L. Clark, P. von der Hardt, eds., Supplement to Atomenergie - Kerntechnik Vol 44, Karl Thiemig, München (1984), p. 735.

13. T. Ebisawa, N. Achiwa, S. Yamada, T. Akiyoshi and S. Okamoto, J. of. Nucl. Sci. and Techn., 16:647 (1979).

14. J. Schelten and K. Mika, Nucl. Inst. and Meth., 160:287 (1979).

POSTER SESSION I

STRUCTURAL STUDIES AND FABRICATION

Xe IMPLANTED Si OBSERVED BY MEANS OF EXAFS

G. Faraci, A.R. Pennisi, and A. Terrasi

Dipartimento di Fisica, Università di Catania
I-95129 Catania, Italy

S. Mobilio, A. Balerna

INFN-Laboratori Nazionali di Frascati
C.P. 13, I-00044 Frascati, Italy

Rare gases introduced in a solid matrix by means of ion implantation are not miscible and hence tend to agglomerate in clusters or bubbles. As far as clusters existence and their aggregation state (solid, liquid or gaseous) are concerned, experimental data are ambiguous even because until now used analysis techniques (e.g., TEM or x-ray diffraction) are sensitive to long-range order.

We report for the first time an EXAFS (Extended X-Ray Absorption Fine Structure) measurement performed at Frascati Lab. Naz. for Xe^+ implanted Si(111). This technique is able to determine the local structural parameters of each individual atomic species and it is applicable to crystals, amorphous, liquid and molecular gases.

Our work, on samples with different percentages of Xe, as implanted or annealed after implantation, detected finally solid bubbles into the host matrix mainly when it is implanted at high doses ($\approx 10^{17}$ at/cm^2) and conveniently annealed.

Studies are in progress now for a more detailed knowledge of these phenomena.

TEM METHODS FOR CHARACTERISING THE STRUCTURE OF MULTILAYERED MATERIALS

C.S. Baxter

Department of Materials Science and Metallurgy
Cambridge University, Pembroke Street
Cambridge, CB2 3QZ, U.K.

Detailed knowledge of the structure of multilayers is an essential prerequisite for understanding their properties. We have therefore developed TEM techniques for examining both metallic and semiconducting multilayers in cross-section. These techniques enable the layer-to-layer variations in composition and lattice parameter to be determined as well as characterising larger scale features such as twinning and gross non-uniformities in the layering. This poster will concentrate on the methods used to assess the structure at the atomic level but will also give examples of other structural features, such as the laminar twinning frequently found in fcc metal multilayers. Features of this type are readily characterised by TEM but may not be obvious from x-ray studies alone.

Information from local variations in lattice parameter extends further out in reciprocal space than information from the underlying 'average' lattice. This information may therefore be truncated by the contrast transfer function (CTF) of even a good high resolution microscope, so that any lattice parameter variation may not be properly represented in the image. A consequence of this is that axial lattice images of Cu/NiPd multilayers, which contain large coherency strains, show no variation in fringe spacing. Higher frequency information can be included within the CTF by using non-axial rather than axial imaging conditions. By this means the lattice variation of a Cu/NiPd multilayer can be revealed, although interpretation of the lattice image is only possible by careful image/simulation matching.

Variations in composition of multilayers are on too fine a scale for the application of conventional EM analysis techniques such as EDX and EELS. Ideally methods which utilise the resolution inherent in TEM images should be used, but unfortunately high resolution images generally contain little compositional information. For semiconductor materials such as GaAs/GaAlAs the composition dependence of the (002) structure factor can be utilised for composition measurement via the intensities in (002) dark field images (although precise quantification requires consideration of inelastic as well as elastic scattering), but such a method is not applicable to multilayers based on fcc metals. Features in conventional bright field and dark field images such as apparent layer widths and interface sharpness are a sensitive function of the imaging parameters (e.g., defocus, specimen thickness and orientation) and such images can easily be misinterpreted. However, careful comparison of experimental series of images with computed image simulations can provide high resolution compositional information. For example, for the metal multilayers studied here it was found that intermixing over one atomic plane either side of the interface provided a better description than intermixing over two atomic planes.

Interpretation of results from cross-sectional TEM specimens is complicated by the fact that coherency strains present in the bulk multi-layers can relax at the surfaces of the thin foils. The specimens are therefore not strictly representative of the bulk multilayer and the plane bending near the surfaces can produce strong contrast in TEM images which may swamp other contrast mechanisms. This novel (and confusing!) contrast has been seen particularly clearly in GaAs/GaInAs multiple quantum wells.

TEM techniques are now available for the atomic level quantification of composition and strains in modulated structures: unfortunately their application is neither quick nor trivial!

CHARACTERIZATION OF R.F. SPUTTERED OXIDE

APPLICATIONS OF MULTILAYERED STRUCTURES IN OPTICS

P. Caburet

MATRA - Thin Films Department

17, rue Paul Dautier, F-78140 Velizy, France

Multilayered structures have been used for many years in the ultra-violet, visible and infrared range. Many components are now available such as mirrors, antireflection coatings, polarizing filters, broad- or narrow-band filters, etc. But one should not consider these structures to be perfect. Ideally, an optical layer should have:

- stable and reproducible refractive index
- low optical losses (absorption, scattering)
- good homogeneity and isotropy
- good mechanical properties (adherence, hardness, resistance to abrasion

Electron-beam evaporation, the most widely used technology for optical coatings, does not fulfill all these requirements. That accounts for the fact that we have come to use R.F. magnetron sputtering as an alternative technology for evaporation.

Experiments on SiO_2, Ta_2O_5 and TiO_2 in the 400 - 1000 nm wavelength range clearly demonstrate the interest of cathode sputtering in optics.

SiO_2 and Ta_2O_5 layers have a stable, well-defined and reproducible refractive index as well as a low absorption and a good homogeneity. Both of them can be used in optical multilayered structure.

TiO_2 depends more on the sputtering conditions, but we are inclined to think that this material is also suitable for optical coatings if these conditions are properly optimized.

Adsorption measurements of moisture on thick single layers proved all these materials to be more compact than the evaporated ones. More-over, spectrophotometric measurements have been performed in air and in vacuum for two mirrors: the negligible shift in the spectral profile of the mirror made by sputtering compared to the important shift of the mirror made by evaporation is also an evidence that the multilayered structures made by sputtering have highly stable properties.

If one considers that sputtered layers usually have particularly high mechanical properties, there is little doubt that cathode sputtering is a major improvement in multilayered structures for ultraviolet, visible and infrared optics.

ION BEAM MIXING IN METALLIC SUPERLATTICES

P. Milołajczak and G. Gładyszewski

Department of Experimental Physics, Institute of Physics

University of Marie Curie-Skłodowska, 20-031 Lublin, Poland

Ion beam mixing has become an important technique in physical metallurgy. This technique is capable of producing metastable crystalline and amorphous alloys as well as equilibrium alloys. The basic physical mechanisms of ion beam mixing in layered solids are not well understood. Unfortunately, we cannot study multilayers which consist of very thin layers by RBS, SIMS or electron microscopy. Our main purpose is to present θ-2θ X-ray diffraction measurements of implanted superlattices as a tool for ion beam mixing study.

Satellite peaks intensity of superlattice varies rapidly with interface thickness (number of mixed layers; see Fig.1). Using computer simulation of X-ray diffractoin spectra it is possible to determine the concentration of elements as a function of a depth after each sequential implantation. This method is very sensitive because extension of interface thickness of about few angstroms is clearly visible on the X-ray diffraction spectra (Fig.2).

This poster included:

- introduction (the model of a superlattice structure, the idea of ion beam mixing, exerimental procedure),
- analysis of influence of superlattice parameters on the X-ray diffraction spectra,
- comparison of experimental and simulated X-ray spectra before and after ion beam mixing in Ag-Cu and Bi-Sb superlattices.

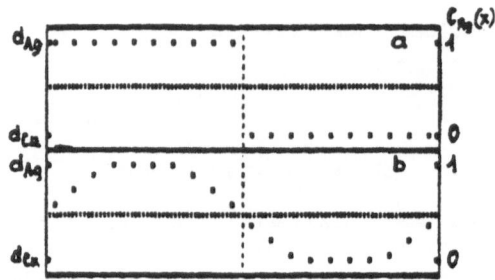

Fig.1 The model of Ag-Cu superlattice
unit cell before (a) and after (b)
ion beam mixing using low ions dose

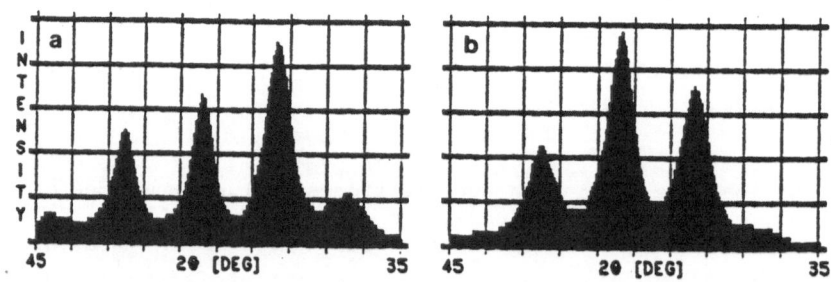

Fig.2 Simulated X-ray diffraction spectra of superlattice Ag-Cu in the
cases as in Fig.1

THE INITIAL STAGES OF OVERLAYER GROWTH STUDIED USING HIGH ENERGY ELECTRON

FORWARD SCATTERING: Cr/Ag(100)

A.D. Johnson

Department of Physics, University of Leicester

Leicester, U.K.

It has recently been shown that both Auger electrons and photoelectrons with kinetic energies above a few hundred eV are very strongly scattered in a forward direction by overlying atoms [1-3]. Thus there are large electron intensity maxima along the major crystal axes. By monitoring these maxima as a function of overlayer deposition it is possible to determine the growth mechanism of the overlayer without the need for any complex theory. Also, by measuring the shift of these peaks, one can directly measure the elastic strain and relaxation at the interfaces.

Polar angle electron intensity distribution scans have been measured for various layers of Cr/Ag(100). The lattice parameter of Cr almost perfectly matches the nearest neighbour distance of Ag so one would expect the Cr to grow in the bulk bcc phase but rotated 45° to the Ag substrate lattice. By measuring the intensity of both the Ag MNN (356 eV) and the Cr LMM (489 eV) Auger lines as a function of polar angle along the Ag[110] and Ag[100] axes, the growth mechanism has been deduced and shown to be in the bulk bcc phase.

The bcc Cr grows as single layer platelets until a coverage of 0.4 ml, when it starts to grow bilayer platelets. This growth continues until 1.6 ml when triple layer (and higher) islanding occurs. Thus high energy electron forward scattering is a very simple but effective method of studying interface formation and the results obtained agree well with previous work done on Cr/Ag(100) [4].

REFERENCES

1. W.F. Egelhoff Jr., Phys. Rev. B 30, 1052 (1984)
2. S.A. Chambers et al.: Phys. Rev. B 35, 2592 (1987)
3. E.L. Bullock, C.S. Fadley: Phys. Rev. B 31, 1212 (1985)
4. D.A. Newstead, C. Norris, C. Binns and P.C. Stephenson: J. Phys. C
   (in print)
5. S.A. Chambers et al.: Phys. Rev. B33, 8810 (1986)

MONTE CARLO CALCULATIONS OF SPUTTERING PROCESSES

T. Krist

Hahn-Meitner-Institut, Glienicker Straße 100

D-1000 Berlin 39

Computer simulations of sputtering processes can roughly be divided into two groups: molecular dynamics calculations and binary collision approximations [1]. The latter group, which is often referred to as Monte Carlo calculations, is dominated by the MARLOWE program for crystalline and TRIM for amorphous targets. Here we shall review some papers concerning the application of TRIM offsprings to sputtering, such as TRIM.SP [2] or TRIDYN [3].

TRIM.SP calculates the scattering angle and the elastic energy loss for each collision of an incoming ion with a target atom using the Kr-C potential and the inelastic energy loss by using semiempirical values. If the energy transferred to a target atom exceeds the bulk binding energy, this atom is treated as a recoil. This recoil and the recoils produced by it are followed until their energy falls below a cut-off value. Afterwards the ion's next collision is calculated. Whenever a recoil transmits the surface a planar attractive potential is applied, its strength being chosen as the heat of sublimation. If the recoil leaves the solid it is counted as sputtered and its angle and energy are stored as well as its origin.

The results presented here refer to: contributions of various sputter mechanisms, sputtering yield as a function of energy and angle of incidence, angular and energy distribution of sputtered particles and their escape depth.

Further extensions of the program to two-component target sputtering and dynamic composition changes are mentioned.

REFERENCES

1. H.H. Andersen: Nucl. Instr. Meth. B18, 321 (1987)
2. J.P. Biersack and W. Eckstein: Appl. Phys. A34, 73 (1984);
   W. Eckstein and J.P. Biersack: Appl. Phys. A37, 95 (1985)
3. W. Möller and W. Eckstein: Nucl. Instr. Meth. B2, 814 (1984)

X-RAY DIFFRACTION STUDIES OF Nb/Ta SUPERLATTICES

F. Lamelas, Hui He and R. Clarke

Department of Physics, Randall Laboratory

The University of Michigan, Ann Arbor, MI 48109, USA

We have grown high crystal quality niobium/tantalum multilayer super-lattices in an MBE system. The films were deposited from electron beam hearths operating at 3-4 kilowatts onto rotating sapphire substrates held at a variety of temperatures ranging from 775°C to 1000°C. With liquid nitrogen filled cryopanels in the deposition chamber, pressures were held at 10^{-9} to 10^{-8} Torr during growth.

X-ray diffraction studies were performed on a 4-circle diffractometer mounted on a 12 kilowatt rotating anode x-ray generator fitted with a molybdenum target. With this setup we are able to perform in-plane x-ray scans without removing the films from the sapphire substrates, due to the penetrating nature of the molybdenum radiation.

Two-dimensional x-ray intensity contour maps generated from the in-plane x-ray scans clearly reveal the in-plane crystalline order of the films and their epitaxial relationship with the sapphire substrates. Out-of-plane scans such as 11ℓ show that the stacking coherence of the films is high. Simple geometrical diagrams are used to explain the orien-tational relationship between the unit cells of sapphire and niobium. In addition, we present evidence for what appears to be a metastable hexagonal phase nucleated at the layer interfaces of the samples.

INTENSITY OSCILLATIONS IN REFLECTION HIGH ENERGY ELECTRON DIFFRACTION

DURING MOLECULAR BEAM EPITAXY OF METALS ON METALS AND THE FABRICATION

OF METALLIC SUPERLATTICES

G. Lilienkamp, C. Kozioł* and E. Bauer

Physikalisches Institut, Technische Universität Clausthal

D-3392 Clausthal-Zellerfeld, FRG

The intensity of the specular reflected RHEED-beam during molecular beam epitaxy of metals (Ni, Cu, Nb, Mo) on W(110) was studied. The preparation was performed under extremely clean conditions, the pressure during growth was below 10^{-10} mbar. After the first few deposited layers intensity oscillations with a constant period corresponding to one monolayer were visible. For Ni on W(110) polar and azimuth angle dependence of the observed oscillations were examined. Near the Bragg condition there are nearly no oscillations visible; out of phase scattering from terraces with one monolayer difference in height and very flat incidence are the best conditions for obtaining many oscillations.

The observed damping of the oscillations is attributed to the formation of small islands, which are not completely smoothed out by more deposited material, so that a quasi-stationary roughness of the surface develops after some time. This is strongly influenced by the temperature of the sample and the residual gas pressure.

RHEED intensity measurements are also a good tool for controlling the fabrication of metallic superlattices. This is shown for the examples Ni/Cu and Nb/Cu superlattices.

---

*Permanent address: Institut of Experimental Physics, University of Wrocław, ul. Cybulskiego 36, PL-50 205 Wrocłoaw, Poland

CRYSTALLIZATION AND MELTING IN MULTILAYERED STRUCTURES

W. Sevenhans, J.P. Locquet, H. Vanderstraeten, and Y. Bruynserade

Laboratorium voor Vaste Stof-Fysika en Magnetisme
Katholieke Universiteit, Leuven, B-3030 Leuven, Belgium

H. Homma and I.K. Schuller

Materials Science Division, Argonne National Laboratory
Argonne, IL 60439, USA

Pb/Ge multilayers were prepared in a load-locked MBE apparatus equipped with two electron beam guns. The evaporation rates were controlled using a quadrupole mass spectrometer in feedback mode. Pb/Ge multilayers were prepared using a microprocessor controlled shutter, to alternately expose the liquid nitrogen cooled sapphire substrates to the flux of evaporant.

Small angle x-ray diffraction spectra from the multilayers show up to the 13th order small angle diffraction peak. The existence of such a large number of small angle peaks implies that the layered structure is well developed. At large angles the continuous variation of the thickness of the amorphous Ge broadens the multilayer peaks which in turn limits the perpendicular coherence length to one Pb layer. In this fashion, the high angle diffraction pattern exhibits one broad peak in conjunction with the secondary fringes.

We have undertaken a series of detailed experiments (temperature dependent x-ray diffraction, TEM and ED) in order to investigate the crystallization of Ge and melting behaviour of Pb in Pb/Ge multilayers.

The peak positions of the wide angle reflections show that the Pb grows along the [111] direction. The mosaic spreadth, determined by rocking, decreases with decreasing Pb thickness and slightly increases with increasing Ge thickness. From electron diffraction the strong amorphicity of the Ge unambiguously has been determined, which is not surprising on view of the low substrate temperature.

During crystallization, the x-ray diffraction peaks of the modulated structure disappear and the Pb texture improves. It is shown that the crystallization temperature decreases with decreasing amorphous Ge thickness and increases with decreasing thickness of the metallic component. These results imply enhanced kinetics of the diffusion at the Pb/Ge interface induced by the metallic character of the latter.

During the diffusion of the Ge in the Pb, the multilayered structure completely disappears causing the small angle x-ray peaks to vanish progressively. Due to the low solubility of Ge in Pb, the diffused Ge will precipitate in crystals giving rise to stress into the Pb lattice. To releave the stress the Pb domains grow allowing the lattice to relax to its state of minimal energy. This evolution was clearly seen by x-ray diffraction, electron diffraction and transmission electron microscopy.

Due to the broken up structure of the Pb/Ge multilayers at $T < 200°C$, no 2D melting could be observed. Recent 2D melting observations can be explained by oxidation of the Pb during heating of the Pb/Ge multilayer.

This work was supported by the Office of Naval Research contract number N00014-83-F-0031, the U.S. Department of Energy, Basic Energy Sciences, under contract number W-31-109-ENG-38, and the Belgian Institute for Nuclear Sciences. International travel was provided by NATO grant number RG85-0695 and the Belgian National Science Foundation.

THE STRUCTURE OF Mo/Ta SUPERLATTICES*

J.L. Makous, and C.M. Falco

Department of Physics
University of Arizona, Tucson, AZ 85721, USA

We grew superlattices of Mo and Ta on sapphire substrates using magnetically enhanced dc triode sputtering, where a rotating substrate table
alternately passes over the targets to create multilayers. Microprocessor
control of the sputtering process allowed us to keep layer thicknesses
constant to ±0.3%. The wavelengths were modulated by an integer number, n,
of atomic planes of Mo and Ta, where n ranged from a value of 153
(λ = 700 Å) to the monolayer limit of 1 (λ = 4.57 Å).

Characterization of the superlattices included the use of a number of
x-ray diffraction techniques and Rutherford backscattering spectroscopy.
We found that these superlattices exhibit long range structural coherence
normal to the plane over the entire wavelength range. Spectra of θ-2θ
x-ray diffraction of Cu-Kα radiation indicate a very narrow Bragg peak
linewidth for all n. Using the Scherrer equation, we converted these linewidths into perpendicular coherence lengths and found that they all have
values between 250 Å and 300 Å down to n = 2, with a slight drop down to
$\approx$165 Å for the n = 1 monolayer wavelength. In addition, resistivity
measurements from 300 K to 7.5 K indicate metallic behavior for all Mo/Ta
superlattices. The temperature dependence of resistivity remains positive
over the entire wavelength range, and the absolute values lie well within
the metallic regime as defined by the Ioffe-Regel condition.

In summary, we have shown that sputtered superlattices of Mo/Ta
remain metallic in both their structure and transport properties for all
superlattice wavelengths.

---

*Supported by the Materials Science Division, U.S. DOE under Contract
DE-FG02-87ER45297

INVESTIGATION OF THE STRUCTURAL PROPERTIES OF THE CdTe-ZnTe

STRAINED-LAYER SUPERLATTICE SYSTEM

G. Monfroy and J.P. Faurie

University of Illinois at Chicago, Department of Physics,

P.O. Box 4348, Chicago, IL 60680

The CdTe-ZnTe strained-layer superlattice (SLS) system was initially proposed as a buffer layer for growing high crystalline quality $Hg_{1-x}Cd_xTe$ epitaxial layers on GaAs substrates (mismatch = 13.7%). This system by itself raised a lot of interest of a fundamental nature: Would these SLS's have to be considered as a new system with constituent layers which are not independent?

A first series of CdTe-ZnTe superlattices (SL) having various buffer layers (BL), constituent layer thicknesses, and number of periods, has been grown on (001) GaAs substrates, and investigated using the photoluminescence and the X-ray diffraction techniques. The first technique indicated a luminescence response much higher than for the corresponding $Zn_xCd_{1-x}Te$ alloys. The response also led to the conclusion that CdTe-ZnTe was a free-standing SL (Miles et al., Appl. Phys. Lett. 48, 1383 (1986)). The X-ray $\omega/2\theta$ scans along the growth direction revealed the high crystalline quality of the samples (Monfroy et al., Appl. Phys. Lett. 49, 152 (1986)). The periods of the SLS's, as well as their average lattice parameters ($<a_\perp>$) along the growth direction were calculated very accurately using an indexing method which takes into account all the reflections along the considered direction, whether they are satellite or central peaks (Knox et al., Acta Crystallog. A, to be published). The free-standing assumption was then checked against the X-ray data. The calculated and experimental lattice parameter (LP) values did not show good agreement. This could be due to the assumptions used to carry out the calculation and/or to the difficulties in estimating the separate LP's, along the growth axis for CdTe and ZnTe, from $<a_\perp>$.

Miles et al. also performed $\omega/2\theta$ X-ray diffraction, and they report that some CdTe-ZnTe SLS's show evidence of large strain fields, while others indicate only very low levels of residual strain. They also suggest that, in this latter case, the constituent LP's returned to their bulk values (Surf. Sci., to be published).

In order to investigate this problem a second series of SLS's has been grown on (001) GaAs substrates. The SLS's were made of 200 periods, each of them consisting of 25 Å of CdTe and 25 Å of ZnTe, and having a

2 μm thick BL of either CdTe, ZnTe, or $Zn_xCd_{1-x}Te$. In-situ electron diffraction (RHEED) experiments have been performed during the growth, and
the evolution of the in-plane LP's followed as a function of the period
number (Monfroy et al., Appl. Phys. Lett., to be published). The results
indicated that, within the experimental accuracy, CdTe and ZnTe have
the same in-plane LP's. Furthermore, two regimes have been observed in
the superlattices: a "transient SL" close to the substrate, and a
"stable SL" after the growth of the first 30 to 50 periods. Finally, an
attempt was made to relate this latter observation to the critical
thicknesses of the SLS's.

ELABORATION AND CHARACTERIZATION OF METALLIC MULTILAYERS IN THE
CNRS - ST. GOBAIN COMMON LABORATORY

J.C. Ousset, M.F. Ravet, M. Maurer, and J. Durand

Presentation of the Laboratoire ixtre CNRS
Saint Gobain, UM 37

Laboratory name: Laboratoire Mixte CNRS - SAINT-GOBAIN - UM 37

Address: CRPAM - BP 28 - 54703 PONT-A-MOUSSON Cédex

Phone: 83.83.24.24

Birth date: 1 January 1986 (official)

            1 October 1986 (effective launching)

Director: Jacques DURAND*

     André THOMY

Staff: - to date 7 scientists:
       M. Maurer, J.C. Ousset, M. Piecuch, M.F. Ravet,
       A. Thomy, D. Girardin, C. Tête

    - 1988: 15 scientists expected

Research topics: - Amorphous metallic alloys
                - Superlattices and multilayered metallic systems
                - Micronic and submicronic metallic powders
                - Laser processing on metallic materials
                - High Tc superconductors

*Died in April 1987

PRELIMINARY STUDIES ON METALLIC MULTILAYERS

    Fabrication of metallic multilayers and superlattices is one of the
topics being developed in our new laboratory. These structures will be
prepared in a metal molecular beam epitaxial deposition system and in a
sputtering one. In the first case the epitaxial growth will be monitored
with reflexion high energy electron diffraction. The composition of both
layers and interfaces will be determined by ESCA and AUGER analysis. The
whole system is to be operational in October 1987.

    At present we are elaborating multilayers in a classical deposition
system. From grazing incidence X-ray diffraction experiments in a high
resolution diffractometer, we obtain information concerning the period of
the sample and the profile of the interfaces.

The systems we have studied are:

- polycrystalline multilayers Cu/Gd (evaporation)
- amorphous multilayers (prepared by M. Maurer by sputtering in Leiden)

$$Al_{1-x}Gd_x/Mo_{1-y}Si_y \quad (x \simeq 0.25, \quad y \simeq 0.30)$$

On these multilayered structures magnetic measurements are performed in a SQUID magnetometer.

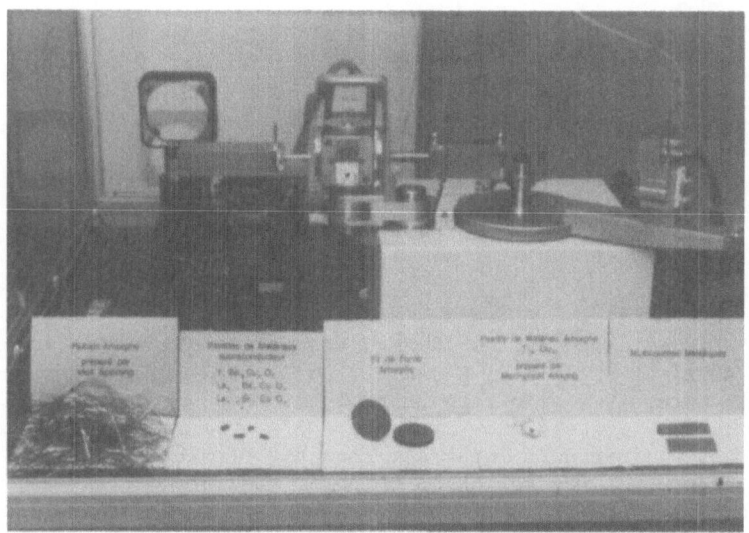

Fig.1 A high resolution x-ray diffractometer: an original device supplied by Philips S.A., including an x-ray generator, a Bartels 4-crystal mono-chromator ($\Delta\lambda/\lambda = 2.3 \times 10^{-5}$, divergence $\Delta\theta = 15"$), a horizontal goniometer (ANVAR - CNRS patent), a spatial localisation detector Raytech (CNRS patent), an adjustable sample holder, automation and piloting by a microcomputer, analysis and calculation software

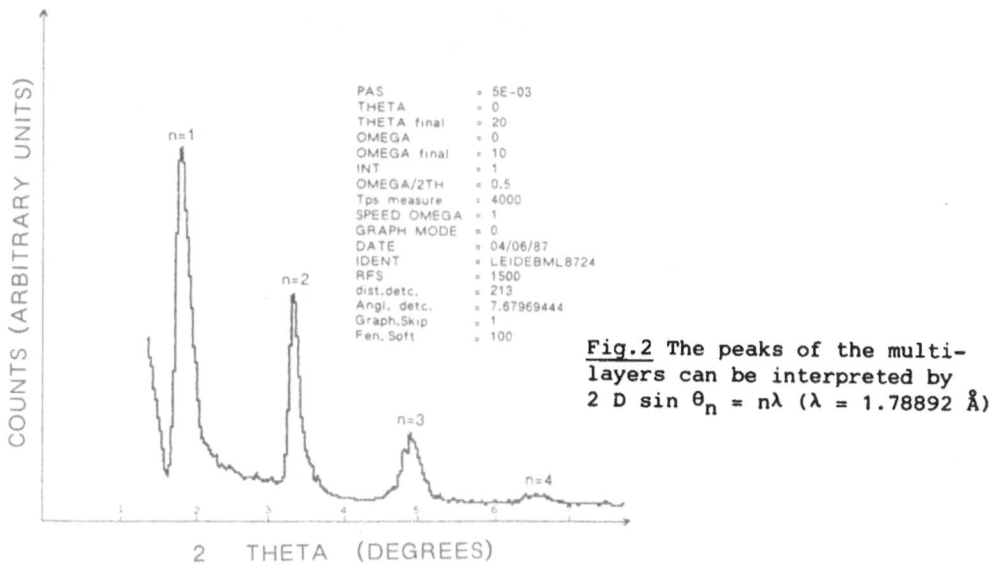

PAS	= 5E-03
THETA	= 0
THETA final	= 20
OMEGA	= 0
OMEGA final	= 10
INT	= 1
OMEGA/2TH	= 0.5
Tps measure	= 4000
SPEED OMEGA	= 1
GRAPH MODE	= 0
DATE	= 04/06/87
IDENT	= LEIDEBML8724
RFS	= 1500
dist.detc.	= 213
Angl. detc.	= 7.67969444
Graph.Skip	= 1
Fen. Soft	= 100

Fig.2 The peaks of the multi-layers can be interpreted by $2\ D \sin \theta_n = n\lambda$ ($\lambda = 1.78892$ Å)

COUNTS (ARBITRARY UNITS)

2 THETA (DEGREES)

HOW TO OBSERVE INTERFACES

P. Houdy and J.A. Sirat

Laboratoires d'Electronique et de Physique Appliquée

3, avenue Descartes, F-94451 Limeil-Brévannes Cédex, France

A diode RF sputtering system is used to realize ultra thin layer stacks ($2d \simeq 40$ Å) of W and C or W and Si, for soft X-ray optics. Regulation of growth parameters allows to control thickness reproducibility with a good accuracy (0.04%). Low pressure ($\simeq 1$ mTorr) minimizes interface roughness. However, <u>intrinsic interface roughness</u> limits multilayer performances as shown by X-ray reflexion computer simulations. In order to understand the interface formation process and structure we have used three methods.

<u>T.E.M.</u>: Smoothing of substrate surface defects for medium lateral size (100 Å) and coalescence of ultra thin layers (5.1 Å) have been observed.

<u>IN SITU KINETIC ELLIPSOMETRY:</u> Three phases have been observed which correspond to interface thicknesses between 2 and 5 Å. Models indicate that carbon and silicon smooth tungsten roughness (at atomic scale) and that tungsten rebuilds silicon surface leading to an asymmetry between W–Si and Si–W interfaces.

<u>S-EXAFS:</u> First experiments indicate that local structure at W–Si interface would correspond to the environment of W in WSi_2.

STUDIES OF THE FABRICATION CONDITIONS OF A VARIETY OF METAL MULTILAYERS

R.E. Somekh

Department of Materials Science and Metallurgy
University of Cambridge, Pembroke Street
Cambridge, CB2 3QZ, U.K.

Over the last few years we have been involved with the fabrication of a range of metal multilayers, ranging from a series of single crystal multilayers, layers for amorphisation studies of mainly Ni/Zr and Co/Ti, some layers for diffusion studies such as a Si/Ge and finally some trial structures for potential X-ray mirror systems such as W/Si and W/C which we extended to W/Ti for modelling purposes. In this paper we will describe the role of the sputtering conditions on the properties of the films produced with particular reference to the problems of producing routine epitaxial films; the role of impurities on the surface mobility of surface atoms and finally the role of reflected neutrals (and sputtered) atoms and their thermalisation as they travel from the target to the substrate.

We use UHV deposition systems capable of routinely achieving (overnight) 0.1 microPa base pressures and which feature a gettering nitrogen cooled can tightly enclosing the two 'lab-made' magnetron sources. We use the power to the magnetrons to control the deposition rates whilst at the same time keeping the sputtering pressure constant.

Our work on single crystal multilayers (mainly Cu/PdNi) confirmed previous work of Schuller and Falco [1] that a relatively high value of PD (the pressure P, target substrate distance D) needs to be quite high (150 - 250 Pa-mms.) to achieve good quality multilayers. This we explain in terms of a detailed calculation of the energy transfer and thermalisation of the reflected neutrals [2]. These with their initially rather high (100 eV plus) energies have much smaller scattering cross-sections than had been previously assumed in earlier calculations of thermalisation.

This type of thermalisation equally plays a role in our understanding of the degree of flatness and amount of amorphisation we observe in Ni/Zr multilayers whose amorphisation has been studied with both DSC and X-ray analysis. We also use this to understand the tensile/compressive stress transition seen in sputtered films [3].

We will also describe some of our preliminary work on structures for X-ray mirrors. We have made a study of the role of oxygen on the quality of some 'model' W/Ti multilayers as defined by the intensity of the superlattice peaks observed to relatively high angle. The addition of oxygen, we believe, reduces the surface mobility of atoms and this may be sometimes beneficial.

REFERENCES

1. K. Meyer, I.K. Schuller, and C.M. Falco: J. Appl. Phys. $\underline{52}$, 5803 (1981)
2. R.E. Somekh: J. Vac. Sci. Tech. A $\underline{2}$, 1285 (1984); Vacuum $\underline{34}$, 987 (1984)
3. D.W. Hoffman: Proc. 7th Int. Conf. Vac. Tech. (Tokyo 1982) p.145

GROWTH AND CHARACTERISATION OF TRANSITION METAL SUPERLATTICES

R.H.M. van de Leur

Delft University of Technology, Delft, The Netherlands

Monocrystalline superlattices of Nb/Ta, Ta/V, Nb/V and Fe/V are grown in an UHV system with two electron beam evaporators. A mass spectrometer is used to control the deposition rate of the constituent metals.

The growth conditions are investigated for Nb/V and then also applied to the other superlattices. To get monocrystalline growth, first a thin layer (≈ 10 nm) of the alloy is deposited on the Al_2O_3 (012) oriented substrate at a temperature of 1070 K. To avoid interdiffusion, the superlattices are grown at the moderate temperature of about 400 K. All the superlattices have a sinusoidal composition variation, imposed during growth, to minimize the elastic energy in the lattice. The modulation period is varied between 0.7 and 13 nm.

The crystallographic structure is studied by X-ray diffraction techniques. Superlattices of good crystallographic quality, as concluded from a Debye-Scherrer kind of photograph, are also investigated on a four-circle diffractometer. X-ray diffraction and model calculations indicate that the superlattices have a long coherence length and that there is no observable interdiffusion.

This work is supported by the Dutch Foundation for Fundamental Research on Matter (F.O.M.)

INTERFACE INDUCED LOSS OF COHERENCE IN SUPERLATTICES

J.P. Locquet, D. Neerinck, H. Vanderstraeten, W. Sevenhans, and Y. Bruynseraede

Laboratorium voor Vaste Stof-Fysika en Magnetisme
Katholieke Universiteit
Leuven, B-3030 Leuven, Belgium

H. Homma and I.K. Schuller

Materials Science Division, Argonne National Laboratory
Argonne, IL 60439, USA

Extending our x-ray diffraction results obtained on amorphous/crystalline multilayers, we model the disorder at the interface of multilayered structures. In an actual x-ray scattering experiment, peak positions are related to atomic and layer periodicities (using Bragg's equation), while linewidths are related (using Scherrer's equation) to the number of scatterers whose positions obey these periodicities. The "coherence length", evaluated for a periodicity, is this number times the periodicity. Assuming a perfect atomic structure of both materials in the multilayer, all loss of range order is induced by fluctuations at the interface. These fluctuations are either continuous variations in the distance between the two materials, and (or) are discrete variations in the number of atoms building up each layer, both giving rise to "roughness". Obviously the latter will mainly influence the long range periodicities, while the small range periodicities (corresponding to high angle diffraction peaks) conserve their coherence length.

Existing models combine continuous and discrete variations as a Gaussian distributed fluctuation on the modulation wavelength λ, and this distribution to fit the experimental data at small and wide angle. For small continuous variations, the low angle linewidth is mainly determined by discrete variations, allowing an independent determination of the continuous and the discrete component. In this paper we study the influence of such a continuous Gaussian distribution at the interfacial distance on the high angle x-ray spectra.

A general relation is obtained which can be applied to crystalline/crystalline as well as crystalline/amorphous multilayers. At each interface the distance between the two blocks of material fluctuates around an average value, following a continuous Gaussian distribution. This fluctuation of "roughness" can strongly reduce the long-range atomic order along the growth direction of the multilayer. Using classical structure factor calculations, we simulate the evolution of x-ray patterns as a function of the fluctuation amplitude, the superlattice wavelength, and the interatomic distances. Applying this model to the crystalline/crystalline case, we fit the experimental linewidths of Pd/Au, Nb/Cu and Pb/Ag multilayers. The fluctuation amplitude obtained at small modulation wavelength λ is of the same order as the difference in lattice parameter. The deduced fluctuation amplitude is 0.1 Å for Pd/Au, 0.4 Å for Nb/Cu and 0.6 Å for Pb/Ag. For crystalline/amorphous systems (Pb/Ge) this amplitude can be significantly larger (1.6 Å).

ACKNOWLEDGEMENT

This work was supported by the Office of Naval Research contract number N00014-83-F-0031, the U.S. Department of Energy, Basic Energy Sciences, under contract number W-31-109-ENG-38, and the Belgian Institute for Nuclear Sciences. International travel was provided by NATO grant number RG85-0695 and the Belgian National Science Foundation.

SOME CHARACTERISTICS OF MULTILAYERS PRODUCED BY MAGNETRON SPUTTERING

DEVICE

B. Vidal*, R. Rivoira*, R. Philip**, and Y. Lepetre**

*CNRS, **Université d'Aix Marseille III,
 Faculté des Sciences de St Jérome
 F-13397 Marseille Cédex 13

High reflectivity multilayers are strongly dependent on the roughness of each interface, which is correlated to the structure of each single layer of the stack. For large values of the period d (d > 40 Å), we generally observe a decrease of the roughness as the stack is increased. In some cases, if the structure of the last additional layer of the stack is good enough, it is possible to smooth the roughness of the previous layer. Using T.E.M. studies of cross-sections of multilayers we have shown some aspects of such phenomena.

In order to study the inner roughness and the stress contribution in the high index material (high density), we fabricated multilayer stacks with a constant thickness of the low index material (typically 200 Å of carbon or silicon) and a decreasing thickness of the high index material (20 Å to 5 Å of W).

By successively tilting the multilayer wedge the T.E.M. observations show the quality of the stacked layers, in particular the granularity of each layer and the roughness of the interface W-C or Si.

We find the following parameters to be strongly related to experimental conditions:

- Control of the substrate temperature during deposition
- Power generators which induce too high an energy ion beam
- Influence of the reactivity of the low index material.

Measurement of the reflectivity of the multilayers at 1.54 Å are usually in good relationship with T.E.M. observations.

NEGATIVE DIFFERENTIAL RESISTANCE (NDR) MEASUREMENTS

Z. Hatzopoulos, A. Georgakilas, L. Fotiadis, and A. Christou

Research Center of Crete, Iraklion, Crete, Greece

We present results on the observation of resonant tunneling negative differential resistance (NDR) through a GaAs/AlGaAs superlattice structure at room temperature and at 4.2 K made by molecular beam epitaxy. The conduction mechanism for the superlattice is miniband conduction and the entire structure behaves similarly to the normal two-barrier structure. Sequential tunneling was not observed.

THE SOLID STATE REACTION IN MULTILAYERS STUDIED BY XAS

C. Brouder, G. Krill, E. Dartyge*, A. Fontaine*, G. Tourillon*,
and P. Guilmin, and G. Marchal

Laboratoire de Physique du Solide, Université de Nancy I, B.P. 239
F-54506 Vandoeuvre lès Nancy Cédex, France

*LURE (Laboratoire d'Utilisation du Rayonnement Electromagnétique)
Université de Paris-Sud, F-91405 Orsay Cédex, France

This poster presents some applications of the x-ray absorption spectroscopy (XAS) to the study of the solid state reaction in multilayers.
Three systems were studied: Nd/Fe, Co/Sn and Ce/Ni. All these samples were
evaporated onto a cooled (77 K) substrate and kept in liquid nitrogen.

- Nd/Fe (171 bilayers of Nd (20 Å) and Fe (40 Å)). No diffusion occurs in
this system before 200°C. However, the spectra are strongly modified when
the temperature is increased from 77 K to room temperature. At low temperature, the Fe layers are very disordered and as temperature is increased,
microcrystals grow so that the spectra look more and more like a spectrum
of bulk bcc iron.

- Co/Sn (150 bilayers of Co (16 Å) and Sn (48 Å)). In this system, an in-
layer structural reorganization occurs between 77 K and 170 K. Diffusion
takes place at room temperature, where the spectra exhibit the presence of
more and more Sn atoms in the neighbourhood of the Co atoms [1].

- Ce/Ni (130 bilayers of Ce (35 Å) and Ni (30 Å)). This system was chosen
because the Ce L_{III}-edge is known to be very sensitive to the local environment through the mixed-valent properties of the Ce atoms. With this probe,
we found that the Ce was almost completely mixed with Ni during deposition,
forming an homogeneous amorphous Ce-Ni interlayer. As temperature is increased, this amorphous interlayer grows to absorb the remaining Ni. It
appears that the interlayer stays homogeneous during the whole process,
showing very few concentration gradients [2].

REFERENCES

1. C. Brouder, G. Krill, E. Dartyge, A. Fontaine, G. Marchal and
   P. Guilmin: In Proceedings of EXAFS and Near Edge Structure IV, ed. by
   P. Lagarde, D. Raoux, J. Petiau (Editions de Physique, Paris 1986)
   p. C8-1061
2. C. Brouder, G. Krill, E. Dartyge, A. Fontaine, G. Tourillon, P. Guilmin,
   G. Marchal: submitted for publication to Phys. Rev. B

GROWTH, PROPERTIES AND CHARACTERISATION OF METAL MULTILAYERS

P. Orozco, H. Bouzar, J.R. Jepson, D.J. Dingley
and N.M. Jennett

H.H. Wills Physics Laboratory, University of Bristol
Bristol, BS8 1TL, UK

Two evaporation units have been constructed for preparation of metal multilayers using thermal and electron bombard sources. CuNi, CuV and CuFe multilayers have been produced with the explicit intention of correlating physical properties with structure as determined by Transmission Electron Microscopy (TEM). Low temperature resistivity and interdiffusion measurements have been performed upon a series of CuNi foils over the modulation wavelength range 1.6 nm to 10 nm. An anomalously high residual resistivity and interdiffusion coefficient were obtained for a modulation wavelength of 2.8 nm to 3.0 nm, the resistivity peak persisting after annealing of the foils for 100 hrs at 380°C even though the x-ray satellite reflections had vanished. TEM, however, indicated that interfacial defects were retained during this treatment. Annealing the foils at 480°C for three hours resulted in a loss of the anomalous resistivity peak, x-ray satellite reflections and interfacial structure.

Superconducting transition effects have been explored for CuV foils of bilayer and multilayer type with modulation wavelengths 25 nm to 100 nm. The bilayer material behaved normally with an onset Tc at 4.1 K but the multilayered foils of both 25 nm and 100 nm showed a loss of superconductivity. TEM of the latter foils showed anomalous behaviour of the diffraction patterns.

Multilayered CuFe grew with [111] fcc parallel to [111] bcc Fe for all modulation wavelengths in the range 0.6 nm to 10 nm. Electron diffraction from the inclined planes cutting through the layers of CuNi material showed a splitting of the diffraction lines indicating a bending of the planes due to interfacial strain. Measurement of the degree of splitting of the deficiency line can be used to determine the strain. These measurements are being repeated on CuFe foils.

POSTER SESSION II

DESIGN, TESTS AND MISCELLANEOUS

APPLICATIONS

X-RAY OPTICS IN LANGMUIR-BLODGETT FILMS

J.J. Benattar*, L. Bosio**, and F. Rieutord*

*DPHG/SPSRM, CEN Saclay, F-91191 Gif-sur-Yvette Cedex, France

**ESPCI, 10 rue Vauquelin, F-75231 Paris, France

The Langmuir-Blodgett (LB) films are multilayered systems of amphi-philic molecules. These films are obtained by successive depositions of monomolecular layers onto a substrate. The molecules are first spread onto a water bath and laterally compressed into a solid-like monolayer (Langmuir film). The film is then transferred by dipping the substrate through the compressed monolayer. This technique allows the construction of numerous interesting physical systems by varying the number of layers and the chemical components.

We present four studies using X-ray reflectivity which show that this technique is well suited for investigations of thin layered films:

1. We present the first X-ray reflectivity experiment on a monolayer onto a water subphase. This experiment opens new prospects to measure the roughness of solid films on water.

2. We show that the combination of X-ray reflectivity and diffraction in transmission allows the direct determination of LB film structures.

3. We have studied the projection of the electronic density along the normal to the layers of differently sequenced LB films. Using both optical formalism and kinetical theory for X-rays, we have analyzed the various interference phenomena.

4. Using the combination of electron microscopy and X-ray reflectivity we have studied the defects and the roughness at the surface of fatty acid multilayers.

FIGURED MULTILAYER STRUCTURES APPLIED TO A LASER PRODUCED PLASMA

F. Bijkerk

LAICA (Laser Application and Information Centre Amsterdam)
c/o Edisonbaan 14, NL-3439 MN Nieuwegein, The Netherlands

A. Verheul, W.J.J. Wolfis, P.F.M. Delmee, G.E. van Dorssen,
W. Kersbergen, E.J. Puik, H. van Brug and M.J. van der Wiel

Association EURATOM, FOM-Institute
for Plasma Physics 'Rijnhuizen', Edisonbaan 14
NL-3439 MN Nieuwegein, The Netherlands

A prime application of figured multilayer structures as X-ray optics is the combination with a laser produced plasma. As a source of soft X-rays, a laser-plasma has proven its attractive properties: a high instantaneous brightness, good shot-to-shot stability and nearly pointlike dimensions. The combination of laser-plasma and X-ray optics would give improved performance in e.g. plasma diagnostics, rapid detection of low Z elements in solid samples, pulsed laser irradiated materials and lithographic imaging of submicron patterns. In the latter application, a parabolic multilayer structure has the potential to convert the emission of the point source into a large parallel and homogeneous beam to prevent several distortions in the lithographic imaging.

To enable the fabrication of highly curved multilayer structures, an electron beam evaporating system with a rotating adjustable substrate holder is being constructed. Layers with laterally graded thicknesses can be deposited on substrates with various types of curvature. In the dual vacuum system, the layer growth is in situ monitored by three detectors, including an X-ray reflection system.

LOCALIZATION EFFECTS IN MULTILAYERS OF METALLIC GLASSES

J.M. Broto

Service des Champs Intenses
Université Paul Sabatier (Toulouse)

and

Laboratoire de Physique des Solides
Université Paris-Sud (Orsay)

The localization effects of disordered metals are due to quantum interferences between conduction electron waves scattered by a random potential. The coherence length of the electron waves is fixed by the inelastic scattering processes and amounts to few hundred Å in metallic glasses at 4.2 K. An applied magnetic field gives rise to a magnetoresistance due to the reduction of the coherence length in the direction perpendicular to the field.

Multilayers of metallic glasses $Cu_{20}Mg_{80}/Cu_{50}Y_{50}$ have been prepared by alternate sputtering and characterized by an energy scanning X-ray diffraction technique. Their magnetoresistance has been measured in pulsed field up to 30 Teslas, and for fields parallel and perpendicular to the layers. In the helium temperature range typical localization effects are observed. When the thickness of the layers is smaller than the coherence length, $\ell_{in} \approx 200$ Å, the magnetoresistance is anisotropic. $\rho(H_{\parallel}) \neq \rho(H_{\perp})$. The difference $\rho(H_{\parallel}) - \rho(H_{\perp})$ is maximum when the magnetic coherence length $\sqrt{2\hbar/eH}$ is about 1.6 t, where t is the thickness.

NUMERICAL DETERMINATION OF THE REFLECTIVITY OF MULTILAYERED NEUTRON MIRRORS

D. Clemens

Hahn-Meitner-Institut

Berlin, FRG

Adapting the theory of electrodynamics in the case of reflection and transmission of electromagnetic waves due to interfaces in multilayered structures, the reflectivity of a given structure in a neutron-optical application can be calculated. In their fundamental paper J. Schelten and K. Mika describe a method of calculating the reflectivity from a product of transfer matrices, set up for each interface separately. This matrix formalism also takes into account refraction as well as absorption effects which are of importance even in ideal structures.

In the presented work, additionally a so-called static Debye-Waller-factor, representing the interface roughness, was integrated into the matrix formalism. The calculations were made with a FORTRAN program running on a SIEMENS 7890 computer, which also provides an option to prepare graphical reflection spectra. In these calculations, that were mainly prepared for so-called "supermirrors", i.e., neutron interference mirrors, giving a high reflectivity (near to 100% when magnetized) up to some multiples of the critical angle, the dominating influence of interface roughness on the reflectivity is proved. The data were in satisfactory good agreement with experimental results.

The program checks neutron-optical quality of given multilayer structures, such as supermirrors, monochromators and reflecting coatings and could be extended to an optimization program for supermirror layer thicknesses.

REFERENCES

1. F. Mezei: Comm. Phys. <u>1</u>, 81 (1976)
2. F. Mezei, P.A. Dagleish: Comm. Phys. <u>2</u>, 41 (1977)
3. J. Schelten, K. Mika: Nucl. Inst. Meth. <u>160</u>, 287 (1979)
4. T. Megademini, B. Pardo: J. Optics <u>16</u>, 289 (1985)

NEW FAST TECHNIQUE FOR ELECTRONIC STRUCTURE CALCULATIONS OF COMPLEX LAYERED MATERIALS

S. Crampin

The Blackett Laboratory, Imperial College

London SW7 2BZ, UK

A method for performing self-consistent calculations of the electronic properties of layered materials is described within the muffin tin approximation. In principle no restriction is placed upon the complexity of material which may be handled in the z direction, while full utilisation is made of the 2-dimensional symmetry within the layers (x-y plane). This leads to a scheme capable of dealing with, for example, interfaces, surfaces, grain boundaries and large period or semi-infinite superlattices. Another possible application ideally suited to the approach is a study of high T_c superconductors (which have a layered structure) and the role of defects and substitution. Since there is no restriction in the formalism upon the form of the potential the method should be extendable to full potential calculations necessary for semiconductor materials.

The heart of the method lies in the Green function scattering theory and a partitioning of the scattering events into intra- and inter-layer contributions. The scattering properties of the atoms within a given layer are used to construct the isolated layer Green function; the Green functions for all the layers are then coupled together to form the Green function for the infinite solid from which the electronic properties may be deduced. It is this last step, the stacking together of the layers, which allows for the unlimited variation in the z direction and leads to an especially efficient technique through the use of layer doubling algorithms.

SUPERCONDUCTIVITY OF REGULAR AND FIBONACCI Mo/V SUPERLATTICES

C.W. Hagen, R. Griessen, and D.G. de Groot

Vrije Universiteit, Amsterdam, The Netherlands

Ø. Fischer and M.G. Karkut

Université de Genève, Genève, Switzerland

Regular and Fibonacci Mo/V superlattices were epitaxially grown by magnetic sputtering under ultra high vacuum conditions on (11$\bar{2}$0) Al_2O_3 substrates held at $\approx$1000 K during deposition [1,2]. The upper critical field determined from magnetic torque as well as resistivity measurements are investigated as a function of temperature and magnetic field orientation and for various modulations of the proximity-coupled superconducting multilayer. For small modulation lengths Λ (Λ = 19.1 and 16.3 nm for the regular and Fibonacci sequence, respectively) a three-dimensional behavior of the superconducting multilayer is observed with a zero-temperature coherence length $\xi_{\parallel} \approx$ 21 nm and an anisotropy $H_{c2\parallel}/H_{c2\perp} \approx$ 2. For multilayers with large modulation lengths ($\Lambda \approx$ 64.4 nm) a peculiar temperature dependence is revealed in $H_{c2\parallel}$, which is attributed to dimensional crossovers similar to those described theoretically by Takahashi and Tachiki [3]. Due to the complexity of our multilayer structure dimensionality features occur at different temperatures. If $\xi_{\perp}$ equals the thinnest V layer thickness, a commensurability between the vortex lattice and the multilayer exists.

The strong pinning of the vortices by the multilayer interfaces in a parallel field gives rise to a drastic difference in the anisotropy of the upper critical fields determined by torque measurement as compared to that determined from the resistance. The Fibonacci superlattice and the corresponding periodic superlattice show the same angular dependence of H_{c2} determined from the resistance, but torque measurements give lower values for H_{c2} in the case of the Fibonacci superlattice.

In general, flux jumps are observed at low temperatures (1.4 K) and small angles of the field with respect to the multilayer surface (<4°).

REFERENCES

1. M.G. Karkut, D. Ariosa, J.-M. Triscone, Ø. Fischer: Phys. Rev. B **32**, 4800 (1985)
2. M.G. Karkut, J.-M. Triscone, D. Ariosa, Ø. Fischer: Phys. Rev. B **34**, 4390 (1986)
3. S. Takahashi, M. Tachiki: Phys. Rev. B **33**, 4620 (1986)

RADIATIVE AND NON-RADIATIVE POLARITON STRUCTURE OF SUPERLATTICES

A. Dereux, J.-P. Vigneron, P. Lambin, and A.A. Lucas

Département de Physique, Facultés Universitaires
Notre-Dame de la Paix
61, rue de Bruxelles, B-5000 Namur, Belgium

Optical properties of multilayered materials, including synthesized superlattices received special attention in the last few years. To some extent, reflectance and absorption properties of these materials are adjustable: the polariton structure depends sensitively on the layer geometry.

Though the approach considered here is applicable to any stratified structure, we emphasize on the polaritons of semi-infinite superlattices, that is superlattices which exhibit a free surface. In these thin-film structures, the accumulation of interfaces gives rise to peculiar electromagnetic eigenmodes, distributed as continuous frequency bands. The termination of the superlattice at the surface modifies the density of these modes, as compared to the mode density of an infinite, truly periodic superlattice. In particular, one notices the appearance of isolated branches analoguous to surface polariton modes of semi-infinite homogeneous materials.

We approach the description of radiative and non-radiative modes, including finite life-time effects, from the point of view of scattering theory, using a Green's function technique. The local density of states of the polariton modes is calculated and provides a complete information on allowed electromagnetic excitations, as a function of frequency and wavelength, at any depth in the superlattice. These results will be used to discuss recent infrared optical experiments performed on thick semiconducting superlattices.

ELECTRONIC STRUCTURE OF STRAINED Si_n/Ge_n(001) SUPERLATTICES

S. Ellialtioglu
Physics Department, Middle East Technical University
Ankara 06531, Turkey

O. Gülseren and S. Ciraci
Department of Physics, Bilkent University
P.O. Box 8, Maltepe, Ankara 06572, Turkey

Using the empirical tight binding method we have investigated the electronic properties of the Si_n/Ge_n(001) strained superlattices as a function of the superlattice periodicity and the band misfit. For $n \geq 3$, we have found some conduction band states localized in Si, and some in Ge, presenting a band discontinuity quite different from the staggered type-II. The hole states localized in Ge appear for $n \geq 4$. The difference between the direct and indirect band gaps is reduced from 0.39 eV for $n = 3$ to 0.035 eV for $n = 6$ which can be considered to be quasi-direct.

We have examined the effect of the band lineup on the direct and indirect band gaps by calculating the electronic structure of Si_4/Ge_4(001) for various values of ΔE_v, the difference between the two valence band maxima.

ΔE_v	-0.66	-0.46	-0.26	-0.06	0.14	0.34	0.54	0.74	0.94	Si%
Γ_c-Γ_v	1.56	1.65	1.66	1.63	1.58	1.51	1.43	1.34	1.24	80
Z_c-Γ_v	1.49	1.50	1.49	1.46	1.41	1.35	1.28	1.19	1.10	80
R^*_c-Γ_v	0.97	1.05	1.13	1.18	1.22	1.25	1.25	1.23	1.19	40
X_c-Γ_v	0.82	0.92	1.02	1.10	1.17	1.23	1.27	1.29	1.30	20
Δ_{di}	0.74	0.73	0.64	0.53	0.41	0.28	0.19	0.15	0.14	-

The position of the CB minimum changes from a Ge-like state at X-point for $\Delta E_v < 0.5$ eV to a delocalized state at a k-point R*, along ΓR, in the range $0.5 < \Delta E_v < 0.6$ eV, and to a Si-like one at the Z-point for $\Delta E_v > 0.6$ eV. The range $0.5 < \Delta E_v < 0.6$ eV is very interesting in the sense that several symmetry points in the BZ have energies very close to the CB minimum.

To explore the effect of superlattice periodicity we have considered the cases for $n = 1$ to 8, with ΔE_v taken to be 0.54 eV [1]. For $n = 1$ and 2, Si_n/Ge_n(001) is still a crystal rather than a heterostructure because the bulk-like regions are merged in the interface [2].

Periodicity n + n	Δ_{direct} [eV]	$\Delta_{indirect}$ [eV]	Δ_{di} [eV]	Si %	k-point for CB minimum	
2 + 2	1.643	1.462	0.181	59	along ZR close to R	
3 + 3	1.576	1.189	0.387	35	ΓX	X
4 + 4	1.432	1.245	0.185	43	ΓR	R
5 + 5	1.355	1.186	0.169	29	ΓX	X
6 + 6	1.245	1.210	0.035	20	at R	R
8 + 8	1.169	1.063	0.106	21	along ΓR	

For n = 3 to 6, the superlattice can support the electron confined states, the localization of which increases with increasing n. However, it is possible to find states localized either at Si or at Ge in the same energy range. The hole confined states are found for n = 4 and n = 6. The number as well as the localization of the hole states increase with n. The band gap is found to be indirect for superlattice periodicities n = 3 to 8, however, the difference between the direct and indirect band gaps reduces to 0.035 eV for n = 6 which suggests a quasi-direct band gap. Conduction band states localized at different sides of the heterostructures, but in the vicinity of the same energy are found. It may be due to a quantum size effect and reveal a novel feature of the band lineup in the strained superlattices with small periodicity. No direct gap as small as 0.76 eV is found to support the direct transition obtained from the recent electroreflectance spectroscopy [3].

REFERENCES

1. C.G. Van de Walle and R.M. Martin: Phys. Rev. B34, 5621 (1986)
2. S. Ciraci and I.P. Batra: Phys. Rev. (in print)
3. T.P. Pearsall et al.: Phys. Rev. Lett. 58, 729 (1987)

ELECTRONIC PROPERTIES OF CuO_2 LAYERS IN HIGH T_C MATERIALS

S. Ellialtioglu and M. Durgut

Physics Department, Middle East Technical University

Ankara 06531, Turkey

The high T_C materials of La-Sr-Cu-O and of Y-Ba-Cu-O structures are both investigated by the use of a simple LCAO model. The local densities of states have shown both two-dimensional and one-dimensional behavior within different energy regions. Around Fermi level the dimensionality and the type of van Hove singularities become important for the enhancement of the critical temperature. The BCS parameters like the superconducting gap, critical temperature and the Debye frequency were calculated and their dependence on the density of states has been illustrated.

ELECTRICAL CHARACTERISATION OF THIN INSULATING LANGMUIR BLODGETT FILMS INCORPORATED IN METAL-INSULATOR-METAL STRUCTURES

N.J. Geddes, W.G. Parker, J.R. Sambles and N.R. Couch*

Thin Film and Interface Group   *Long Range Research Group
University of Exeter   GEC Hirst Research
Department of Physics   East Lane
Stocker Road   Wembley
Exeter, UK   London, UK

Renewed interest in the use of LB films as the passivating layer in electronic devices has led to our study of M-I-M structures incorporating such multilayers. Our initial studies concentrated on the Ag/LB/Ag system, silver being chosen for the electrodes due to its oxide free surface after evaporation and ω-tricosenoic acid for the LB material due to its ease of deposition [1]. In these structures the LB film was found to be non-insulating giving junction resistances typically of the order of ohms (even with 40 layers of ω-tricosenoic acid) [2,3]. The conduction mechanism through these junctions was caused by damage incurred during the evaporation of the top silver electrodes. To obviate this problem a metal for the top electrode with a much lower evaporation temperature was required. Magnesium was chosen, since its evaporation temperature of 300°C coupled with its low mass number made it ideal for the purpose. The result from Ag/40 layers LB/Mg were very promising giving fully insulating junctions. The LB film thickness was then reduced progressively down to 10 layers of ω-tricosenoic acid (300 Å thick) with no loss in its integrity. Electrical characterisation of these junctions with Ag bottom electrodes, Mg top electrodes and LB film thickness from 10-2 layers of ω-tricosenoic acid has been undertaken. I/V characteristics of these junctions yield information on the capacitance, resistance and conduction processes through the LB films in the absence of both conducting filaments or on insulating oxide (present when Al is used as a base electrode as has often occurred) [4]. The essential features of these new data will be presented.

REFERENCES

1. G. Veale, I.R. Girling and I.R. Peterson: Thin Solid Films $\underline{127}$, 293 (1985)
2. N.R. Couch, C.M. Montgomery and R. Jones: Thin Solid Films $\underline{135}$, 173 (1986)
3. N.J. Geddes, J.R. Sambles, D.J. Jarvis and N.R. Couch: Proc. of 3rd MED Conf., Washington (1986)
4. R.H. Tredgold and C.S. Winter: J. Phys. D: Appl. Phys. $\underline{14}$, L185 (1981)

LATTICE DEFORMATION AND MISORIENTATION OF GaAs LAYERS GROWN ON (100) Si
BY MOLECULAR BEAM EPITAXY

F.E.G. Guimaraes and K. Ploog

Max-Planck-Institut für Festkörperforschung
D-7000 Stuttgart 80, FRG

GaAs epitaxial layers of thicknesses ranging from 0.5 to 4 µm were
grown by molecular beam epitaxy (MBE) on (100) oriented Si substrates
(2° off along the <011> direction) using the two-step growth method. The
structural properties of the grown layers have been determined by measuring
the symmetric (400) and the asymmetric (511) and ($\bar{5}$11) reflections with a -
high-resolution double-crystal x-ray diffractometer.

The GaAs layers are deformed due to a biaxial tensile stress and are
misoriented with respect to the <100> growth direction. The misorientation
corresponds to a tilt of the <100> GaAs plane towards the <011> direction.
We measure the values of the lattice constant parallel and perpendicular
to the substrate surface ($a_\parallel$ and $a_\perp$) and we find that the strain level of
the GaAs layers does not depend on the layer thickness, if the GaAs layers
are grown at the same temperature. A halfwidth of 0.04° is observed for the
(400) GaAs reflection, indicating a good crystal quality. The understanding
of the structural properties of the GaAs layers on the Si is fundamental
for the interpretation of the optical properties of these layers.

The structural properties of the GaAs layers are explained by the
accommodation of the large lattice mismatch at the growth temperature due
to the formation of misfit dislocations and lattice misorientation. As a
result, the compressive stress is relieved at this temperature. The final
stress state of the layers is in fact developed during the cooling of the
samples to room temperature due to the different thermal expansion coefficients of GaAs and Si.

X-UV POLARIZER BASED ON MULTILAYERED MIRRORS

A. Khandar and P. Dhez

UA775 and LURE

Université Paris XI, F-91405 Orsay, France

Possibilities to otain X-UV polarizers are demonstrated. Through an historical survey of polarization in the visible and x-ray range it appears that scattering perpendicular to the incident direction will be more useful for the X-UV than dichroism and birefringence.

The usual characterizaton of the matter-wave interaction from a macroscopic (refractive index) or microscopic (form factor) approach from visible to X rays is given by using Maxwell's equations. This justifies the choice of polarization by reflection in the X-UV.

By using the calculation adapted for X-UV interference mirrors it is shown that good reflectivity and high polarization rate at 45° incidence can be obtained simultaneously. This situation is equivalent to the Brewster incidence angle on the simple diopters in the visible. However, these cannot be used in the X-UV since the reflectivity is very small.

Two successive 45° reflections were used on identical X-UV interference mirrors optimized for 304 Å. These were set in the crossed polarizer scheme and a cosine square variation in the intensity versus rotation angle was observed. This behavior is similar to that expected from Malus' law in the visible. From this variation one can evaluate the polarizing rate for the mirrors.

An X-UV polarimeter based on this scheme was developed and used following a monochromator connected to the synchrotron radiation source ACO at LURE to measure the polarization rate at 154 Å and 304 Å. This source is highly polarized in the elctron orbit plane. The polarimeter consists of a single X-UV multilayer receiving the radiation at 45°. For the measurement the plane of incidence of the multilayer is turned to vary the angle between this plane and the principal plane of polarization. From the observed modulated intensity the polarization rate of incident radiation is deduced.

(This work is a part of A. Khandar's thesis, June 19, 1987, Université Paris XI)

STRAIN MEASUREMENTS IN $Si/Si_{0.5}Ge_{0.5}$ AND W/Mo SUPERLATTICES BY He ION CHANNELING

S. Mantl, K. Kasper*, H.J. Jorke* and K. Reichelt

Institut für Festkörperforschung, KFA Jülich

Postfach 1913, D-5170 Jülich, FRG

He ion channeling experiments have been performed on molecular beam grown $Si/Si_{0.5}Ge_{0.5}$ superlattices and n-channel $Si/Si_{0.5}Ge_{0.5}$ MODFET structures. Angular yield scans provide directly the tetragonal strains in the layers. The strain fields in $Si/Si_{0.5}Ge_{0.5}$ superlattices grown directly on (100) Si and on $Si_{1-y}Ge_y$ buffer layers were compared experimentally. In the first case only the $Si_{0.5}Ge_{0.5}$ layers are tensilly strained whereas on a $Si_{0.71}Ge_{0.29}$ buffer layer with a thickness of 230 230 nm almost equal but opposite strains have been found. Strain symmetrization yields the minimum elastic energy and thus to the most stable structure. An example of strain measurements on an n-channel MODFET structure will be shown. Ion channeling is probably the only technique which allows strain measurements in structures, like the MODFET structure investigated, where the thicknesses vary from layer to layer.

The W/Mo superlattices were prepared by UHV evaporation on (100) MgO substrates. Channeling measurements proved that the lattice mismatch of 0.6% is accommodated completely by elastic strain in the superlattice. On a Mo buffer layer only the W layers are compressively strained. Perfect agreement between the experimentally determined strain value and theoretical estimates implies a pseudomorphic structure of the W/Mo superlattice.

Complete strain relaxation in the superlattice was observed after 2 MeV Mo irradiation with a dose of $2 \cdot 10^{15}$ Mo/cm^2. The structure remains single crystalline during irradiation. The tetragonal strain relieves linear with irradiation dose. From Rutherford backscattering analysis we conclude that the strain relaxation during the irradiation occurs due to the production of dislocations, e.g. misfit dislocations, and not due to layer mixing.

REFERENCES

1. S. Mantl, E. Kasper and H.J. Jorke: MRS, Heteroepitaxy, Anaheim 1987 (to be published)

2. E. Kasper: This conference and references therein
3. S. Mantl, D.B. Pker and K. Reichelt: Nucl. Instr. Meth., Proceedings
   IBMM '86, Catania, Italy

---

* AEG Telefunken Forschungsinstitut, Postfach 1730, D-7900 Ulm, FRG

REFLECTION OF VERY LOW ENERGY ELECTRONS BY INDIUM/GOLD BILAYERS

C. Marliere and J.P. Chauvineau

Institut d'optique, L.A. CNRS no.14, Bât. 503

B.P. 43, F-91406 Orsay Cedex, France

A 30 nm gold layer is deposited on a floatglass substrate at room temperature and annealed at 450 K. The film is polycrystalline but each crystal has its [111] axis (Au: f.c.c) perpendicular to the substrate. Then an indium overlayer is evaporated at 77 K. It was proved earlier that, under these conditions, indium grows one gold monoatomic layer after the other.

That bilayer is bombarded with electrons in normal incident whose energy can be varied between few eV to 50 eV. The total electron current transmitted by the sample is then detected and measured, during the variations of electron energy if the sample thickness is fixed, or during the indium deposit if the energy is constant. By that mean we try and study quantum size effects already observed with electrical resistance and work function variation. First results will be presented.

MBE GROWTH OF MAGNETIC MULTILAYERS

F. Nguyen van Dau, P. Etienne, R. Bisaro, J. Chazelas, F. Keller, and G. Creuzet

Thomson-CSF - L.C.R. - B.P. 10, F-91401 Orsay, France

J. Massies

C.N.R.S. Sophia-Antipolis, Rue Bernard Gregory
F-06560 Valbonne, France

A. Fert

Université Paris XI - Laboratoire de Physique des Solides
F-91405 Orsay, France

We present an original MBE chamber, especially conceived to study the growth of magnetic multilayers. We equipped this new chamber with in-situ analysis instruments such as quadrupole mass spectrometer, RHEED, with the possibility of analysing the diffraction intensities during growth (RHEED oscillations).

We have developed a new kind of cells, with electronic bombardment heating, in order to evaporate more refractory materials such as iron.

Preliminary results concerning epitaxial growths of Al/GaAs(001) and Ag/GaAs(001), Fe/GaAs(001) and multilayer Fe/Ag//GaAs(001) are presented: RHEED patterns, RHEED oscillations during GaAs buffer layer growth, ex-situ analysis (STEM, X-ray diffraction patterns).

Work supported by the French M.R.E.S.

REFLECTIVITY AND ROUGHNESS OF MULTILAYER MIRRORS

A. Renwick

Kings College, London

The influence of boundary roughness on the reflectivity of multilayer x-ray mirrors is discussed.

The roughness of interfaces is usually taken into consideration by multiplying the reflection coefficient of a boundary by an exponential term containing the r.m.s. roughness. This is sufficient when the r.m.s. roughness is small in comparison with the wavelength of the radiation but at shorter wavelengths it is not sufficient to describe the performance of a multilayer.

Another approach to this problem, proposed by A.V. Andreev, uses the concept of a transition layer and solves the Helmholtz equation to calculate the specularly reflected and the non-specularly scattered intensities. This approach has been extended to multilayer structures and a brief summary of the theory is given. Calculations of reflectivity using this theory are shown and compared with calculations using the existing theory. These show that in the limit of a smooth interface both theories predict the same reflectivity curve, but as roughness increases there are differences in the predicted reflectivites particularly for higher order Bragg reflections.

ELECTRIC FIELD DEPENDENT EXCITON ENERGY; PHOTOLUMINESCENCE QUENCHING AND NONLINEAR ABSORPTION IN GaInAs/InP QUANTUM WELLS

M.G. Shorthose

Clarendon Laboratory

University of Oxford, Oxford, UK

Optical properties of GaInAs/InP quantum wells at room temperature have been studied. Electric field induced exciton energy shifts, photoluminescence quenching and saturation of optical absorption have been observed.

Photocurrent and photoluminescence spectra from 100 Å wells contained within p^+ and n^+ InP layers in a conventional pin structure have been obtained; reverse bias voltages of up to 12 V were applied. The exciton peaks in the photocurrent spectra are seen to broaden and shift to lower energy; the photoluminescence peak, which is due to $n = 1$ heavy-hole exciton recombination, also shifts to lower energy and is completely quenched at high voltages. These results are similar to those reported previously for GaAs quantum wells [1] and ascribed to the so-called quantum-confined Stark effect. The overall shift of the $n = 1$ exciton peak is 47 meV, which compares to the value 20 meV observed in GaAs quantum wells. The decrease in the photoluminescence intensity with field is somewhat larger than expected purely on the basis of the measured carrier sweep-out by the field, and is most likely due to a decreasing oscillator strength caused by a field-induced reduction in the overlap of electron-hole wavefunctions.

A model of the quantum confined Stark effect has been developed in which R matrix propagation techniques are used to calculate the quasi-stationary subband energy levels of a quantum well system in an electric field. The lifetime of each state is obtained by using boundary conditions which assume that carriers within the wells eventually escape confinement by tunnelling.

Screening of the electron-hole interaction has been studied by measuring the effect on the transmission of light tuned to the exciton peak in a GaInAs/InP multiple quantum well structure, of pump laser light tuned both resonantly and nonresonantly with the exciton. Bleaching of the $n = 1$ heavy-hole exciton has been observed at incident light intensity of less than 20 Wcm^{-2}. This is lower than the saturation intensity of 70 Wcm^{-2} reported by Fox et al. [2]. It has been found that the wavelength dependence of the saturation intensity closely follows the absorption spectrum for the sample, indicating that the mechanism for screening the electron-hole interaction is the same whether pumping resonantly or non-resonantly with the exciton. This is evidence that very few excitons are present at room temperature, being rapidly ionized to free carriers by interaction with phonons.

REFERENCES

1. M. Yamanishi, Y. Kan, T. Minami, I. Suemune, H. Yamamoto, Y. Usami: Superlattices and Microstructures 1, 2, (1985) p.111
2. A.M. Fox, A.C. Maciel, M.G. Shorthose, J.F. Ryan, M.D. Scott, J.I. Davies J.R. Riffat: Appl. Phys. Lett. (accepted for publication)

ENHANCED HOLE MOBILITY IN Si DOPING SUPERLATTICES

D.W. Smith, R. Biswas, R. Houghton, E.H.C. Parker,
and T.E.Whall

Advanced Semiconductor Research Group
Dept. of Physics, University of Warwick
Coventry, England

The construction of a superlattice in a homogeneous semiconductor by modulation impurity doping is a recent development. One of the likely advantages of such a structure is enhanced charge carrier mobilities in the layers.

This investigation to establish the conduction mechanisms is concerned with a so-called p-i-p-i, or high-low, multilayer in Si.

Boron doped, twenty layer structures of period 300 A have been grown by MBE with doping levels of 5×10^{17} cm^{-3} in the p(i) regions and 10^{19} cm^{-3} in the p$^+$. The doping profiles have been extensively studied by SIMS and an electrolytic Schottky barrier CV technique, the results of which are in good agreement.

The serious problem of making ohmic contacts to all layers whilst electrically isolating the Si substrate has cast doubt upon the accuracy of the previously reported mobility measurements in such structures [1]. This problem has been solved by a bevel and stain technique preceding the evaporation of Al contacts and the alloying of these above the eutectic temperature.

Preliminary Hall measurements, over the temperature range 77 K to 300 K, on such structures show significant mobility enhancement as compared to uniformly doped epitaxial layers and, indeed, bulk Si.

This enhancement is thought to be due to charge carrier spillage into the low doped region, thereby reducing subsequent impurity scattering. The results are not conclusive and the mechanism may be more involved. The possibility of a modification to the effective mass has also been suggested [2]. Further detailed investigations on optimised structures are in progress.

REFERENCES

1. Y. Shiraki: Private communication (1987)
2. K. Nakagawa and Y. Shiraki: Surface Science **174** (1986)

STRUCTURAL INVESTIGATIONS OF Mo-Ge MULTILAYERS BY X-RAY ANOMALOUS
SCATTERING AND EXAFS

Lane Wilson and Arthur Bienenstock

Department of Applied Physics, Stanford University

Stanford, CA 94305, USA

Studies of atomic arrangements in amorphous and crystalline, low
period Mo-Ge multilayers using synchrotron radiation X-ray diffraction
(transmission and reflection), supplemented by EXAFS, are described.
Differential anomalous scattering and EXAFS were utilized to determine the
environment of each species. Intermixing, as well as a BBC epitaxial Ge
structure, are among the observed structural characteristics.

A wide range of structural variations is exhibited by the Mo and Ge
layers and interfacial regions as layer thicknesses are varied. This
indicates that specification of the composition profile or the layer
thicknesses alone is insufficient for characterization of the multilayers. Thick layers of Mo ($\geq$30 Å) exhibit bulk BCC crystalline structure
while thick layers of Ge ($\geq$30 Å) exhibit 4-fold coordinated amorphous RTN
structure. An amorphous Mo-Ge alloy interfacial region between the thick
layers is suggested by the Ge edge Differential Distribution Function. As
the Ge layer is made thinner (10 Å) while the Mo remains at least relatively thick ($\geq$20 Å), the Ge structure is transformed into a novel form, a
metallic, strained BCC structure coherent with the Mo BCC structure. If,
however, the Mo layer thickness is decreased ($\leq$10 Å) while the Ge layers
remain thick, the Mo is no longer crystalline but is incorporated in an
interfacial amorphous Mo-Ge alloy separating layers of a-Ge. If both
layers are decreased (Mo $\leq$ 20 Å, Ge $\leq$ 20 Å), with the Ge thickness equal
to or greater than the Mo thickness, only an amorphous Mo-Ge alloy with a
layered concentration gradient remains. The amorphous alloys show striking
similarities with homogeneously sputtered samples on the scale of local
atomic structure. Indeed, data taken with the scattering vector in the
layer plane and perpendicular to the layer plane indicate the same distribution function. This demonstrates that the tendency to alloy is strong
and bonds are similar in all directions even though the concentration
modulation is strong enough to yield multilayer Bragg reflections. There
is no evidence of an amorphous Mo structure of concentration greater than
75% in any of the samples.

THE MOLECULAR BEAM EPITAXIAL GROWTH OF GaAs ON Si

D. Woolf

Department of Physics, University College Cardiff

P.O. Box 78, Cardiff, CF1 1XL, Wales, U.K.

The aim of this project is not only to grow high quality GaAs onto an Si substrate, but also to gain an understanding of the growth process and the physics of the GaAs/Si interface.

Specifically, the project divides into the following sections:

a. Investigating the cleaning of Si(100) stepped surfaces, employing surface science techniques.

b. The investigation of the adsorption of As, Ga and combinations of As and Ga on clean Si(100) stepped surfaces.

c. To explore the abruptness, crystallography, chemical reactions, band discontinuities, etc. of the As/Si, Ga/Si and GaAs/Si interfaces and their dependence on preparation conditions. The As/Si, Ga/Si and GaAs/Si surfaces and interfaces will be examined using a range of surface science techniques including high resolution XPS, AES, LEELS, LEED, ARUPS, etc.

d. To grow high quality GaAs on Si and to characterise the layers.

This project, so far, has involved the commissioning of the VG V8ØH MBE system, and also investigating the cleaning of Si.

POSTER SESSION III

MAGNETIC MULTILAYERS

MAGNETIC ANISOTROPY IN MODULATED Nd-Fe MULTILAYERS

L.T. Baczewski*, M. Piecuch, J. Durand+, P. Delcroix, and G. Marchal

Laboratoire de Physique du Solid (U.A. au C.N.R.S. no.155)
Université de Nancy I, B.P. 239, F-54506 Vandoeuvre les Nancy Cédex
France

1. INTRODUCTION

Rare-Earth-Transition Metal (RE-TM) amorphous thin films have recently attracted considerable attention [1,2,3] not only because of basic physical interest but also as a technologically useful material for perpendicular magnetic recording. On the other hand, Nd-Fe based alloys are the best permanent magnets up to date and their main phase $Nd_2Fe_{14}B$ has a layer-like structure. So compositionally modulated Nd-Fe films can help to understand the basic mechanisms responsible for the strong perpendicular anisotropy found in Nd-Fe thin films.

In this paper we present the results of magnetic properties investigation of modulated Nd-Fe multilayers in comparison with amorphous Nd-Fe thin films.

2. EXPERIMENTAL

The samples were prepared by alternate evaporation of Nd and Fe layers in an ultra-high vacuum chamber onto mica substrates at 410 K. Then the films were transferred to kapton polyimide. The quality of the samples regarding crystals dimensions, periodicity and interface sharpness has been checked by electron microscopy and X-ray diffraction. All the films were continuous and crystals of iron and neodymium - 200 Å and 40 Å in width, respectively - were found.

Magnetic properties were studied by means of [57]Fe Mössbauer spectroscopy and vibrating sample magnetometer (VSM).

3. RESULTS AND DISCUSSION

Samples of the following iron layer thicknesses were studied: 31 Å, 22 Å, 16 Å and 13 Å while the Nd layers thickness was kept constant at about 38 Å except for the sample with 16 Å of Fe for which the Nd layer

---

*On leave from Institute of Physics, Polish Academy of Sciences, Warsaw, Poland

thickness was 32 Å. These values would correspond to the compositions for homogeneous alloys: $Nd_{29}Fe_{71}$, $Nd_{37}Fe_{63}$, $Nd_{40}Fe_{60}$, $Nd_{50}Fe_{50}$, respectively. Magnetic hysteresis loops were obtained for each sample at room temperature (RT) and 4.2 K for the direction of applied magnetic field (up to 20 kOe) parallel and perpendicular to the sample plane.

In Fig.1 the hysteresis loops at 4.2 K of the $Nd_{50}Fe_{50}$ sample for both directions of the applied field are shown.

For all investigated samples strong perpendicular magnetic aniso-tropy was found at cryogenic temperature.

However, at room temperature all the samples behave like soft magnetic materials with weak perpendicular anisotropy which decreases with the increasing of Fe content, vanishing for 71 at % of Fe where the magnetic easy axis is in plane of the sample.

Temperature dependence of perpendicular anisotropy was confirmed also by Mössbauer spectroscopy and the temperature measurements of saturation magnetization, but a detailed discussion will be submitted elsewhere [5].

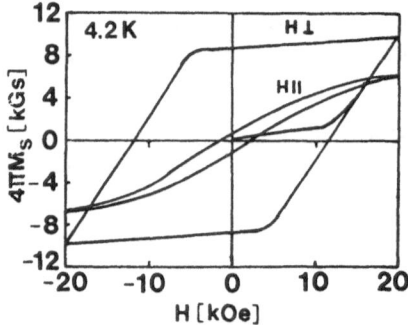

Fig.1 Hysteresis loops of an $Nd_{50}Fe_{50}$ (13 Å Fe and 38 Å Nd) multilayers sample for the direction of applied magnetic field parallel and perpen-dicular to the sample plane at 4.2 K

Saturation magnetization $4\pi M_S$ at H = 20 kOe, T = 4.2 K for the investigated multilayers is presented in Fig.2. The results are compared with those obtained by TAYLOR et al. [4] for amorphous Nd-Fe thin films. The values of $4\pi M_S$ are always higher for multilayers but the difference largely increases with the increasing of the iron content. Under the assumption that Fe atoms have the magnetic moment of 2.2 μ_B/at like in pure iron, the magnetic moment of Nd atoms at 4.2 K was calculated. This assumption is certainly oversimplified regarding the structure of the Nd-Fe interface, but can be useful as a first-order approximation. For the samples $Nd_{29}Fe_{71}$, $Nd_{37}Fe_{63}$, $Nd_{40}Fe_{60}$ and $Nd_{50}Fe_{50}$ magnetic moments of Nd atoms in μ_B/at are the following: 2.52, 1.3, 1.39 and 1.48, respectively. It can be seen that high content of Fe in the bcc alpha phase induces a large magnetic moment in Nd. Thinner iron layers have much weaker influence on Nd atoms.

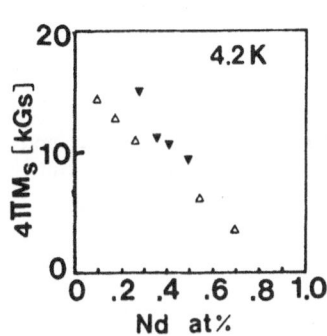

Fig.2 Saturation magnetization $4\pi M_S$ at 4.2 K for Nd-Fe multilayers as a function of Nd content (▼). The data for amorphous thin films (△) are from Ref.4

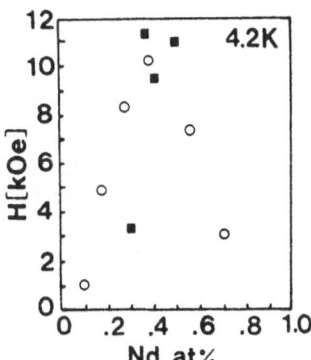

Fig.3 Coercivity H_C of Nd-Fe modulated multilayers at 4.2 K (■) as a function of composition. The data for amorphous thin films (o) are taken from Ref.4

For explaining a bigger difference in $4\pi M_S$ for high Fe content (shown in Fig.2) we assume that ferromagnetic coupling occurs between Nd and Fe layers. For high iron content ($\omega \geq 25$ Å of Fe) we have seen in Mössbauer spectra pure bcc alpha iron and for lower Fe layers width - amorphous-like structure. It seems that for bcc iron the magnetic coupling between Nd and Fe is much stronger than for iron in amorphous structure.

The coercive force at 4.2 K for Nd-Fe modulated multilayers as a function of Nd content in comparison to the data for amorphous thin films obtained by e-beam evaporation are shown in Fig.3

It could be seen that for both systems a significant maximum of H_C exists near the composition of 40 at % of Nd. A similar behaviour can be seen for amorphous Nd-Fe thin films [1,4] and melt-spun amorphous Nd-Fe alloys [6] where the maximal values were found also for the compositions between 40 and 50 at % of Nd.

From the shape of M (H) curves (abrupt increase of M near H = H_C) for Nd-Fe multilayers it seems that the domain wall pinning process can be regarded as coercivity mechanism.

REFERENCES

1. T. Suzuki: J. Magn. Mag. Mater. $\underline{50}$, 265 (1985)
2. T. Morishita, Y. Togami, K. Tsushima: J. Phys. Soc. Jpn. $\underline{54}$, 37 (1985)
3. N. Sato: J. Appl. Phys. $\underline{59}$, 1942 (1986)
4. R.C. Taylor, T.R. McGuire, J.M. Coey, A. Gangulee: J. Appl. Phys. $\underline{49}$, 2885 (1978)
5. L.T. Baczewski, P. Piecuch et al.: to be published
6. J.J. Croat: J. Appl. Phys. $\underline{53}$, 3161 (1982)

+We would like to dedicate this paper to our colleague and friend
Prof. J. DURAND who died suddenly during the final redaction of this text.

STRUCTURE AND MAGNETIC PROPERTIES OF COBALT ULTRATHIN FILMS

P. Beauvillain, P. Bruno, C. Chappert, H. Hurdequint*, K. Le Dang,
C. Marlière**, D. Renard**, and P. Veillet

Institut d'Electronique Fondamentale
Laboratoire de Physique des Solides**
Institut d'Optique**, Laboratoires associès au CNRS
Université Paris-Sud, F-91405 Orsay, France

Sample Preparation

Au/Co/Au and Au/Cu/Co/Cu/Au sandwiches are prepared by slow evaporation of the metals from tungsten crucibles in a ultrahigh vacuum (P < 10^{-9} Torr) deposition unit. The first gold layer is annealed at about 440 K. This substrate is polycrystalline, textured, with a mean lateral size of the crystallites around 2000 Å. Its surface is oriented (111) and atomically flat. Thicknesses of Au, Cu and Co layers are respectively 250 Å, 30 Å and from 2 to 80 Å, and are controlled by a quartz oscillator. The growth is followed by in-situ resistivity measurements [1].

Structural Characterization

The sandwiches are monocrystalline through the total thickness and have a textured polycrystalline structure. The mean lateral sizes of the crystallites are 2000 Å and 500 Å, respectively for Au/Co/Au and Au/Cu/Co/Cu/Au, as shown by Transmission Electron Microscopy (TEM) images and diffraction patterns [1]. Grazing X-ray reflectivity and X-ray Bragg diffraction allows direct thickness measurements, quartz calibration and gives some information on the interface roughness [1]. The Au/Cu/Co/Cu/Au sandwiches interfaces are rougher than the Au/Co/Au ones.

TEM diffraction patterns show that the sandwich follows the polycrystalline structure of the gold substrate with a good crystalline coherence through the whole thickness. In the Au/Co/Au sandwiches, these patterns indicate that the cobalt exhibits a hcp structure. This fact is supported by nuclear magnetic resonance (NMR) [2,3]. On the other hand, from NMR studies, the cobalt film appears to have a fcc structure in Au/Cu/Co/Cu/Au sandwiches. This fcc growth might be induced by the larger roughness of the Cu substrate surface, Cu and fcc Co having nearly the same lattice constant.

Magnetic Properties

Magnetization experiments are performed with a home-made fully automated low field SQUID magnetometer [4] with a resolution in magnetic moment up to $2 \cdot 10^{-10}$ emu cgs. We have studied the remanent magnetization (RM) of the cobalt sandwiches. On the contrary to the nickel films, the cobalt remains ferromagnetic down to one atomic layer [5]. Moreover, as shown in Fig.1, the RM after saturation in a magnetic field applied parallel or perpendicular to the film plane reveals that the easy magnetization direction switches from in-plane to out-of-plane when the Co thickness is decreased [5].

The magnetic anisotropy energy can be written $E_a = (K^{eff} - 4\pi M^2/2) \sin^2 \theta$ (Eq.1). $-1/2\ 4\pi M^2$ is the shape anisotropy and K^{eff} is the magnetocrystalline anisotropy: $K^{eff} = 1/V\ (V \times K^V + 2\ S \times K^S) = K^V + 2\ K^S/e$ (Eq.2), where K^V and K^S are respectively the bulk and interface anisotropies and e the cobalt thickness. For each cobalt film, the magnetic anisotropy constant K^{eff} is deduced from the angular variation of the resonance field in FMR experiments [3]. The magnetic anisotropies K^V and K^S are then deduced from Eq.2 as shown in Fig.2. We have obtained:

$K^V = 3 \cdot 10^6$ erg cm^{-3} for Au/Co/Au
$K^V \simeq 0$ for Au/Cu/Co/Cu/Au
$K^S = 0.5$ erg cm^{-2} for both samples

The large value of the bulk anisotropy (not far from the bulk value $K^V = 4.2 \cdot 10^6$ erg cm^{-3}) of Au/Co/Au and the zero bulk anisotropy of Au/Cu/Co/Cu/Au are respectively consistent with the hcp and fcc dominant structures. The positive surface anisotropy is responsible of the easy direction reversal at low thicknesses.

This value of K^S is in good agreement with a Neel's model calculation [6]. In this localized moments model with nearest neighbours interactions we can calculate K^S in connection with the magnetostriction constants.

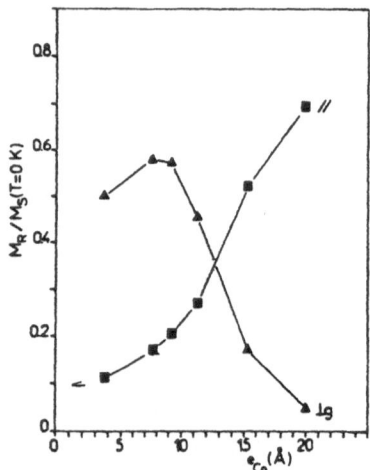

Fig.1 Dependence versus the Co thickness e of the parallel and perpendicular remanent magnetization normalized to the expected bulk saturation magnetization

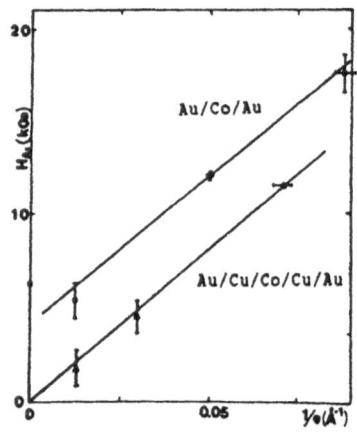

Fig.2 Plot of the K^{eff} anisotropy constant measured at room temperature on our Co sandwiches

REFERENCES

1. D. Renard, G. Nihoul: Philos. Mag. B55, No.1, 75 (1987)

2. K. Le Dang, P. Veillet, C. Chappert, P. Beauvillain, D. Renard,
   J. Phys. F: Met. Phys. 16, L109 (1986)
3. C. Chappert, K. Le Dang, P. Beauvillain, H. Hurdequint, D. Renard:
   Phys. Rev. B 34, No.5, 3192 (1986)
4. P. Beauvillain, C. Chappert, J.P. Renard: J. Phys. E: Sci. Instr. 18,
   839 (1985)
5. C. Chappert, D. Renard, P. Beauvillain, J.P. Renard, J. Seiden: J. Mag.
   Mag. Mat. 54-57, 795 (1986)
6. P. Bruno et al.: to be published

MAGNETIC PROPERTIES OF Pd/Co MULTILAYERS

H.J.G. Draaisma

Eindhoven University of Technology, Department of Physics

NL-5600 MB Eindhoven, The Netherlands

Multilayers consisting of alternatingly ferromagnetic and nonmagnetic layers may have interesting magnetic properties, induced by the limited thickness of the layers and the great number of interfaces. The anisotropy of the magnetic moment, which determines the preferred direction of the magnetization, is one of the properties which can be influenced by the multilayered structure. Of specific interest are systems in which the preferred direction is oriented perpendicular to the interfaces and thus the film plane. This has its application in so-called perpendicular recording, which might increase the information density in recording, and is also of importance in magneto-optical data storage.

We have prepared Pd/Co multilayers by e-beam evaporation in UHV on glass at room temperature. The films are polycrystalline with a [111] (fcc) texture perpendicular to the film. Decreasing the Co layer thickness we found an increasing tendency towards perpendicular anisotropy, resulting in square perpendicular magnetization curves for Co layers of 2 Å. This can be ascribed to the interface between Pd and Co. The Pd thickness has influence on the magnetic field at which perpendicular saturation is reached. This saturation field is analyzed on the basis of an extension of a domain model with dipolar interactions between the layers only.

The interface between Pd and Co has to be sharp: Pd/Co multilayers produced by ion beam sputtering did not show this anisotropy probably due to the mixed interface between Pd and Co. Ion beam mixing and annealing of the vapour deposited multilayers also lowered the anisotropy.

REFERENCES

H.J.G. Draaisma, W.J.M. de Jonge, F.J.A. den Broeder
J. Magn. Magn. Mat. 66, 351 (1987)

F.J.A. den Broeder, H.C. Donkersloot, H.J.G. Draaisma, W.J.M. de Jonge
J. Appl. Phys. 61, 4317 (1987)

PREPARATION AND MAGNETIC PROPERTIES OF RARE-EARTH-TRANSITION METAL
THIN FILMS AND MULTILAYERS: A STARTING PROJECT

D. Givord and J.P. Rebouillat

C.N.R.S. - 166 X, F-38042 Grenoble Cédex, France

Objective of the Project

-       Understanding of basic magnetic properties in ultra-thin films (some atomic layers)

-       Magnetic interactions between layers exhibiting magnetism of different nature (3d and 4f magnetism, ferro-antiferromagnetism)

-       Magnetic anisotropy in thick films and multilayers (in the range of 1000 Å) exhibiting controlled microstructures (for magnetic recording)

Film Elaboration Technique

-       Pulsed laser evoporation under ultra high vacuum: a Nd-YAG laser beam is focused on a target, the very high electromagnetic radiation (P $\simeq$ 10^{12} W/cm^2) causes instantaneous sublimation of the target material and particles are projected on a facing substrate.

Advantages of the Method

i)   The target remains at low temperature (incident mean power < 1 watt) and consequently:

- no deterioration of the vacuum due to evaporation
- no reactivity of the target with residual gazes or with crucible

ii)   No limitation for target composition: any solid element (even refractory) compound or alloy
- evaporation is congruent
- no need for multisources monitoring

iii)   Evaporated particles are excited
- increased mobility on the deposit
- epitaxial growth easier
- lowering of the substrate temperature

Structural and Magnetic Characterization

- RHEED, LEED, AUGER spectroscopy
- SQUID magnetometer
- Kerr Effect (transverse and longitudinal) magnetometer
- Mössbauer spectroscopy (detection of conversion electrons)

REFERENCES

1. J. Desserre and J.F. Eloy: Thin Solid Films $\underline{29}$, 29 (1975)
2. S.V. Gaponov, S.A. Gusev, B.M. Luskin, N.N. Salashchenko, and E.S. Gluskin: Opt. Commun. $\underline{38}$, 7 (1981)
3. R.N. Sheftal and L.V. Cherbakov: Cryst. Res. Tech. $\underline{16}$, 887 (1981)
4. J.T. Cheung: Appl. Phys. Lett. $\underline{43}$, 255 (1983)
5. K.H. Müller: Surf. Sci. $\underline{184}$, L375 (1987)

ELECTRODEPOSITED Co-P AMORPHOUS MULTILAYERS:

A STUDY OF MAGNETOELASTIC PROPERTIES

G. Rivero, M. Liniers, E. Ascasibar and J.M. Gonzalez*

Laboratorio de Magnetismo
U. Complutense, E-28003 Madrid, Spain

*Instituto de Ciencia de Materiales de Madrid
 C.S.I.C., E-28006 Madrid, Spain

Tube-shaped samples were electrolytically obtained by using a cylindrical distribution of electrodes and square pulsed electrolytic current density [1,2].

The composition and structural parameters of the resulting multilayered structure can be controlled through the variation of the amplitude and width of the positive and negative current pulses.

In this work, negative pulses were adjusted in order to achieve a high concentration of P ($\approx$30% at P), whereas we used four different amplitudes of the positive pulses. Therefore, the layers corresponding to the negative pulses were nonferromagnetic, thus making it possible to control the shape anisotropy associated with the perpendicular magnetization configuration in the ferromagnetic layers.

Samples show magnetic anisotropy with azimuthal easy axis. The anisotropy field is about 16 Oe and the coercive field along the azimuthal direction 0.2 Oe (at 20 Hz). Typical M_z-H_z and M_ϕ-H_ϕ hysteresis loops are shown in Figs. 1a and 1b.

ΔE effect has been measured by using a magnetoelastic resonance technique [3]. Figure 2 shows the variation of the magnetoelastic resonance frequency with the longitudinal DC bias field.

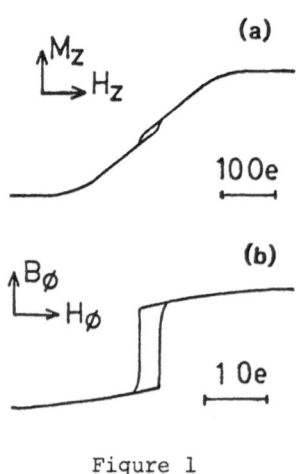

Figure 1

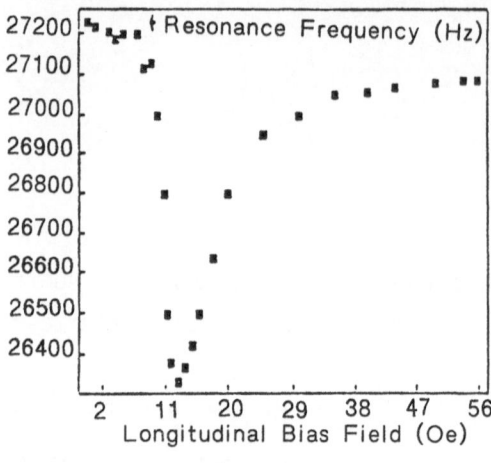

Figure 2

REFERENCES

1. G. Rivero: Ph.D. Thesis, U. Complutense (1985)
2. G. Rivero: Communication to this Conference
3. A. Hernando, V. Madurga, J.M. Barandiaran, M. Liniers: J. Magn. Magn. Mat. <u>28</u>, 109 (1982)

INTERACTIONS IN TWO-LAYER MAGNETIC FILMS

H. Niedoba

Laboratoire de Magnétisme et d'Optique des Solides

C.N.R.S., 1, place Aristide-Briand, F-92195 Meudon Cédex

    Some properties of sandwich films consisting of two magnetic layers
either in direct contact or separated by a nonmagnetic layer are reviewed.
First, we consider various interaction phenomena which are observed:
direct exchange coupling, magnetostatic couplings, as well as interactions
between domain walls in different films. Interaction phenomena have a
number of effects on the wall structures and energies, on the coercivity
and mobility of the walls. The use of Lorentz microscopy in the study of
magnetic domains and walls in two-layer films is illustrated by several
examples.

STATIC AND DYNAMIC CHARACTERIZATON OF INTERFACIAL ANISOTROPY   IN MAGNETIC
SUPERLATTICES*

M.J. Pechan

Department of Physics, Miami University
Oxford, OH 45056, USA

I.K. Schuller

Materials Science Division, Argonne National Laboratory
Argonne, IL 60439, USA

We have observed large discrepancies between the DC magnetization and
ferromagnetic resonance (FMR) magnetic anisotropy in Mo/Ni superlattices.
We show that higher order anisotropy is present in both measurements and
develop an analysis by which the first order and second order energies can
be treated independently. The low temperature results show a systematic
divergence of the first order FMR and DC anisotropy as the Ni layer thickness decreases. It is shown that the source of the divergence is an enhanced surface anisotropy measured by FMR. Possible sources of this enhancement are discussed.

ACKNOWLEDGEMENTS

This work was supported in part by the U.S. Office of Naval Research
Contract number N0014-83-F-0031 and the U.S. Department of Energy,
BES-Materials Sciences Contract numbers W-31-109-ENG-38 and
DE-FG02-86ER45281.

*Full report presented in Physical Review Letters (in press)

CRYSTAL STRUCTURE AND MAGNETIC CHARACTERISTICS OF Tb/Fe MULTILAYERS*

D.C. Person, W.R. Bennett, and C.M. Falco

Optical Sciences Center and Department of Physics

University of Arizona, Tucson, AZ 85721, USA

We have grown Tb/Fe multilayers with modulation wavelengths in the range of 10 to 200 Å by alternately sputtering these components. The individual layer thicknesses were kept constant to within ±0.3% over the entire sample thickness of ≈0.2 µm by feedback control of the sputtering rates and microprocessor control of the substrate rotation. Composition of the samples was determined from calibrated sputtering rates, and was confirmed using Rutherford backscattering spectrometry. The layered structure was demonstrated by low angle x-ray diffraction. High angle x-ray results show that the Tb and Fe layers are polycrystalline for high modulation wavelengths, but are amorphous for low modulation wavelengths. A vibrating sample magnetometer was used to measure the coercivity and saturation magnetization over the temperature range 200 to 430 K. Comparing results for pure Fe, pure Tb, and Tb/Fe multilayers demonstrates that the Tb and Fe multilayers couple to form an antiferromagnetic film.

*Research supported by the Optical Data Storage Center (ODSC) at the University of Arizona

THEORY OF FERROMAGNETIC MULTILAYERS

F. Fishman, F. Schwabl and D. Schwenk

Physik-Department der Technischen Universität München

D-8046 Garching, FRG

We present a theory of ferromagnetic periodic multilayers, consisting of two alternating ferromagnetic materials. The basis of our theory is an inhomogeneous Ginzburg-Landau (G.L.) functional, where the G.L. coefficients are chosen to model the alternating layers and interface interactions. We study the transition temperature of the composite material as a function of the material parameters. Further, we calculate the static magnetisation profile for different temperatures. We find that in the vicinity of the lower transition temperature Tl the penetration of the magnetisation into the low temperature ferromagnet is characterised by a 1/x law, while far from Tl the penetration is described by an exponential function. The resulting average magnetisation is also evaluated. The spin dynamics is studied by using the Bloch equation. Here we consider some limiting cases as well as the general situation, where both dipolar and exchange interactions are included. We examine the magnon dispersion relation. A prominent feature is that band gaps vanish at certain values of the wave vector k. We also compute the cross section for inelastic scattering of neutrons by the magnons.

REFERENCES

H. Schmidt and F. Schwabl: Z. Phys. B 30, 197 (1978)
F. Fishman, F. Schwabl and D. Schwenk: Phys. Lett. A 121, 192 (1987)

Research supported in part by the Bundesministerium für Forschung und Technologie and the Deutsche Forschungsgemeinschaft

PREPARATION AND CHARACTERIZATION OF METALLIC MAGNETIC SUPERLATTICES

L. Smardz

Institute of Molecular Physics, Polish Academy of Sciences

Smoluchowskiego 17/19, PL-60179 Poznań, Poland

The Cu/Ni, Cu/Fe, Cu/Co and Pd/Ni superlattices composed of alternating regions of m atomic planes of Cu(Pd) and n atomic planes of Ni(Fe, Co) were prepared by the sputtering technique [1]. A layer 200 – 500 Å of Cu was first deposited onto the mica substrate to establish a strong texture. The samples prepared had thicknesses between 0.5 and 1 μm, wavelengths of modulation between 20 and 150 Å, and average compositions which varied from 10 to 80 at. % of ferromagnetic component.

The structure of the samples was examined by standard X-ray diffraction with Cu-Kα radiation. The wavelength of modulation was determined from the location of the satellite peaks around the central Bragg peak and was consistant with the value obtained from total thickness, divided by the number of repetitions.

The magnetic properties of the samples were investigated at R.T. using a magnetic balance and the FMR method, and at 4.2 K by a SQUID magnetometer and a vibrating sample magnetometer. It was found that the average magnetization of the Ni fraction, calculated per unit volume of Ni fraction in the magnetic alloy CuNi, is independent of the thickness of the Cu sublayers and decreases rapidly for Ni thickness less than 8 atomic planes [2]. The magnetization of Fe and Co sublayers was calculated assuming a rectangular profile of modulation. For Fe(Co), thickness greater than 20 Å, magnetization was equal to the bulk value [3,4]. From this we conclude that the interfaces between Fe(Co) and Cu sublayers seem to be sharp.

REFERENCES

1. L. Smardz, J. Baszyński and J. Dubowik: Acta Phys. Pol. (to be published)
2. L. Smardz and J. Baszyński: J. Appl. Phys. (to be published)
3. L. Smardz, J. Baszyński, V.D. Kuznetsov and I.E. Kuznetsov: Acta Magnetica, Supplement '87 (1987)
4. L. Smardz, J. Baszyński and A.N. Bazhan: Acta Magnetica, Supplement 87 (1987)

Materials Index

Subject Index